Sustainable Land Management

Selim Kapur · Hari Eswaran · W. E. H. Blum
Editors

Sustainable Land Management

Learning from the Past for the Future

Springer

Editors
Dr. Selim Kapur
Department of Soil Science &
Archaeometry
University of Çukurova
01330 Adana, Turkey
kapurs@cu.edu.tr

Dr. Hari Eswaran
United States Department of Agriculture
Natural Resources Conservation Service
PO Box 2890
Washington, DC, USA
hari.eswaran@wdc.usda.gov

Dr. Winfried E. H. Blum
Institute of Soil Research
Department of Forest and Soil Sciences
University of Natural Resources
and Applied Life Sciences (BOKU)
Vienna, Austria
herma.exner@boku.ac.at; winfried.blum@boku.ac.at

ISBN 978-3-642-14781-4 e-ISBN 978-3-642-14782-1
DOI 10.1007/978-3-642-14782-1
Springer Heidelberg Dordrecht London New York

Library of Congress Control Number: 2010938610

Cover design: deblik, Berlin

Printed on acid-free paper

Springer is part of Springer Science+Business Media (www.springer.com)

Dedicated to the Dear Memory of Prof. Dr. Cemil Cangir (1946–2010)

Cemil Cangir the warrior of soil protection in Thrace, Turkey, passed away suddenly in Tekirdağ on May 10th 2010, at the age of 64, now lying in a village cemetary, whom he sought to be buried as a site overlooking the view of the soils which he dedicated his life in search of his unique protection strategies. The ever-threatened fertile soils of Thrace by the inappropriate growth of the industry and urbanisation are now seeking for the decendants of the "Cangir School" to take over the task and proceed the fight for their protection. The incomparable community and NGO solidarity Cemil had achieved as his life-long dream was based on a simple key, namely the "Integrated Sustainable Basin Management of the land".

Selim Kapur and Koray Haktanır

Preface

Inappropriate use of land and water is still on-going, despite all efforts undertaken by numerous institutions dealing with land and soil protection, as for example the European Soil Protection Strategy, aiming at a common legal instrument for soil protection in Europe.

One of the failures lies in conceptual problems, because the protection of land and water have to be done at river basin scale.

The "Anthroscape"-approach presented in this book is thought as a contribution to improve the sustainable management of land and water at river basin scale, based on an analysis of the past developments, in order to shape the contemporary and future use of land in a sustainable way.

Moreover, this book contains information about traditional techniques and knowledge, shown at specific examples of the "Anthroscapes", which were developed by an integrated management, with a holistic perspective, to meet the needs of humans in a landscape, like water, food, biodiversity and culture.

Therefore, the title of this book is "**Sustainable Land Management: Learning from the Past for the Future**".

Each chapter of the book has been discussed on several occasions in the field as well as during international conferences and meetings related to the topic of "Anthroscape development". Moreover, the chapters were arranged in such a way that they are complimentary, regarding specific aspects of the "Anthroscape" concept.

Chapter 1 introduces into the book, describing the key components of sustainable land and water management at river basin scale, which was specifically developed by H. Eswaran who has devoted the many efforts to develop a conceptual framework on the basis of his long-standing and world wide experience on soils and landscapes. Chapter 2 is mainly designated to the recent development of the EU Soil Thematic Strategy. Chapter 3 reflects the major benefit of studies of land/soil at river basin scale, disregarding frontiers and regions. Chapter 4 describes the major soils of Mediterranean Anthroscapes, with special reference on renovated traditional land use techniques in the Apulia region of southern Italy. Chapter 5 discusses the presently observed side-effects of touristic activities in the Italian Alps, including major threats to European water sources. Chapter 6 deals with North European Anthroscapes, and negative impacts, by measures within a particular region in South East Norway. Chapter 7 describes an expensive and time consuming practice, partially discussed very controversially, in the use of contemporary technology, for using soils with calcrete (petrocalcic) horizons in Sardinia, Italy, for increasing the productivity of vineyards. Chapter 8 deals with the use of soils on mine waste deposits in southeast Spain and Chap. 9 describes important Asian Anthroscapes, created by anthropogenic activities during thousands of years, for the cultivation of rice.

In Chap. 10, Anthroscapes from Morocco were described, under the aspect of the use of exotic plants to rehabilitate degraded rangeland and to mitigate desertification.

Chapter 11 deals with anthropogenically re-shaped land and its historical and cultural background in the Adana region in Turkey.

Chapter 12 explains the impact of the ancient practice of shifting agriculture to the forest ecosystem and the low-input agricultural system to the Anthroscapes in Sarawak, Malaysia and Chap. 13 refers to impacts by improper land use in the Mediterranean area, and especially attempts to re-use abandoned mining landscapes in Murcia, southern Spain. Chapter 14 deals with possible future impacts of climate change on Mediterranean coastal areas, including the Anthroscape in Adana, Turkey.

In Chap. 15 field trials for the evaluation of indigenous techniques for soil and water conservation in Niger (Africa) were discussed, searching less costly and labor-intensive conservation techniques.

Similarly, Chap. 16 deals with local knowledge in protecting landscapes in Japan, with special attention to the cultural background and wisdom of the eco-life.

Chapter 17 is very specific and deals with an experimental study, aiming to recover ancient natural landscapes in the Adana region, reconstructing lost parts of the landscape and regenerating biodiversity.

Economic and social problems in degraded landscapes, including the evaluation of the costs of land degradation are discussed in Chap. 18.

We specially thank the everlasting encouragement and patience of Drs. Hans Guenther Brauch and Christian Witschel, from Springer Editors. Our thanks are also due to Dr. Ismail Çelik, Dr. E. Akça, Dr. Kemal Gülüt, Dr. M. Dingil and Mr. K. Y. Koca for their support in editing this book.

We finally express our sincere gratitude to all the authors of this book.

Adana, Turkey

Selim Kapur
Hari Eswaran
Winfried E. H. Blum

Contents

The Anthroscape Approach in Sustainable Land Use 1
H. Eswaran, S. Berberoğlu, C. Cangir, D. Boyraz, C. Zucca, E. Özevren, E. Yazıcı, P. Zdruli, M. Dingil, C. Dönmez, E. Akça, I. Çelik, T. Watanabe, Y.K. Koca, L. Montanarella, M. Cherlet, and S. Kapur

Anthroscapes in the Light of the EU Soil Thematic Strategy 51
Luca Montanarella

Soil Erosion-Desertification and the Middle Eastern Anthroscapes 57
Uriel N. Safriel, Pedro Berliner, Ariel Novoplansky, Jonathan B. Laronne, Arnon Karnieli, Itzhak Moshe, A. Kharabsheh, A. Ghaleb Mohammad, and G. Kusek

Soils of the Mediterranean Region, Their Characteristics, Management and Sustainable Use 125
Pandi Zdruli, Selim Kapur, and Ismail Çelik

Mountain Anthroscapes, the Case of the Italian Alps 143
Franco Previtali

Development and Challenges of the Anthroscapes in the Clay Soil District of Eastern Norway 163
Arnold Arnoldussen

Anthroscapes and Anthropogenic Soils in North-Western Sardinia: The Soils with Calcrete Horizon in the Alghero Area (Italy) 179
S. Madrau and C. Zucca

Stimulated Soil Formation in a Degraded Anthroscape: A Case Study in Southeast Spain 193
J.M. Arocena, J.M. van Mourik, and Á. Faz Cano

Asian Anthroscapes: China and Taiwan 205
Zeng-Yei Hseu, Chen-Chi Tsai, Heng Tsai, Zueng-Sang Chen, and Hari Eswaran

An Anthroscape from Morocco: Degraded Rangeland Systems and Introduction of Exotic Plant Material and Technology 243
C. Zucca, F. Previtali, S. Madrau, E. Akça, and S. Kapur

The Historical Anthroscape of Adana and the Fertile Lands 259
Nurettin Çelmeoğlu

Impact of Shifting Agriculture on the Sustainability of Anthroscapes in Sarawak, Malaysia 285
E. Padmanabhan and H. Eswaran

Roman Mining Landscapes in the Murcia Region, SE Spain: Risk Assessment of Mine Ponds 293
A. Faz Cano, A. Zanuzzi, P. Martinez-Pagan, J. Acosta, D. Carmona, S. Martinez-Martinez, and M. Muñoz

Anthroscape of the Mediterranean Coastal Area in the Context of Hydrogeology: Projected Impacts of Climate Change 311
Katsuyuki Fujinawa

Re-evaluating Indigenous Technologies for Sustainable Soil and Water Management in the Sahel: A Case Study from Niger 333
Takanori Nagano, Haruhiko Horino, and Takashi Kume

Local Wisdom of Land and Water Management: The Fundamental *Anthroscape* of Japan 351
Tsugihiro Watanabe

Reconstructing the Past by Regenerating Biodiversity: A Treatise on Weed Contribution to Soil Quality at a Post-cultivation Succession 363
I. Çelik, K.T. Yilmaz, H. Eswaran, A. Mermut, M. Dingil, Z. Kaya, A. Demirbaş, I. Aksit, I. Ortaş, M. Gök, C. Akpınar, T. Nagano, N. Ae, Y.K. Koca, and S. Kapur

Economic and Social Impact of the Degraded Antroscapes 379
Nejat Erk

Index 411

Contributors

J. Acosta Sustainable Use, Management and Reclamation of Soil and Water Research Group, Agrarian Science and Technology Department, Technical University of Cartagena, Paseo Alfonso XIII, 52, 30203 Cartagena, Murcia, Spain

N. Ae Kobe University, Kobe, Japan, aenoriha@kobe-u.ac.jp

E. Akça Department of Technical Programs, University of Adiyaman, Adıyaman, Turkey

C. Akpinar Department of Soil Science, University of Çukurova, Adana, Turkey, cakpinar@cu.edu.tr

I. Aksit University of Erciyes, Kayseri, Turkey, iaksit@erciyes.edu.tr

Arnold Arnoldussen Norwegian Forest and Landscape Institute, PB 115, 1431 Ås, Norway, Arnold.Arnoldussen@skogoglandskap.no

J.M. Arocena Canada Research Chair in Soil and Environmental Sciences, University of Northern British Columbia, Prince George, BC, Canada V2N4Z9, arocenaj@unbc.ca

S. Berberoğlu Department of Landscape Architecture, University of Çukurova, Adana, Turkey

Pedro Berliner Blaustein Institutes for Desert Research, Ben-Gurion University, Negev, Israel

D. Boyraz Department of Soil Science, Namık Kemal University, Tekirdağ, Turkey

C. Cangir Department of Soil Science, Namık Kemal University, Tekirdağ, Turkey

D. Carmona Sustainable Use, Management and Reclamation of Soil and Water Research Group, Agrarian Science and Technology Department, Technical University of Cartagena, Paseo Alfonso XIII, 52, 30203 Cartagena, Murcia, Spain

Ismail Çelik Department of Soil Science, University of Çukurova, 01330 Adana, Turkey, icelik@cu.edu.tr

Nurettin Çelmeoğlu Greater Municipality of Adana, Adviser to the Mayor, Atatürk Bulv. 67, Kat 8, Adana, Turkey, celmeoglun@gmail.com

Zueng-Sang Chen Department of Agricultural Chemistry, National Taiwan University, Taipei 10617, Taiwan, ROC, soilchen@ntu.edu.tw

M. Cherlet Joint Research Centre of the European Commission (JRC), Institute for Environment and Sustainability (IES), Land Management and Natural Hazards Unit, DESERT Action, Via E. Fermi 2749, TP28, I-21020 Ispra (VA), Italy, michael.cherlet@jrc.ec.europa.eu

A. Demirbas Department of Soil Science, University of Çukurova, Adana, Turkey, ademirbas@cu.edu.tr

M. Dingil Department of Soil Science, University of Çukurova, 01330 Adana, Turkey, mdingil@cu.edu.tr

C. Dönmez Department of Landscape Architecture, University of Çukurova, Adana, Turkey

Nejat Erk Department of Economics, University of Çukurova, 01330 Balcalı, Adana, Turkey, erk@cu.edu.tr

H. Eswaran United States Department of Agriculture, Natural Resources Conservation Service, PO Box 2890, Washington, DC, USA, hari.Eswaran@wdc.usda.gov

Á. Faz Cano Departamento de Ciencia y Technología Agraria, Universidad Politécnica de Cartagena, Paseo Alfonso XIII, 52, 30203 Cartagena, Murcia, Spain; Sustainable Use, Management and Reclamation of Soil and Water Research Group, Agrarian Science and Technology Department, Technical University of Cartagena, Paseo Alfonso XIII, 52, 30203 Cartagena, Murcia, Spain, angel.fazcano@upct.es

Katsuyuki Fujinawa Shinshu University, Wakasato 4-17-1, Nagano 380-8553, Japan, fujinawa@shinshu-u.ac.jp

A. Ghaleb Mohammad Hebron University, Hebron, Palestinian National Authority

Haruhiko Horino Graduate School of Life and Environmental Sciences, Osakafu University, 1-1 Gakuen-cho, Sakai-shi, Osaka 599-8231, Japan

Zeng-Yei Hseu Department of Environmental Science and Engineering, National Pingtung University of Science and Technology, Pingtung 91201, Taiwan, ROC

Selim Kapur Department of Soil Science, University of Çukurova, 01330 Adana, Turkey; Department of Archaeometry, University of Çukurova, 01330 Adana, Turkey, kapurs@cu.edu.tr

Arnon Karnieli Blaustein Institutes for Desert Research, Ben-Gurion University, Negev, Israel

Z. Kaya Department of Soil Science, University of Çukurova, Adana, Turkey, zkaya@cu.edu.tr

A. Kharabsheh Al-Balqa'Applied University, Al-Salt, Jordan

Y.K. Koca University of Çukurova, Department of Soil Science, 01330 Adana, Turkey, ykkoca@cu.edu.tr

Takashi Kume Research Institute for Humanity and Nature, 457-4 Motoyama-cho, Kamigamo, Kita-ku, Kyoto 603-8047, Japan

G. Kusek Agricultural Rural Services Directorate, Malatya, Turkey

Jonathan B. Laronne Department of Geography, Ben-Gurion University, Negev, Israel

S. Madrau DIT (Department of Territorial Engineering), University of Sassari, V.le Italia 39, 07100 Sassari, Italy; DIT (Department of Territorial Engineering), University of Sassari, Via de Nicola, 9, 07100 Sassari, Italy; Nucleo Ricerca Desertificazione, University of Sassari, V.le Italia 57, 07100 Sassari, Italy, madrau@uniss.it

S. Martinez-Martinez Sustainable Use, Management and Reclamation of Soil and Water Research Group, Agrarian Science and Technology Department, Technical University of Cartagena, Paseo Alfonso XIII, 52, 30203 Cartagena, Murcia, Spain

P. Martinez-Pagan Sustainable Use, Management and Reclamation of Soil and Water Research Group, Agrarian Science and Technology Department, Technical University of Cartagena, Paseo Alfonso XIII, 52, 30203 Cartagena, Murcia, Spain

A. Mermut Department of Soil Science, University of Harran, Şanlıurfa, Turkey, mermut@harran.edu.tr

Luca Montanarella European Commission, Joint Research Centre, Institute for Environment and Sustainability, 21020 Ispra, Italy, luca.montanarella@jrc.it

Itzhak Moshe Keren Kayemet Le'Israel, PO Box 283, Jerusalem, Israel

M. Muñoz Sustainable Use, Management and Reclamation of Soil and Water Research Group, Agrarian Science and Technology Department, Technical University of Cartagena, Paseo Alfonso XIII, 52, 30203 Cartagena, Murcia, Spain

Takanori Nagano Kobe University, Kobe, Japan; Graduate School of Agricultural Science, Kobe University, 1-1 Rokkodai-cho, Nada-ku, Kobe 657-8501, Japan, naganot@ruby.kobe-u.ac.jp

Ariel Novoplansky Blaustein Institutes for Desert Research, Ben-Gurion University, Negev, Israel

I. Ortas Department of Soil Science, University of Çukurova, Adana, Turkey, iortas@cu.edu.tr

E. Özevren Department of Desertification, Ministry of Environment and Forestry, Ankara, Turkey

E. Padmanabhan Department of Science and Mathematics, Curtin University of Technology, Sarawak, Malaysia, eswaran_padmanabhan@petronas.com.my

Franco Previtali Department of Environmental and Territorial Sciences, University of Milano-Bicocca, Piazza della Scienza, 1-20126 Milano, Italy, franco.previtali@unimib.it

Uriel N. Safriel Blaustein Institutes for Desert Research, Ben-Gurion University, Negev, Israel, uriel36@gmail.com

Chen-Chi Tsai Department of Natural Resources, National I-Lan University, I-Lan 26501, Taiwan, ROC

Heng Tsai Department of Geography, National Changhua University of Education, Changhua 50018, Taiwan, ROC

J.M. van Mourik IBED-Earth Surface Processes and Materials, Universiteit van Amsterdam, Nieuwe Achtergracht 166, 1018 WV Amsterdam, The Netherlands

Tsugihiro Watanabe Research Institute for Humanity and Nature (RIHN), 457-4 Motoyama Kamigamao, Kita-ku, Kyoto 603-8047, Japan

K.T. Yilmaz Department of Landscape Architecture, University of Çukurova, Adana, Turkey, tuluhan@cu.edu.tr

A. Zanuzzi Sustainable Use, Management and Reclamation of Soil and Water Research Group, Agrarian Science and Technology Department, Technical University of Cartagena, Paseo Alfonso XIII, 52, 30203 Cartagena, Murcia, Spain

Pandi Zdruli International Centre for Advanced Mediterranean Agronomic Studies, Mediterranean Agronomic Institute of Bari, Italy Via Ceglie 9, 70010 Valenzano (BA), Italy, pandi@iamb.it

C. Zucca Department of Territorial Engineering and Desertification Research Group (NRD), University of Sassari, Via de Nicola, 9, 07100 Sassari, Italy; Nucleo Ricerca Desertificazione, University of Sassari, V.le Italia 57, 07100 Sassari, Italy, clzucca@uniss.it

The Anthroscape Approach in Sustainable Land Use

H. Eswaran, S. Berberoğlu, C. Cangir, D. Boyraz, C. Zucca, E. Özevren, E. Yazıcı, P. Zdruli, M. Dingil, C. Dönmez, E. Akça, I. Çelik, T. Watanabe, Y.K. Koca, L. Montanarella, M. Cherlet, and S. Kapur

Abstract The final outcomes of the approach of the anthroscape, seeking to develop a powerful tool for use by the local communities/administrations and relevant bodies for a bottom to top approach, as a quantified entity, means to direct the future land and water use decisions to be taken at lower levels – as farm domains etc. – leading to the development of an "Anthroscape Land Quality Class" map and the relevant "Ideal Land Use Patterns" of the Seyhan Anthroscape. These two final

H. Eswaran (✉)
United States Department of Agriculture, Natural Resources Conservation Service, Washington, DC, USA
e-mail: hari.eswaran@wdc.usda.gov

S. Berberoğlu and C. Dönmez
Department of Landscape Architecture, University of Çukurova, Adana, Turkey

C. Cangir and D. Boyraz
Department of Soil Science, Namık Kemal University, Tekirdağ, Turkey

C. Zucca
Department of Territorial Engineering and Desertification Research Group (NRD), University of Sassari, Sassari, Italy

E. Özevren and E. Yazıcı
Department of Desertification, Ministry of Environment and Forestry, Ankara, Turkey

P. Zdruli
International Centre for Advanced Mediterranean Agronomic Studies, Mediterranean Agronomic Institute of Bari, Via Ceglie 9, 70010 Valenzano (BA), Italy

M. Dingil, I. Çelik, and Y.K. Koca
Department of Soil Science, University of Çukurova, 01330 Adana, Turkey

E. Akça
Department of Technical Programs, University of Adiyaman, Adıyaman, Turkey

T. Watanabe
Research Institute for Humanity and Nature (RIHN), 457-4 Motoyama Kamigamao, Kita-ku, Kyoto 603-8047, Japan

L. Montanarella
European Commission, Joint Research Centre, Institute for Environment and Sustainability, 21020 Ispra, Italy

S. Kapur et al. (eds.), *Sustainable Land Management*,
DOI 10.1007/978-3-642-14782-1_1, © Springer-Verlag Berlin Heidelberg 2011

products of quantification are sought to be capable in revealing the magnitude and the distribution of the degradation of the selected area, as well as allocating the ideal land use types given for the percentages of the distributions of land except their specific location. In this context, the downstream part of the map will show the abundant degradation arising via the intensive cultivation practices where the class stated in the map reveals the urgent need of an integrated SLWM Programme to revert the lower class C and D ALQCs to higher ALQCs to meet the requirements of the sustainable use of the land. Whereas, the higher ALQC land in the transition or upstream zone stands for higher resilience and lower input requirements to meet the ideal use of the land. The Net Primary Production (NPP) – The remainder of photosynthesis and respiration – which has significance in the global and regional carbon budgets and is a function of the standing biomass (an important component of the carbon cycle and a key indicator of ecosystem performance) was utilized via land cover and management to serve as a supplementary indicator of the *Anthroscape* and the *Anthroscape Land Quality*.

Keywords Anthroscape · Seyhan Basin · Sustainable use · Anthroscape Land Quality Class · Net primary production

1 Introduction

The approach is endorsed in developing sustainable land management programs and the introduction of the "Anthroscape Land Quality Classes" to replace the classical land capability classes of land use. The "Anthroscape" context of the Seyhan Basin primarily requires the integration of the topographic, vegetative, land use, demographic and socio-economic attributes together with the information on indigenous technologies for the development of a holistic sustainable land management program. The attributes of the factors said above are included in the three subregions within the basin as delineated below by Atalay's (2002) in the lately developed Eco-regions of Turkey (Fig. 1):

(a) Downstream (Adana-southern, Karataş) (Fig. 2)
(b) Transition zone (Adana-northern, Karaisalı, Pozantı) (Fig. 2)
(c) Upstream (Saimbeyli, Tufanbeyli, Feke, Çamardı-Mediterranean highlands; Ulukışla, Pınarbaşı, Sarız, Tomarza, Develi and Yahyalı-Margins of the Inner/Central Anatolian Plateau adjacent to the Mediterranean) (Fig. 2)

M. Cherlet
Joint Research Centre of the European Commission (JRC), Institute for Environment and Sustainability (IES), Land Management and Natural Hazards Unit, DESERT Action, Via E. Fermi 2749, TP28 I-21020 Ispra (VA), Italy

S. Kapur
Department of Soil Science, University of Çukurova, 01330 Adana, Turkey
and
Department of Archaeometry, University of Çukurova, 01330 Adana, Turkey

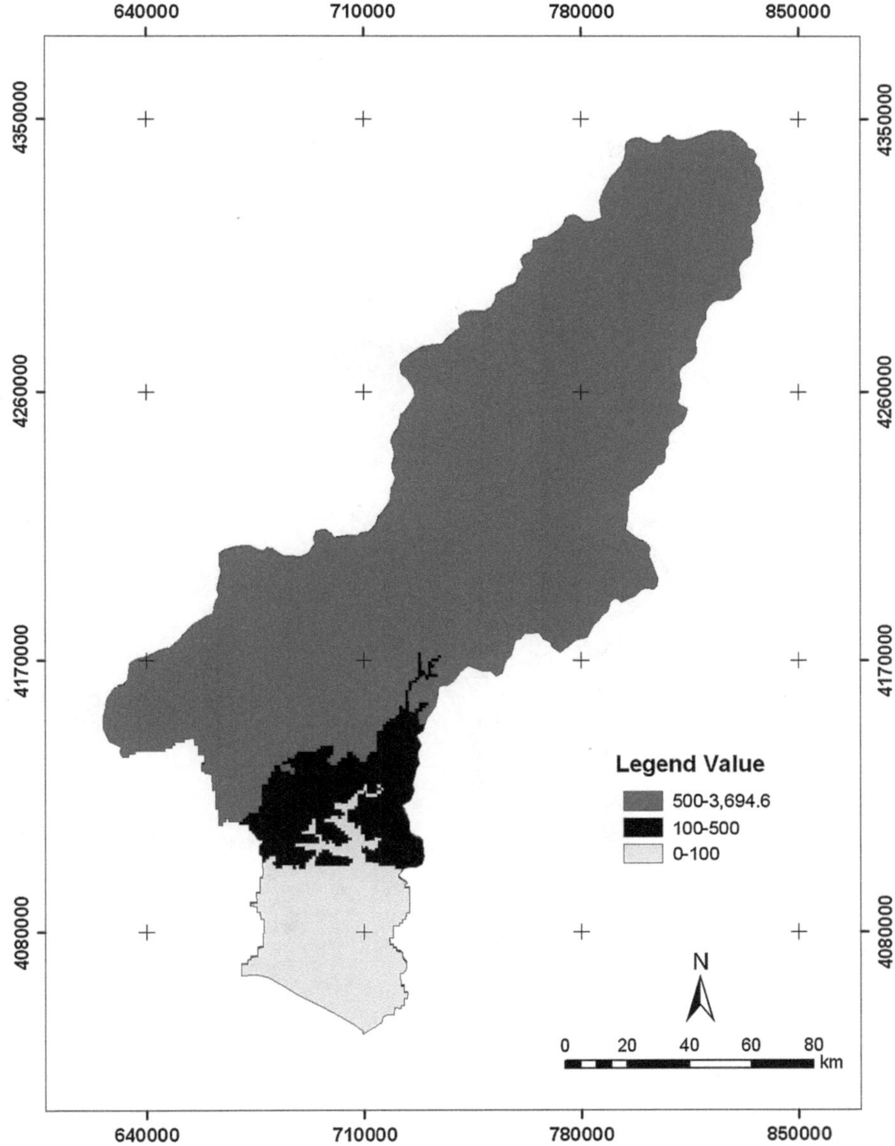

Fig. 1 The sub-regions of the Seyhan Basin (0–100 m downstream, 100–500 transition, 500–3694.6 m upstream)

Impacts of misuse of the natural resources (land, water, vegetation) and loss of the heritage in land management (indigenous crops, cultivation and animals) are mainly related to the population pressure and the migration of people from rural areas to the urban i.e. from the upstream to the downstream in an intra basin as well as an inter

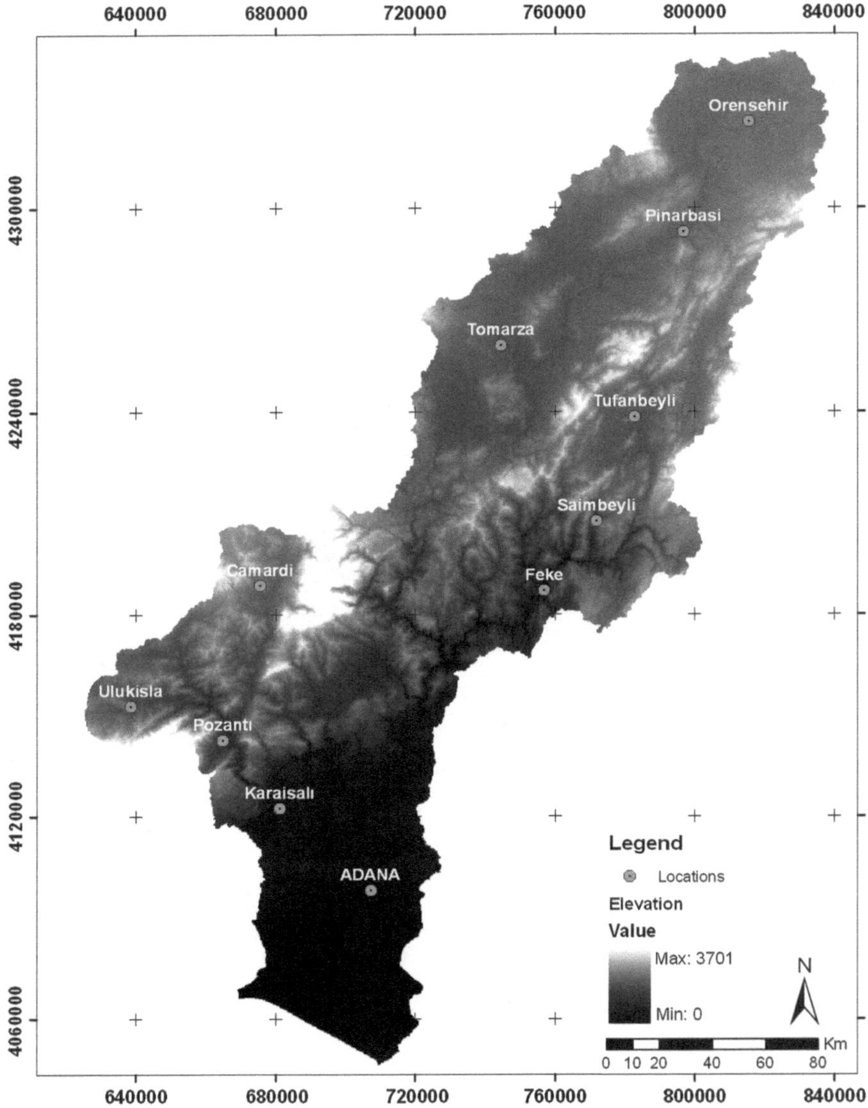

Fig. 2 Town centers/Townships of the Seyhan/Adana Basin (Karataş, Karaisalı, Pozantı, Çamardı, Ulukışla, Yahyalı, Develi, Sarız, Saimbeyli, Tufanbeyli, Feke)

provincial scale. This process leads to the degradation of the prime lands adjacent to the settlement areas and the rural areas by abandonment (Cangir et al. 2000; Zucca et al. 2006).

The "Anthroscape" context, as it embraces the components of the integrated environment, bears significance in assessing human-induced land degradation in variable basin scales, where the major negative changes (the driving forces-D of the matrix developed by the EEA (2002), i.e., the Driving force-Pressure-State-

Response matrix: DPSIR) lead to declines in land quality. Thus, understanding the soil-landscape relationships on Anthroscapes is a prerequisite for addressing land degradation and desertification, especially in line with the "Strategy" requirements and operational objectives of the United Nations Convention to Combat Desertification (UNCCD) and its bodies, namely the Science and Technology Committee and the Global Mechanism, within the framework of the 10-year Strategy Plan sought for the implementation of Sustainable Land Management. Conventional descriptions and analyses of soils may not demonstrate the subtle changes in soil properties and functions and consequently, the concept of Anthroscapes is generally confined to situations where marked differences or deviations from the normal, natural landscapes are observed.

The interim report of this project illustrated that the anthroscape model is feasible in understanding of the underlying processes and the data leading to further stages of the implementation of this context to the spatialized quantification of the causes and effects of improper use of resources. The anthroscape based approach, which is the main issue of this chapter, is expected to link land use change impacts according to agricultural land use management practices with potential causes and their magnitude of policy drivenness. We, thus, foresee having a thematic and spatial analysis frame, where the anthroscape will become a tool for optimized assessment and planning.

2 Materials and Methods

2.1 The Study Area

The Seyhan Basin (SA) is 2.7% (2,106,304 ha) of the total land area of Turkey and is located between 36°33′–39°12′N longitudes and 34°24′–36°56′E latitudes (Fig. 3). The basin is surrounded by Konya, Kızılırmak, East Mediterranean, Ceyhan and the Euphrates basins. The climate of the area is Mediterranean with mild rainy winters and dry hot summers with minor continental areas in the north margins of the basin.

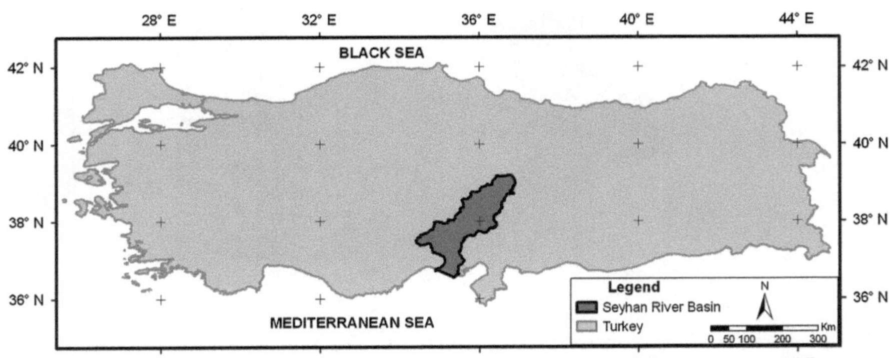

Fig. 3 The Seyhan River Basin/Anthroscape

The annual precipitation varies from 350 to 1,500 mm with a mean annual temperature of 13.1°C. Cultivated crops dominate the plains, whereas the highlands are mainly allocated for forestry (Fig. 4) and animal husbandry. The upstream of the

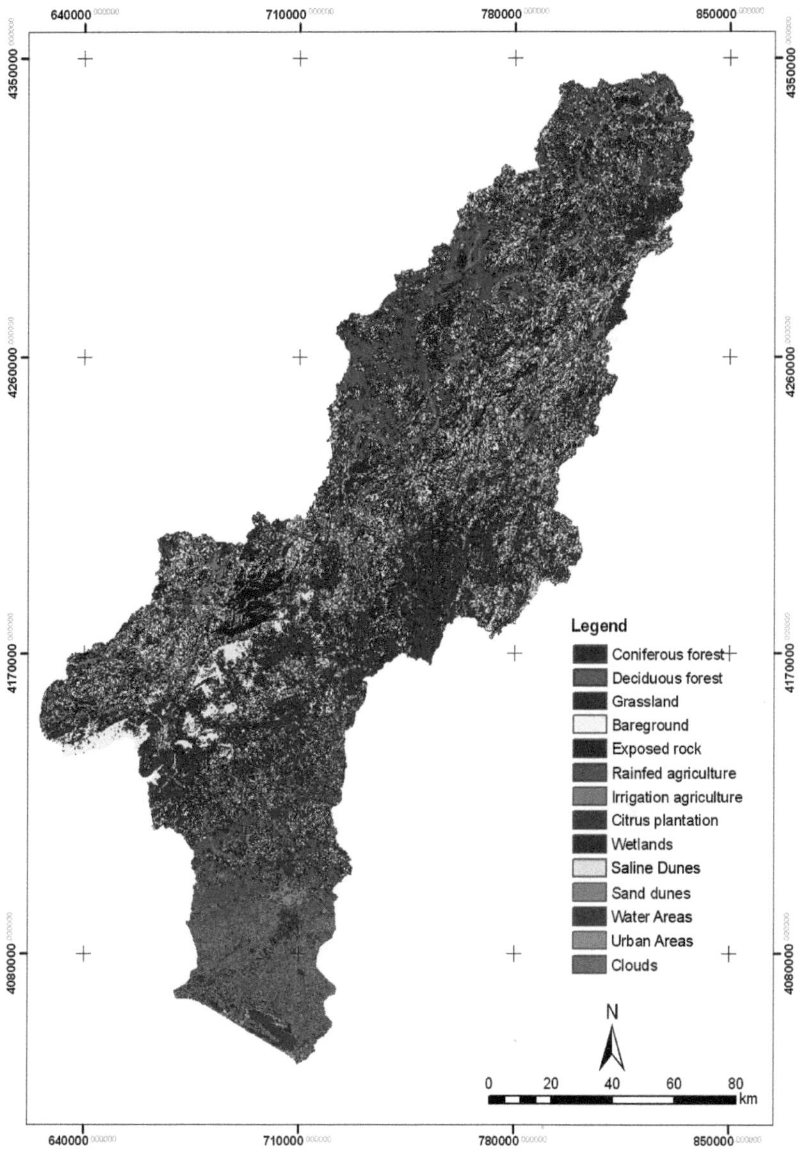

Fig. 4 Vegetation of the Seyhan Basin (http://www.wrrc.dpri.kyoto-u.ac.jp/~tanaka/ICCAP/SurfaceParameter.html)

basin is a mountainous area and an undulating terrain ranging from 1,000 to 3,000 m (Fig. 5), which is responsible for the complex river network (Fig. 6).

The length of the Seyhan River, flowing to the Mediterranean Sea, is 560 km, comprising a number of tributaries. There are a number of water bodies (five Hydroelectric power plants/dams and ten irrigation ponds) along the main course

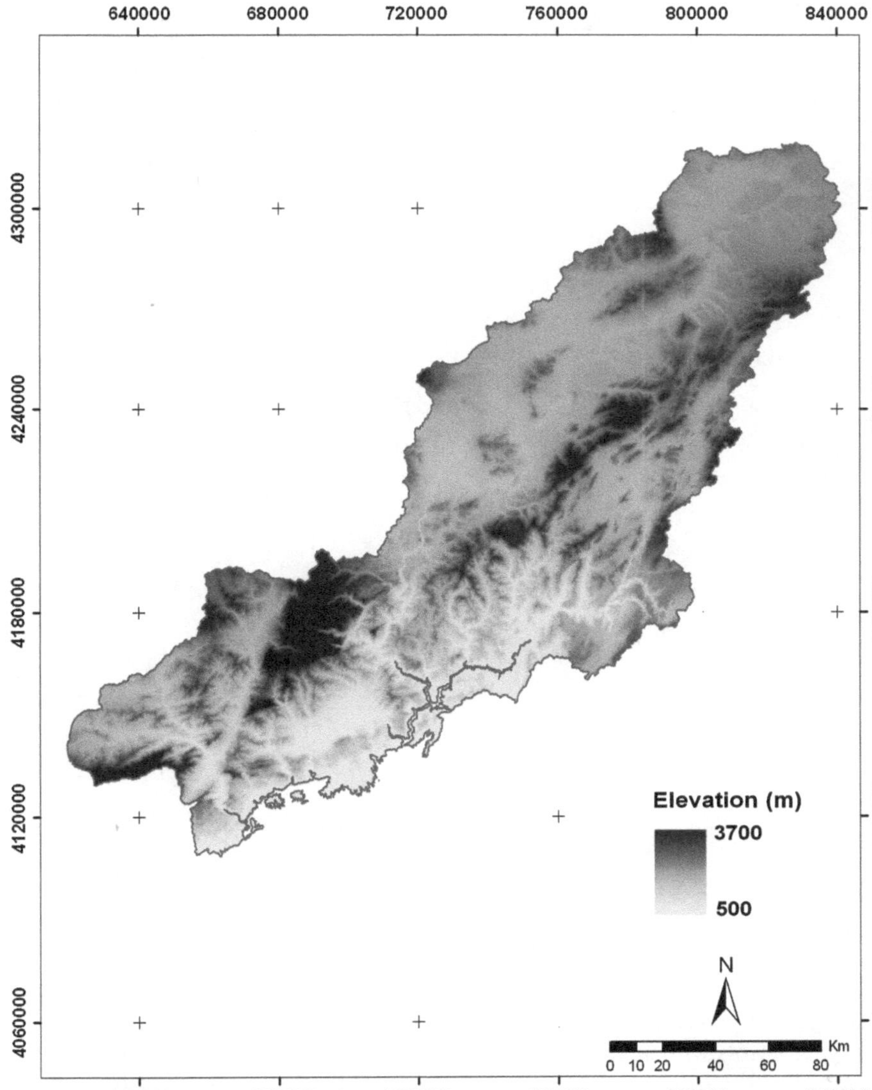

Fig. 5 The mountainous area and an undulating terrain of the study area

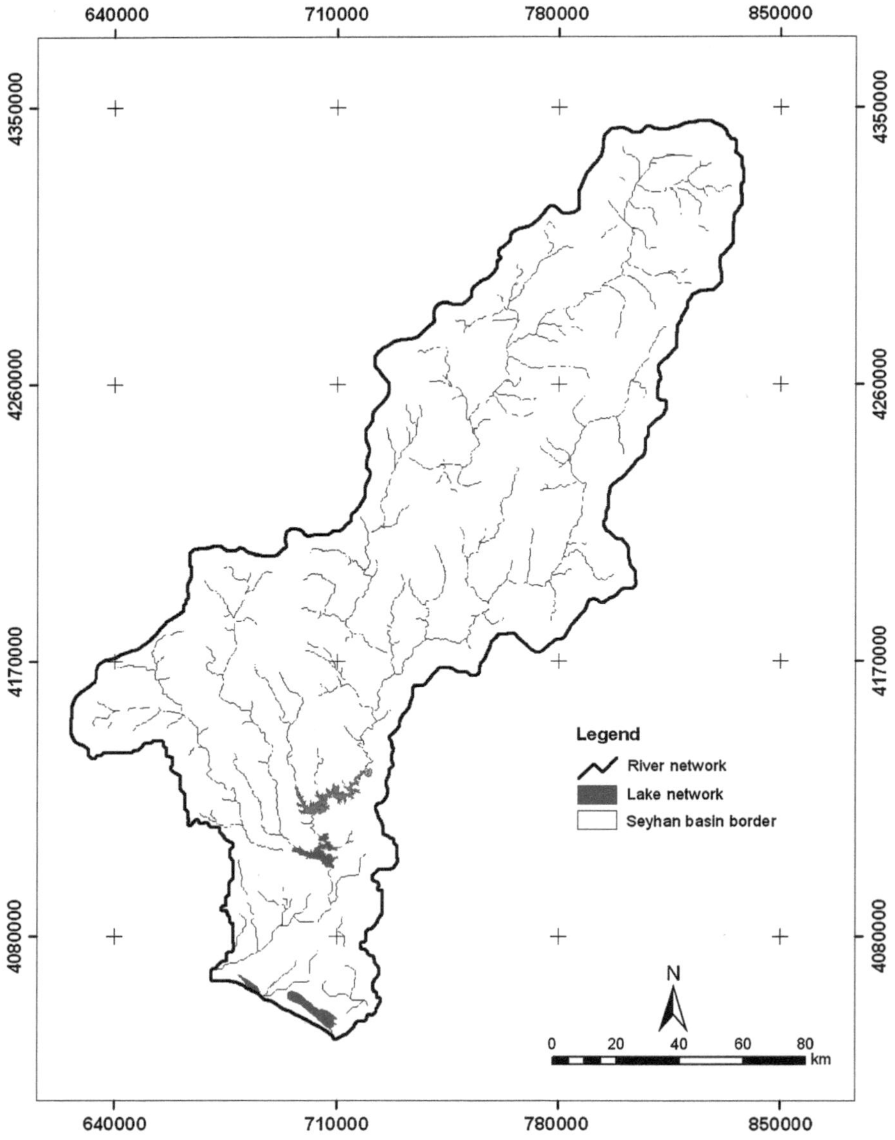

Fig. 6 The River Network of the Seyhan Basin (SHW 2005)

of the Seyhan River and its tributaries. The soils of the Seyhan Basin are dominantly Leptosols with natural vegetation and forests in the upstream and Fluvisols/ Vertisols with various crops in the downstream (Fig. 7). Many soil types were developed in the basin according to different parent materials and geo-morphological surfaces as well past and present climatic conditions.

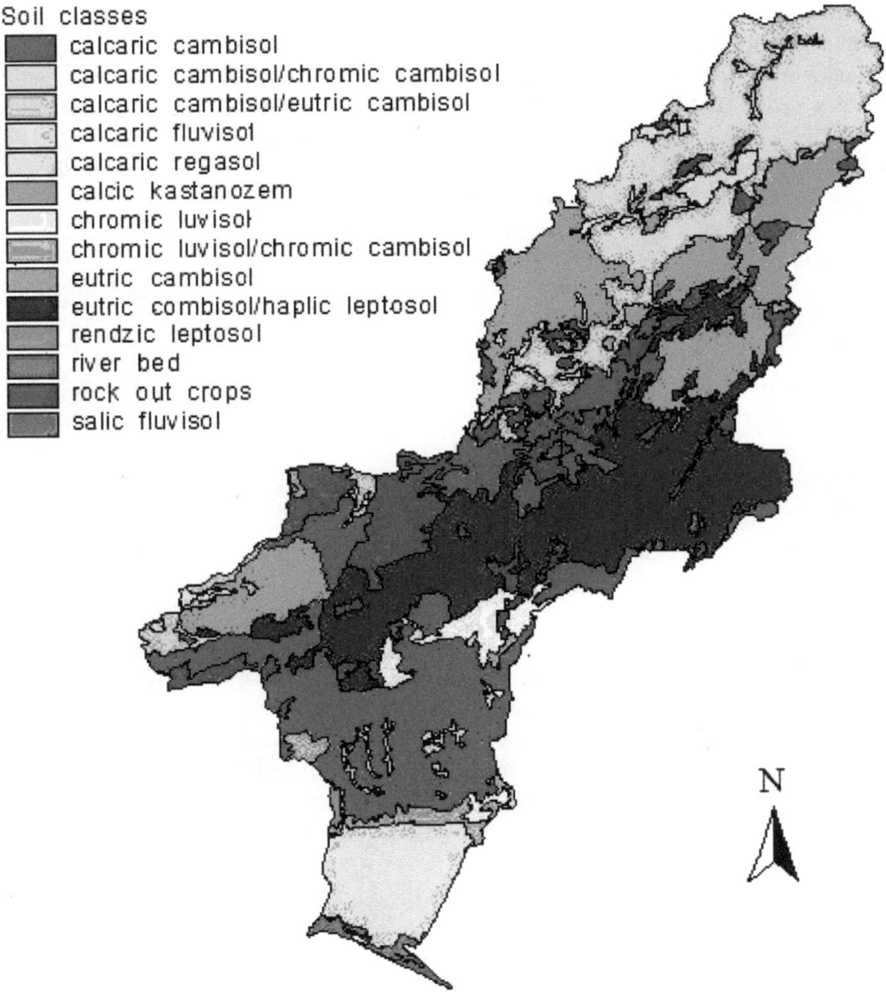

Fig. 7 The soils of the Seyhan Basin (1:200,000) (Toprak Su, 1974)

2.2 Driving Forces and Impacts of Misuse

The driving forces and their impacts determined in the region will be employed to the development of the thematic maps designating the allocated sites for different land uses including agriculture, forestry, biodiversity, and integrated settlement areas as satellite towns with specific functions in production, e.g. a satellite town, since the Roman period, allocated for olive groves and for olive oil production (the example of Salento in Italy-sample site for an EU Project), but both connected and integrated for mutual income generation via a common and well selected market initially local shifting over to the national. This system of satellite towns seeks administrative and production complementarity's (the Multi-integrated satellite system developed by the ancient Romans in the Mediterranean coastal/karstic terrain zone – MEF 2005) that can be developed for different crops concerning the phases from the initial until the final phase-the manufactured goods. However, the determination of the ultimate settlement sites of the satellite towns should primarily follow a phase of a buffer settlement zone in proximity to metropolitan areas, which would merely be based on the original settlement sites preferred by the migrants based on their urgent needs. These sites would help in determining the permanent satellite towns yielding information on preferences of the migrants considered before and during the course of the migration process. (a) The problems of the town centers as the driving forces, to serve as the initial given data to activate solutions, are illustrated in Table 1 – in order of importance – and (b) the analysis of the problems/drivers/threats complemented by the opportunities available in Table 2 as SWOT (Strengths, Weaknesses, Opportunities, and Threats) analysis, to be implemented, for sustainable management programs developed for local conditions.

2.3 Socio-Economy

The majority of the inhabitants of the Seyhan Basin relying upon agriculture and middle to low standard jobs in industry,-apart from the appreciably high population bound to the high standard employment of the private enterprise and large-scale agricultural ownership, struggle with problems of low/insufficient income which was basically due to the high inflation and fluctuating economic conditions of the past. The economic fluctuations are more significance than the former following the recent economic policies (the shift to the new national currency and the following economic policies that sought to decrease the high rate inflation) in the country (Table 1).

The SWOT analysis as the layout of the drivers for the improper use of the land points out to the necessity of solving the specific problems particular to the Seyhan Basin with the available opportunities as the remedies. However the analysis lacks the particularities and the modes of implementation for the said precautions (Table 2). Thus, the approach of designating the LQCs, concerning the variable facets of sustainability as the base material in initiating an implementation program, for a management plan to complement the SWOT analysis will bear the power stated in the legislations to be mentioned in Sections 2.5 and 2.7.

2.4 Migration

The high migration from the east of Turkey, to the Seyhan Basin has drastically increased at a rate of 48% (Table 3) from 1960 to 1970 and 76% of the total

Table 1 Magnitude of problems in the Seyhan Basin

Issues	Downstream								Transition								Upstream							
	L. Ruminants	S. Ruminants	Cereals	Indigenous crops	Fodders	Apiculture	Fisheries	Vegetables–fruits	L. Ruminants	S. Ruminants	Cereals	Indigenous crops	Fodders	Apiculture	Fisheries	Vegetables–fruits.	L. Ruminants	S Ruminants	Cereals	Indigenous crops	Fodders	Apiculture	Fisheries	Vegetables–fruits.
Socio-economic																								
Migration	3	3	1	1	1	2	1	1	3	3	1	2	1	3	1	2	4	4	4	4	2	3	2	2
Lack of capital	4	3	2	2	1	2	2	1	4	4	2	2	2	3	2	1	5	5	4	4	3	4	4	3
Education	2	2	1	1	1	1	2	1	3	3	2	2	1	2	2	2	4	4	2	3	2	3	3	2
Low income	3	3	2	2	2	3	3	2	3	3	2	2	2	3	3	3	4	4	3	1	1	1	1	3
Production																								
Animal husbandry																								
Animal breeding	3	2	0	0	0	2	2	0	4	4	0	0	0	3	1	0	4	4	0	0	0	3	3	0
Studs	3	2	0	0	0	3	3	0	4	3	0	0	0	3	2	0	5	5	0	0	0	4	4	0
Artificial insemination	2	0	0	0	0	0	0	0	2	0	0	0	0	0	0	0	3	0	0	0	0	0	0	0
Shelters	3	2	0	0	0	0	0	0	3	3	0	0	0	0	0	0	5	4	0	0	0	0	0	0
Lack of fodder supply	2	2	0	0	3	0	0	0	2	2	0	0	2	3	0	0	2	2	0	0	1	0	0	0
Lack of grazelands	3	3	0	0	0	0	0	0	4	4	0	0	0	0	0	0	2	2	0	0	0	0	0	0
Lack of shepard	2	4	0	0	0	0	0	0	3	4	0	0	0	0	0	0	2	2	0	0	0	0	0	0
Lack of veterinary services	1	1	0	0	0	0	0	0	1	1	0	0	0	0	0	0	2	3	0	0	0	0	0	0
Low productivity	2	2	0	0	2	2	3	0	3	3	0	0	0	1	2	0	4	3	0	0	0	2	2	0
Lack of labor	2	2	0	0	1	1	0	1	3	2	0	0	0	2	2	1	3	3	0	0	1	1	2	0
Crop production																								
Low productivity	3	3	0	0	0	1	1	0	3	2	0	0	0	1	2	0	2	2	4	3	1	2	2	2
Low input	4	3	2	2	2	3	2	2	4	4	2	2	2	3	4	3	5	4	3	2	1	3	3	3
Mechanization	1	1	2	1	1	3	2	2	2	2	2	2	2	3	2	2	4	4	4	4	5	4	5	3
Seedling	0	0	0	0	0	2	1	0	0	0	0	0	1	2	2	1	0	0	2	2	1	3	2	1
Unused land	0	0	0	0	1	1	0	0	0	0	0	0	0	1	1	0	0	0	3	3	2	2	2	2
Erosion	0	0	0	0	0	0	0	0	0	0	1	1	1	1	0	0	0	0	3	3	3	2	3	2
Insufficient irrigated land	0	0	1	1	0	0	0	0	0	0	1	1	1	0	0	0	0	0	3	3	3	1	1	3
Land fragmentation	0	0	2	2	2	1	1	1	0	0	2	2	2	1	1	1	0	0	4	4	2	2	2	2
Water resources																								
Drainage	0	0	3	3	3	0	0	3	0	0	3	3	3	0	0	2	0	0	0	0	0	0	0	0
Inadequate use of water resources	2	2	1	1	1	2	3	1	2	2	1	1	1	2	3	1	4	3	3	3	3	3	4	3

(*continued*)

Table 1 (continued)

Issues	Downstream								Transition								Upstream							
	L. Rumi-nants	S. Rumi-nants	Cer-eals	Indi-genous crops	Fod-ders	Api-cul-ture	Fish-eries	Vege-tables–fruits	L. Rumi-nants	S. Rumi-nants	Cer-eals	Indi-genous crops	Fod-ders	Api-cul-ture	Fish-eries	Vege-tables–fruits.	L. Rumi-nants	S Rumi-nants	Cer-eals	Indi-genous crops	Fod-ders	Api-cul-ture	Fish-eries	Vege-tables–fruits.
Lack of regulation for water use	3	3	2	2	2	2	3	2	3	3	3	2	3	3	4	2	5	5	4	4	4	4	4	4
Extension																								
Lack of extension services	4	4	3	3	3	3	3	4	4	4	3	3	3	4	4	3	5	5	4	4	4	3	4	3
Lack of extension to women	3	3	2	2	2	2	3	2	3	3	2	2	2	3	3	2	5	5	5	5	4	5	5	4
Lack of organization																								
Lack of information on organization	1	1	2	2	2	1	1	1	2	2	3	3	3	2	2	2	4	4	3	3	3	2	2	2
Agricultural legislations																								
High credit rates	4	4	2	2	2	2	3	2	4	4	2	2	2	3	2	2	4	4	2	2	1	3	3	2
Lack of subsidies	3	3	2	2	3	2	2	1	3	3	2	2	2	1	2	1	4	4	1	1	1	2	2	1
Marketing																								
Lack of marketing structure	2	2	1	1	1	1	2	1	3	3	1	1	1	0	1	2	4	3	2	3	1	1	2	1
Quality and standards	3	3	2	2	2	2	2	3	3	3	2	2	2	3	2	2	3	3	4	4	2	2	2	1
Lack of product processing units	2	2	1	1	2	3	4	2	3	3	2	3	4	3	4	3	5	5	5	5	5	5	3	4

Numbers indicate the relative risk in a descending order

5: very high, 4: high, 3: moderate, 2: few, 1: NA (Governorate of Adana 2005)

Table 2 SWOT (strengths, weaknesses, opportunities, and threats) analysis of the development plan (Governorate of Adana 2005)

Aims (amaçlar)	Strategy (ne olmalı nasıl olmalı)	Current state and advantages (mevcut durum ve üstünlükler)	Bottlenecks (darboğazlar-zorluklar)	Opportunities (fırsatlar)	Threads (tehditler)
Productivity: animal production	Selection of indigenous animals and breeding	Basin is ecologically suitable for various type of animal husbandry	Low insurance cover	Basin highlands are suitable for grazing	Animal diseases are common
	Improving feeding, maintenance, and nutrition conditions	High demand to animal products	High fodder costs	Livestock exchange market	
		Intensive artificial insemination	High animal product imports	Intensive studies on combating diseases	
	Establishment of effective marketing, organization system	Infrastructure is suitable for mass production	Climatic conditions (humidity and temperature)	Several research projects are undertaken by local universities	
	Development of agriculture based industry	Number of small (ca. 500,000) and large ruminants (ca. 300,000) are high	Lack of animal transportation control	Presence of an Animal Disease Institute in the basin	
	Preventing overgrazing of grazing lands	Suitable coastal zone is for fisheries	Low market rates	Stable market prices	
	Obtaining studs	Capacity of enterprises are low	Habit of feeding with feed grains		
	Increasing number of animals in each establishment	Marketing organizations are not sufficient	Fluctuating milk prices		
	Preventing slaughtering of young animals	Sheltering is not sufficient	High number of family enterprises		
			Insufficient grazing land amelioration studies		
			Fragmentation of grazing lands		

(*continued*)

Table 2 (continued)

Aims (amaçlar)	Strategy (ne olmalı nasıl olmalı)	Current state and advantages (mevcut durum ve üstünlükler)	Bottlenecks (darboğazlar-zorluklar)	Opportunities (fırsatlar)	Threads (tehditler)
Plant production	Seedlings suitable and certified for use in basin	Basin is ecologically suitable for various type of plant production	Drainage and irrigation problems	Great indigenous variety of vegetables and fruits	Natural hazards (floods, mudflows, earthquakes)
	Improper use of agricultural inputs	Marketing problems	Insufficient mechanized agriculture	Micro-ecological maps should be produced	Insects and weeds
	Irrigation of potential irrigable lands and amelioration of lands with drainage problems	Chance of testing new varieties and species	Harmful Insects and weeds	Fodder seeds should be produced	Fluctuating market prices in the country and globally
	Land use:	Production according to EU standards	Low sapling production	Presence of organic agriculture	
	Decreasing fallow lands	Rich natural flora	Marketing and organization problems	Agricultural Research Institute and University undertakes several scientific research	
	Put into use of potential arable lands	Close to export centers	Lack of quality control mechanism	Presence of product processing facilities	
	Preventing land fragmentation	Intensive production of vegetables, citrus and cotton	Excess use of agro-chemicals	IV. Class soils are suitable for agro-forestry and olive production	
		Lack of sufficient education and information on production	Lack of standardization	Rain fed and irrigated agriculture potentials	
		Reducing facility and field capacities	Insufficient legislations		
		Natural flora on coastal zones	Lack of coastal zone destruction control		

		Presence of natural soil productivity enhancers	Low soil organic matter		
			Misuse of I–III. Class soils		
			Land fragmentation		
Fishery products	Education of stake holders for environmental friendly practices	Suitable for marine fishing and ponding	High fodder costs	Water resources are suitable for fishery products	Natural hazards
	Establishment of integrated fishery production and processing facilities	Presence of lagoons, fish traps and dam lakes	Low fish production	Opportunities of cooperation among university, governmental bodies and private enterprises	Environmental pollution
	Decreasing cost of fodders	Local export companies experiences	Insufficient fish processing facilities	Land resources are potentially suitable for various productions	Hunting
	Allocation of water bodies for fishery products	High demand from the country and abroad	Production is not in compliance with EU standards		
		Fisherman shelters (marinas and ports) are not suitable	Pollution of water resources		
		Lack of information and application of new technologies	Marketing		
			Fluctuation of lagoon water levels		
Increasing agricultural income	Improving organization and system of marketing	Presence of numerous crop types	Less number of cooperatives	Product quality is generally suitable for marketing	
		High export potential	Insufficient legislations	Easy cooperation among producers, exporters, university and governmental bodies	
		Presence of stock exchange, agricultural products office and unions	Standardization of production	Transportation and packaging problems will be easily overcome	

(continued)

Table 2 (continued)

Aims (amaçlar)	Strategy (ne olmalı nasıl olmalı)	Current state and advantages (mevcut durum ve üstünlükler)	Bottlenecks (darboğazlar-zorluklar)	Opportunities (fırsatlar)	Threads (tehditler)
		Current animal and crop production and processing facilities	Lack of collaboration among governmental agencies, locals and NGOs		
		Low capacity use of marketing network	Insufficient information on new crops within the farmers associations		
		Lack of expertise services	Lack of information and education services		
		Lack of trademark and promotion studies	Production without planning Production is not globally competitive		
Food security	Improving control and inspection services	Increase of irrigation lands due to high production capacity	Insufficient controlling bodies in food production	Legislation no 2090 provides funds to farmers facing natural hazards	
	Quality and standard production	Limitations on food storage due to the high temperature and humidity	The production insurance system should be extended Supporting the establishment of large production facilities	Suitable lands for the establishment of large production facilities	
Sustainable agriculture	Improving and supporting ecological agriculture	Variety of soil, flora and fauna types	Insufficient information on organic agriculture	High income potential from relatively small lands (<1 ha)	
	Use of lands according to their capabilities	Decrease in the use of agro-chemicals	Marketing and storing problems	Long vegetation period for vegetables	
	Combating erosion	High demand for organic products	Insufficient packing facilities	Potential lands for organic production of a wide variety of crops	

	Improper use of natural resources	Intensive production	Drainage, salinity and irrigation problems	Protection of the environment and the soils should be undertaken
	Pollution due to excess use of agricultural chemicals	Suitable ecological conditions	Expensive investment for purification facilities	New research should be undertaken for alternative products (rape, etc.)
	Exploitation of soils for brick and ceramic production	Presence of NGOs	Land fragmentation	
	Competition among agricultural facilities	High entrepreneurial spirit	Unwillingness of private sector cooperation	
	Encouraging greenhouse production on suitable sites		Lack of legislations Lack of modern technology	
Domestic and external trade	Organization of producers	High demand	Insufficient Organization	High production for external demands
	Standardization of quality	Great variety of plant pattern	Production not standardized	Availability of vegetable suitable for industrial use (canned foods, etc.)
	Supporting high yielding crops	Cropping is possible at all seasons due to climatic conditions	Importing should be limited	Suitable infrastructure for the market
	Supporting export	Means of transportation	Fluctuating product prices	
		Continuous production of MAJOR crops (cotton, olives)	Lack of legislations	
		Suitable conditions for large investments	Insufficient network between producers and consumers	
			Insufficient product processing and packing facilities	

2,000,000 (2000 census) throughout the basin, with the rest 24% living in the rural areas (Fig. 8) due to the mass migration of the poorer inhabitants of the eastern and southeastern rural areas of the country in search of jobs in fertile plain of the irrigated plain (Fig. 1). This has expanded the urban occupation in expense of the middle and lower parts of the fertile soils of the delta. The rush and consequent permanent settlement of the migrants has exceeded "the settlement-urban/rural-farmland and the natural environment" ratios stated in the legislation of settlement and housing. This is the ratio of land occupation by structures vs. natural resources developed for the sustainable management of the land and stands for a horizontal and vertical permit for structures. This permit is for maximum two floors, bound to local settlement habits, where the first floor of the structure is advised to be used as animal shelters and heating purposes via the natural body warmth of the animals

Table 3 The population increase at town centers of the Seyhan Basin from 1960 to 1970

Towns	1960	1970		Change in 10 years
		Towns	Rural	
Adana (Center)	280,658	351,655	415,975	48
Karataş	28,813		39,843	38
Karaisalı	44,531	1,690	48,565	9
Feke	16,054	3,930	20,100	25
Pozantı	12,310	3,675	13,486	10
Tufanbeyli	17,863	3,825	20,465	15
Saimbeyli	11,507	2,676	13,893	21
Çamardı	14,703	2,721	17,563	19
Ulukışla	25,495	5,920	30,208	18
Pınarbaşı	48,868	8,263	51,619	6
Sarız	21,150	3,060	23,486	11
Tomarza	32,660	4,828	38,153	17
Develi	49,749		58,334	17
Yahyalı	23,276		30,600	31

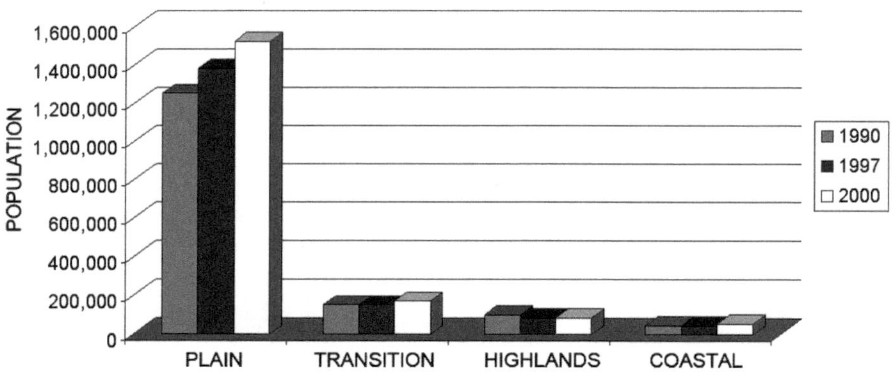

Fig. 8 The population distribution of the Seyhan Basin (TÜİK 2000)

and the second floor is the living premises for people. However the local Municipalities and the Regional Directorates of Settlement and Housing are empowered to determine/modify the ratios stated in the law mentioned above according to local needs – the economic and social trends – for each demographic section of Turkey based on the properties of land, vegetation, crops and industrial trends. This consequently caused the substitution of the locals to the migrants in the plain increasing the degradation of the soil by excess and improper cultivation and irrigation practices. The much lower amounts of increase in population in the highlands documents the contemporary urge, developing among the younger generations of rich farming background, to move to larger western/southern Turkish towns, to invest to the ever growing industries (transportation, textile, and electronic). The recently increased farm inputs, especially for fuel and fertilizers, which jeopardize the economic balance between the inputs and outputs, is also one of the causes of the farmers to sell their land to others which are more eager to cultivate the plain i.e. the landlords as well as the immigrants (Table 3).

2.5 *Fragmentation*

Fragmentation is a direct driver connected to

(a) The conversion of land to smaller ownerships/deeds via heritage to family members
(b) Selling of land to migrants, especially in the downstream part of the basin

And an indirect driver concerning

(a) Soil sealing undertaken as the construction of communal infrastructure as roads, dams and new irrigation network
(b) Energy/power cable network prohibiting cultivation and reforestation in the vicinity
(c) Shifts in river beds due to changes in their flow rates bound to the changes in land use, drainage system, climate change and groundwater depth changes

However the state, as a precaution to land fragmentation and as a regulation under the "Legislation for Land Consolidation" prohibits the concerned land heritage divisions, at lands of prime agricultural value, by advising the parties to turn over their properties to their counterparts, thus keeping land use rights at limited number of ownerships and preventing the abrupt increase in prices of land and decline in total yields on smaller parcels. The upstream land abandonment by the younger generation seeking for better job prospects in the metropolitan area of the basin has, in a way, decreased the pressure of overgrazing of the improper areas in the forests and communal pastures (despite the continuation of the goat grazing activities by the elders at some easily accessible sites by them), whereas it has caused the improper state of land use upon the prime soils adjacent to Adana City (Table 2).

2.6 Natural Resources

The trends concerning the conventional activities inducing the increasing rates of intensive agriculture is the prime factor threatening the sustainable use of the natural resources, and thus, the lack of the paradigm shift to sustainability, which is particular to specific areas and cultures. Accordingly, the SWOT analysis performed in this report has helped in identifying the drivers of mismanagement along with the approach of the DPSIR matrix of the EEA (2002). Some of the drivers employed to prepare the LQC maps of the basin with particular land allocations are mentioned in the following and in Sect. 2.4.

Fire hazards and improper exploitation of the timber products are the main causes for the decreasing of the forest cover in the area. The following threat to forest resources is the tradition of seasonal migration to highlands i.e. cooler spots, in need of extra space in the woodlands for permanent settlement structures.

Degradation of the biodiversity by the unplanned use of the wetland products via cutting down of reeds, extraction of beach and dune sands, fishing, hunting, cultivation of cash-crops (water melon, peanuts, vegetables requiring high water demands and fertilizers especially nitrogen – applied as 300 kg/ha which is about twofold over the current EU level of N use), and building second houses as summer resorts on the beach and the calcrete ridges (that have been allocated for indigenous crops and trees such as olives, vineyards, figs and graze lands over the millennia) located on the mainland boundaries of the lagoons as well as the boundaries of the downstream with the transition to the upstream.

Degradation of the water balance/the hydrologic cycle of the wetlands due to excess water use in the upstream and the downstream together with the increasing amounts of pesticides and fertilizers in the latter along with the improper use of the natural habitat-fauna and flora – around the lagoons and sand dune areas of the delta (Figs. 9 and 10).

2.7 Environmental Problems (Improper: – Land/Soil, and Water Use and – Crop Selection and Animal Production)

The present pollution of the surface waters and the soils caused by the municipal wastes of the Adana city (5 km north of the town center) is an urgent health threatening and environmental conservation issue for the region requiring high priority measures of precaution, which are compulsory for EU accession policies. The dumping ground of the city is located at an improper site, which was selected

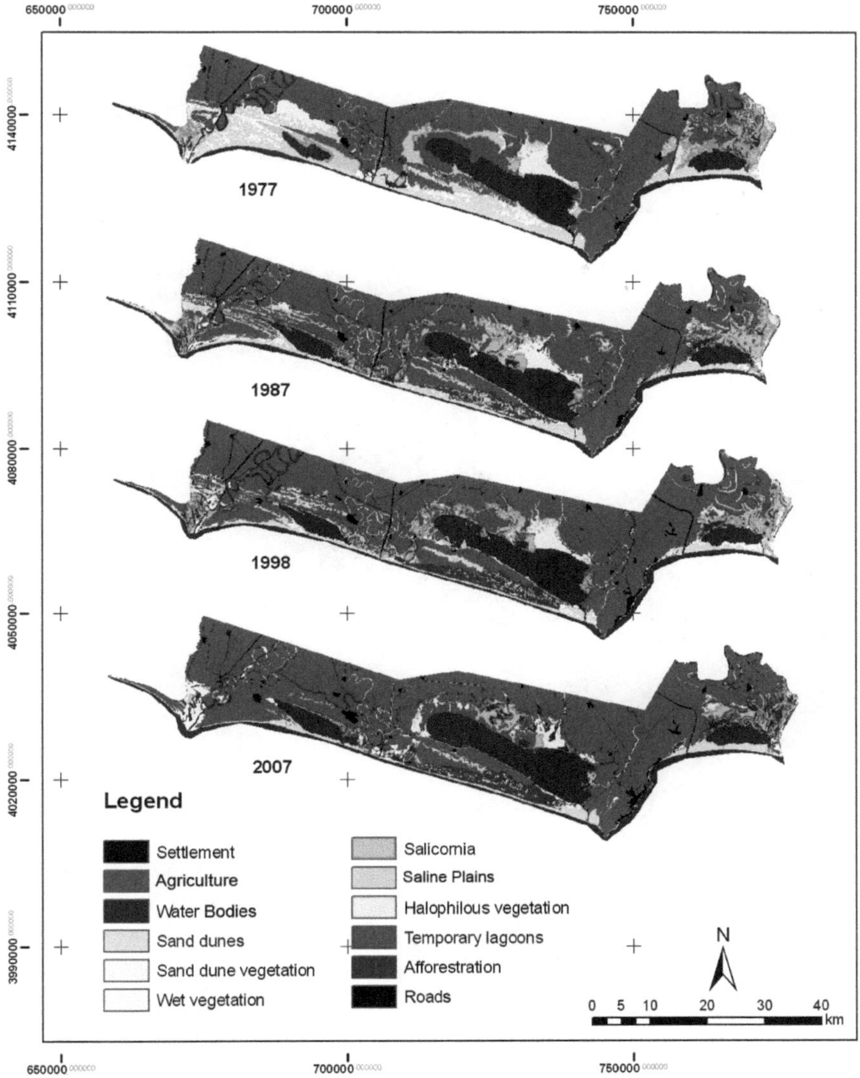

Fig. 9 Land use/cover classes of the Cukurova Delta from 1977 to 2007

only for its high elevation to the town and the so-called underlying porous rock (calcrete), despite the fact that the calcrete and the unique agro-ecosystem it underlies is, and has been since millennia, part of the indigenous (since the Early Roman period) anthroscape allocated for olive/carob/vineyard/fig production.

The pollution of the soil and surface waters caused by the excess use of herbicides, insecticides and fungicides in both the up and downstream at especially

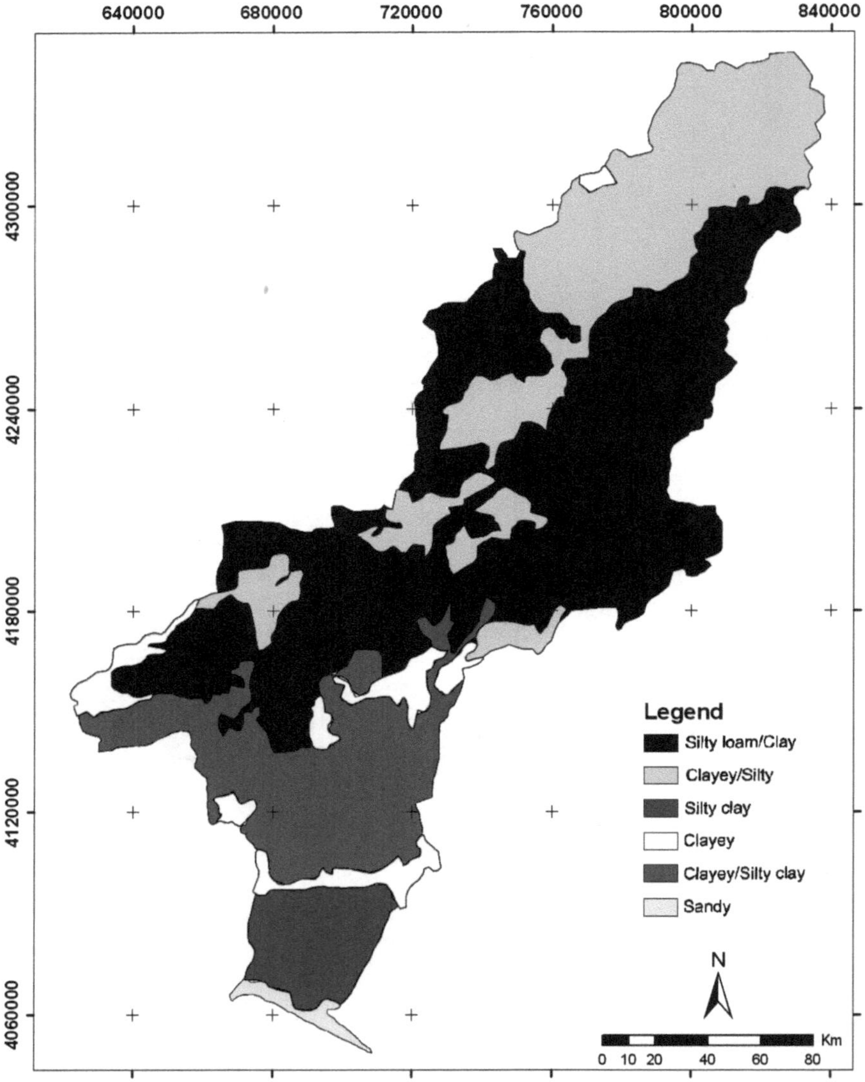

Fig. 10 Soil texture classes of the Seyhan Basin

the maize fields irrigated by shallow to deep groundwater wells in the former and at cotton, maize, wheat, peanut, and water melon fields as well as citrus orchards in the latter, dominant in clayey soils, and irrigated by water distributed from an extensive canal network (Fig. 6).

Sand dunes of the lagoon area in the downstream are leveled for cultivation of cash-crops (water melon, peanut, melon, tomato) requiring high amounts of agricultural chemicals which in turn increase eutrification of the marine ecosystem (Fig. 9).

An equally hazardous driving force to the agricultural chemicals is the excess and improper use of fertilizers especially in the downstream where high plant nutrient demanding crops, such as maize and increasing fields of cotton are widespread. The excess use of "P" fertilizers causes the accumulation of cadmium in the soils, whereas the excess use of the nitrogen fertilizers is responsible for the pollution of the shallow to deep groundwater wells and reservoirs in especially coarse textured soils (Fluvisols and Cambisols) of the south of the Seyhan Basin (Fig. 7).

The lack and/or insufficiency of the waste water purification plants in the coastal part of the basin is responsible for the pollution of the unique Mediterranean Sea leading to the decline of the ecotouristic quality of the coasts and eventually the loss of biodiversity and wildlife along with the conversion of the previously successfully anchored sand dunes to watermelon and peanut fields in need of high amounts of fertilizer and pesticide applications for the former and fertilizers for the latter (Fig. 9).

The unfiltered emissions of the industry as the driver of air pollution as well as the loss of the prime land for their infra and ultra structures, with the parallel development of the ever increasing soil sealing/improper urbanization (Fig. 11), especially in the downstream, due to the migration from the eastern part of the country , seeking for better job prospects are the drivers of the loss of the

Fig. 11 Urban development through the northern fringe of the city (1998) (Alphan 2003)

non-renewable natural resources – the improper use of the soils – and the increasing evapo-transpiration to follow, which in turn is responsible for the short and long term climate changes (Figs. 12–14).

The degradation of the soil structure by improper and excess tillage practices throughout the basin, especially in the downstream, where the so-called cash-crops are cultivated demanding sophisticated mechanization and water are among the major land degradation drivers, which would shortly lead to a status of declining crop yields followed by an unexpected poverty. The urgent need for the preservation of the indigenous cotton and relevant field crops like sesame and orchards of citrus, with sprinklers and drips for irrigation as factors of renovation, as well as allocation of rain fed crop to sites of long standing traditional use of Mediterranean crops, to sites in the transitional zone with calcretes, would be the remedy of land degradation and the declining economical situation.

Finally the lack of the integrated implementation of the available legislations and regulations issued earlier, such as the, "Protection of the soils and land management" – The Ministry of Agriculture and Rural Affairs, "Protection of the

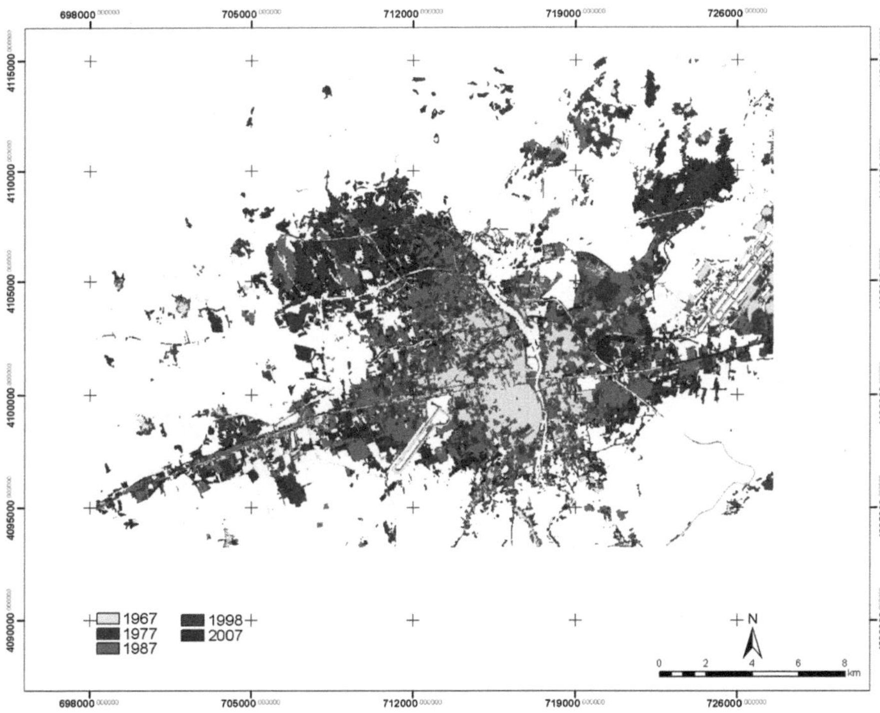

Fig. 12 Urban expansion of Adana and the surrounding areas from 1967 to 2007 (Dinç et al. 2001)

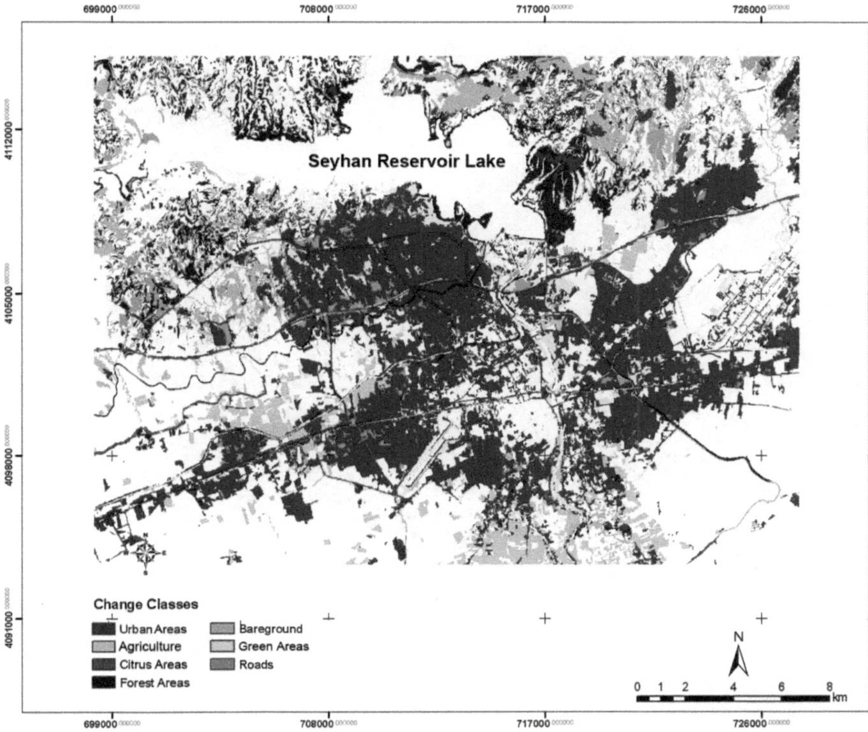

Fig. 13 Land use/cover changes of Adana city from 1967 to 2007

Environment," the "National Action Programme for Desertification," the "Law of Forestry" – The Ministry of Environment and Forestry, the "Law of Public Works and Settlement" – The Ministry of Public Works and settlement, the "Regulation on water pollution and earthquakes," together with the delayed empowerment of the Ministry of Natural Resources and Energy on the legislation concerning the Sustainable Use of Water, i.e., "The Water Law," have been responsible for the major part of the improper use of the natural resources in the country. The relevant shortcomings, are hopefully, sought to be overcome by the development of the recent activities and plans undertaken for the implementation of the European Union and un regulations of the Conventions to Combat Desertification (UNCCD), ON Biological diversity (UNCBD) and Climate Change (UNFCCC) to be adapted all aiming the sustainable use of the land and the water as united entities (efforts on the synergy of the three conventions are to be finalized by January 2010) also by considering the rich heritage of the ancient lands, namely Anatolia and the riparian countries of the Mediterranean Basin.

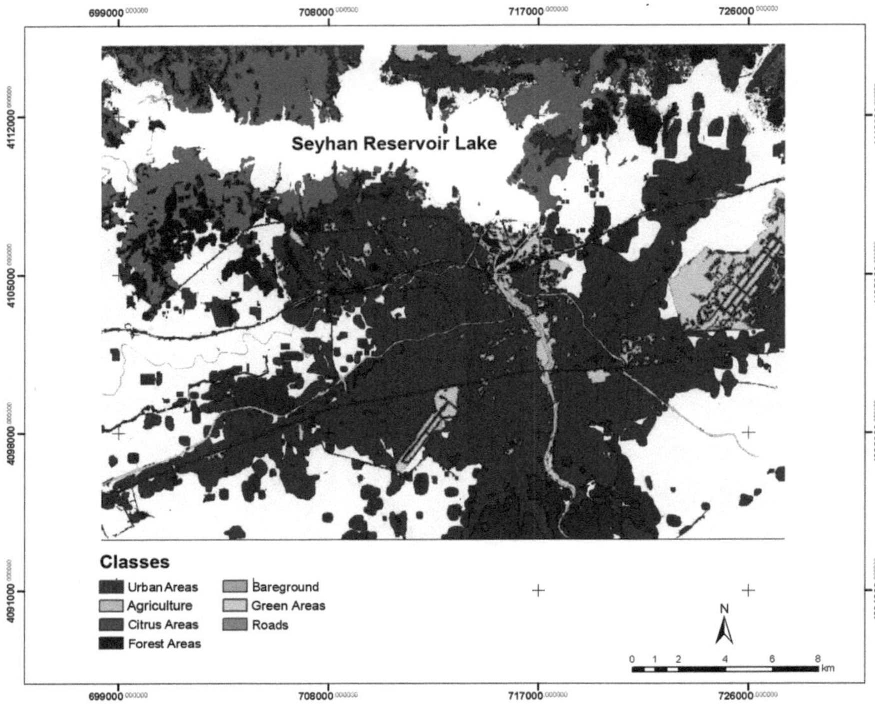

Fig. 14 Future urban development in Adana city by 2023

2.8 *Quantification of the Anthroscape*

The misuse of the land/soils, analyzed by the SWOT and the DPSIR matrix, disregarding their "Land Quality Classes – LQC," which are the modified contemporary versions based on the "Land Capability Classes – LCC" of Klingebiel and Montgomery (1961), are based on 25 stress classes/constraints/drivers (especially the nationwide soil sealing, and improper use/allocation of land including the Adana context) and IX land quality classes of land/soil sustaining "grain production" – the major crop for human sustenance and indicator of potential power to survive against the odds of nature and human induced hazards – as stated by Beinroth et al. (1999) and earlier by Karlen and Stott (1994) (Table 4). Land quality may be defined as the ability of the land to perform specific functions without becoming degraded. It should be noted that this definition contains an element of time as it implies the sustainability of performance functions. This is a marked departure from previous definitions, which consider land quality as a static attribute. The previous understanding, of soils/land as being sites/shelters of mere intensive crop production, has led to the loss of prime soils as well as the agro-ecosystems of unique indigenous value obtained/renovated (the Mediterranean

Table 4 Properties of land quality classes (Beinroth et al. 1999)

Land quality class	Properties
I	This is prime land. Soils are highly productive, with few management-related constraints. Soil temperature and moisture conditions are ideal for annual crops. Risk for sustainable grain crop production is generally <20%
II, III	The soils are good and have few problems for sustainable production. The lower resilience characteristics of Class II soils make them more risky, particularly for low-input grain crop production. However, their productivity is generally very high and consequently, response to management is high. Conservation tillage is essential. Risk for sustainable grain crop production is generally 20–40%
IV, V, VI	If there is a choice, these soils must not be used for grain crop production, particularly soils belonging to Class IV. All three classes require important inputs of conservation management. Lack of plant nutrients is a major constraint and so a good fertilizer-use plan must be adopted. Productivity is not high and so low-input farmers must receive considerable support to manage these soils or be discouraged from using them. Land can be set aside for national parks or as biodiversity zones. Risk for sustainable grain crop production is 40–60%
VII	These soils may only be used for grain crop production if there is a real pressure on land. They are definitely not suitable for low-input grain crop production; their low resilience makes them easily prone to degradation. As in Classes V and VI, biodiversity management is crucial in these areas. Risk for sustainable grain crop production is 60–80%
VIII, IX	These soils belong to very fragile ecosystems or are very uneconomical to use for grain crop production. They should be retained under their natural state. In Class VIII, which is largely confined to the tundra and boreal areas, timber harvesting must be done very carefully with considerable attention to ecosystem damage. Risk for sustainable grain crop production is >80%

agro-ecosystem of the Hittites, Greeks, Romans, Seljuks and the Ottomans) by the numerous societies that have intermingled over the millenia.

The constraint-free classes of I, II and III are suitable for most agricultural uses, whereas classes IV, V and VI require considerable management inputs and conservation practices. Classes VI, VII, VIII and IX include the extremely dry, wet, cold, or steep areas, i.e., the deserts and specific agro-ecosystems, and thus the lands unsuited for sustainable grain production as stated in general by Beinroth et al. (1999). However, these are the threatened land pieces under constraint, that when modified according to the particular attributes of Beinroth et al. (1999) (crops/soils/topographies-geomorphic units/geological-parent materials) together with the concepts of the agro-ecosystem/anthroscape and the indigenous knowledge stated by Eswaran et al. (2005), Kapur and Akça (2002), and Kapur et al. (2002), seeking sustainable land management, are to be accepted as the higher quality natural/human cultivated-managed lands in the upstream, downstream and the transition zone of the SA. An example for this would refer to, the number – the lower numbers – designating the lower soil capability/quality in the hierarchy of the Beinroth et al. (1999) classification (e.g. Classes VIII, VII, and VI), which stand/stands for the higher quality anthroscape (the Anthroscape Land Quality Class – ALQC) – "class A" – in the modified version of this report for the upstream and the transition zone

Table 5 Major land resources stresses or conditions and modification for SA

Major land resources stresses or conditions (Beinroth et al. 1999)				Modification due to different/suitable ecological conditions or stresses of the Seyhan anthroscape	
Stress class	Land quality class based on land capability classes	Major land stress factor	Criteria for assigning stress	Land quality class-Anthroscape Land Quality Class/ landscape unit	Description
25	IX	Extended periods of moisture stress	Aridic SMR,[a] rocky land, dunes	N/A	N/A
24	VIII	Extended periods of low temperatures	Gelisols	N/A	N/A
23	VIII	Steeplands (elevation: 500–2,000 m)	Slopes greater than 32%	VIII/B–C	Sloping land with excess grazing, should be allocated for biodiversity/forestry may require seeding for enhancing vegetation depending on actual cover (high or medium input) (species present and/or recommended: trees – *Prunus dulcis*, *Prunus amygdalus* or *Amygdalus communis*, *Junglans regia*, *Castanea sativa*; shrubs and grasses – *Arceuthos drupace*, *Ceratonia siliqua* (up to 700 m), *Thymus serpyllum*, *Thymus vulgaris*)
22	VII	Shallow soils (elevation: 100–500 m)	Lithic subgroups, root restricting layers <25 cm	VI–VII/C–D	Soils overlying massive rock (limestone, basalt, dolomite, serpentine, etc.) should be allocated for biodiversity/ rangeland/forestry. May require afforestation, fencing and seeding for grazeland development and cash-crop production (species present and/or recommended: trees – *Ceratonia siliqua*, *Laurus nobilis*; shrubs and grasses – *Rosmarinus officinalis*, *Capparis spinosa*, *Capparis ovata*, *Thymus serpylum*, *Thymus vulgaris*)
21	VII	Salinity/alkalinity	Salic, halic, natric categories	IV–VII/C	Introduction of halophytes (species present and/or recommended: Aegilops spp., *Agropyron elongatum*, Bromus, *Echinochloa crusgalli* L., *Prospis farcta*, *Salsola kali* L., *Salicornia europea* L., Tamarix spp.) for income generation and land protection. Local halophytes may not be sufficient for land protection and income generation, necessitating the introduction/adaptation of exotic species

20	VII	High organic matter	Histosols	N/A	N/A
19	VI	Low water holding capacity	Sandy, gravelly, and skeletal families	IV–VI/C	Improper use: Leveled sand dunes (Arenosols) converted to irrigated cultivation with high agro-chemical input (fertilizers, pesticides, etc.). Proper use: Coastal greenbelts (buffer zones between cultivated Arenosols and the immediate coast/beach) with dominant stone pine (*Pinus pinea*) as a cash-crop, and *Acacia cyanopyhlla* L. as a nitrogen fixer in soils. May require afforestation, fencing and seeding of exotic species
18	VI	Low moisture and nutrient status	Spodosols, ferritic, sesquic, and oxidic families, aridic subgroups	N/A	N/A
17	VI	Acid sulphate conditions	"Sulph-" great groups and subgroups	N/A	N/A
16	VI	High P, N, organic compounds retention	Anionic subgroups, acric great groups, oxidic families	N/A	N/A
15	VI	Low nutrient holding capacity	Loamy families of Ultisols, Oxisols	N/A	N/A
14	V	Excessive nutrient leaching	Soils with udic, perudic SMR, but lacking mollic, umbric, or argillic horizons	N/A	N/A
13	V	Calcareous, gypseous conditions (elevation: 0–100 m)	With calcic, petrocalcic, gypsic, petrogypsic horizons; carbonatic and gypsic families; exclude Mollisols and Alfisols	IV–VII/A–B	Calcretes underlying shallow soils (Leptosols) seems to be a morphologic limitation to agriculture, however, the physical and mineralogical properties of the calcrete provide suitable growing conditions for indigenous crops (Class A if land is allocated to indigenous olive, fig, vineyards, almonds and plant species-indigenous cash crops – of the natural vegetation; Species present and/or recommended: trees – *Pinus pinea*, *Ceratonia siliqua*,

(*continued*)

Table 5 (continued)

Major land resources stresses or conditions (Beinroth et al. 1999)				Modification due to different/suitable ecological conditions or stresses of the Seyhan anthroscape	
Stress class	Land quality class based on land capability classes	Major land stress factor	Criteria for assigning stress	Land quality class-Anthroscape Land Quality Class/ landscape unit	Description
					Laurus nobilis; shrubs and grasses – *Rosmarinus officinalis*, *Capparis spinosa*, *Capparis ovata*) in agro-ecosystem management. And, the introduction of gypsum tolerant plants (Class B if land is allocated to Alfalfa, barley, groundnut, lentil, wheat, oats, tomatoes and onions due to tillage and agricultural input requirement)
12	V	High aluminum	pH < 4.5 within 25 cm and Al saturation $>60\%$	N/A	N/A
11	V	Seasonal moisture stress	Ustic or Xeric suborders but lacking mollic or umbric epipedon, argillic or kandic horizon; exclude Vertisols	I–V/C–D	Crops (grain, cotton, citrus, maize) of the downstream may require irrigation even in Late Winter and Early Spring due to global and local climate changes? Priority of water use may shift to household consumption due to the pressure of increasing population and migration
10	IV	Impeded drainage	Aquic suborders, "gloss" great groups	N/A	N/A
9	IV	High anion exchange capacity	Andisols	N/A	N/A
8	IV	Low structural stability and/or crusting	Loamy soils and Entisols except fluvents	IV–VI/B–C	May require high organic matter input and chemical fertilizers for cash crops in the downstream on river banks. And should be covered by natural vegetation for preventing erosion in the transition and upstream zones (expenses for seeding for enhancing vegetation depending on actual natural cover, medium to high input)

7	III	Short growing season due to low temperatures	Cryic or frigid STR[b]	N/A	N/A
6	III	Minor root restricting layers	Soils with plinthite, fragipan, duripan, densipan, petroferric contact, placic horizons, <100 cm	N/A	N/A
5	III	Seasonally excess water	Recent terraces, aquic subgroups	N/A	N/A
4	II	High temperatures	Isohyperthermic and isomegathermic STR excluding Mollisols and Alfisols	N/A	N/A
3	II	Low organic matter	With ochric epipedon	II–IV/B–C	May require high organic matter input and chemical fertilizers for cash crops
2	II	High shrink/swell potential	Vertisols, vertic subgroups	II/C	Cotton, maize, sunflower, wheat, canola, etc. require intensive agricultural practices (tillage, mechanization, fertilizers, pesticides, irrigation, etc.)
1	I	Few constraints	Other soils	A-B-C-D-E	Depends on indigenous management as the major component to determine the anthroscape classes. For example, the shift from the indigenous olive production on deep Calcisols to citrus will reduce the quality of the anthroscape from Class A to C due to the high input requirements of tillage, agro-chemicals, and irrigation

[a], [b] Knowledge of soil moisture and temperature regimes (SMR and STR) for land use and other applications of climatic data. These are useful as air temperature and rainfall information. SMRs and STRs are studied by meteorologists, climatologists, ecologists, agronomists, soil scientists, civil engineers, farmers, and many others who use soils

Table 6 The Anthroscape Land Quality levels

Anthroscape Land Quality Class	Magnitude of the ITK-decreasing/economic input-increasing
A (100–80%)	Management practices (crops – trees, shrubs, herbs, animals – small ruminants of the highlands, land use – tillage, terracing, rotation, use of manure, etc.) except irrigation (drip irrigation of olives on calcretes) are *indigenous. Appropriate C-management level* (*type of management appropriate to the agro-ecosystem*)
B (80–60%)	Management practices *partly* depend on external input i.e. seeding of rangelands, selection and introduction of local/exotic trees, shrubs and grasses (*Laurus nobilis*, *Ceratonia ciliqua*, *Rosmarinus officinalis*, *Capparis spinosa*, Thymus spp), in transitional and highland zones, and halophytes on the coast. *Appropiate C-management level*
C (60–40%)	Management practices *moderately* depend on external input i.e. use of agro-chemicals on local (olives, vineyards and vegetables) and cash crops (cotton, wheat, and citrus) and excess tillage/subsoiling along with cross-breeding of small ruminants. *Partially appropriate C-management level*
D (40–20%)	Management practices *abundantly* depend on external input i.e. fertilisation, pesticides, mechanization and irrigation of cotton, citrus, maize, soybean, canola, low and high tunnels vegetables i.e. cash-crops and large ruminants. *Inappropriate C-management level*
E (20%)	Management practices *dominantly* depend on high inputs e.g. greenhouses, large ruminants. *Inappropriate C-management level* (type of management inappropriate to the agro-ecosystem)

(Tables 5–7). Whereas, for the land of the downstream area (i.e. The flat table lands), which are Beinroths (1999) high LQCs (e.g. Classes I, II, and III), but require high inputs in agricultural technology and practices, stand for the lower/limited class of the anthroscape land quality as the "classes B or C, D to E" (Table 6) i.e. The percentages given in Table 6 are listed in a descending order based on the increasing amounts of economic inputs in agriculture and decreasing ITK. Ultimately, the LQCs/AQCs of the study area are grouped accordingly at Tables 5, 8–11 by delineating the relevant parts of the SA based on the Anatolian Eco-regions determined by Atalay (2002) as below:

(a) The deep to moderately deep Vertisols and Fluvisols of the table land/the downstream zone sought to be allocated to the production of specific crops of cash value such as indigenous cotton (planted since the sixteenth century), citrus and contemporary maize and grain/wheat (Anatolian species adapted since decades) along with olives and carobs at specifically selected areas with Cambisols and Calcisols of the transition zone between the upstream and the downstream, i.e., especially in the northern part of Adana (Fig. 15).
Sand dunes and the wetland Arenosols sought to be allocated to a green belt/buffer strip of local indigenous trees and shrubs of biodiversity together with cash-crops such as Stone Pine (Pinus pinea) to be allocated between the peanut and vegetable/watermelon production belt and the Mediterranean Sea (Kapur et al. 1999; Akça et al. 2004), i.e., the coastal zone.

Table 7 The ideal land use patterns of the Anthroscape Land Quality classes of the SA

	ALQC	Agriculture	Animal husbandry	Forestry	Biodiversity	Urban/industry/ infrastructure	Agro-tourism/ ecotourism	Major use
Karataş/Adana Coastal and Downstream								
I	D	75	10	5	5	5	0	Cash crops
II	D	75	10	5	5	5	0	Cash crops
III	D	75	10	5	5	5	0	Cash crops
IV	C	55	20	10	10	5	0	Cash/indigenous
V	D	15	40	10	10	25	0	Large ruminants
VI	C	60	20	10	5	5	0	Cash crops
VII	C	20	20	30	20	1	9	Halophytes
VIII	C	0	20	50	20	1	9	Biodiversity
Adana-Pozantı-Karaisali transition								
I	C	70	10	5	5	5	5	Indigenous and cash crops
II	C	70	10	5	5	5	5	Indigenous and cash crops
III	C	70	10	5	5	5	5	Indigenous and cash crops
IV	A	60	10	10	5	5	10	Indigenous crops
V	A	60	10	10	5	5	10	Indigenous crops
VI	A	60	10	10	5	5	10	Indigenous crops
VII	B	40	10	30	10	1	9	Indigenous trees, shrubs and grasses
VIII	B	10	0	40	30	1	19	Biodiversity
Saimbeyli, Tufanbeyli, Feke, Çamardı townships-Mediterranean highlands								
I	C	80	5	5	0	5	0	Vegetables, poultry, large ruminants
II	C	80	5	5	0	5	0	Vegetables, poultry, large ruminants
III	C	80	5	5	0	5	0	Vegetables, poultry, large ruminants
IV	C	80	5	5	0	5	0	Orchards, cereals, vegetables, large ruminants
V	B	10	15	50	10	5	10	Orchards, shrubs, grasses, small ruminants
VI	B	10	15	50	10	5	10	Orchards, shrubs, grasses, small ruminants
VII	B	10	10	50	15	5	10	Forestry, shrubs, grasses, apiculture, small ruminants
VIII	A	0	0	70	25	0	5	Biodiversity

(*continued*)

Table 7 (continued)

	ALQC	Agriculture	Animal husbandry	Forestry	Biodiversity	Urban/industry/ infrastructure	Agro-tourism/ ecotourism	Major use
Tomarza, Develi and Yahyalı-Margins of the inner/Central Anatolian Plateau adjacent to the Mediterranean								
I	B	60	15	20	5	5	0	Orchards, large ruminants
II	B	60	15	20	5	5	0	Orchards, large ruminants
III	B	60	15	20	5	5	0	Orchards, large ruminants
IV	B	40	10	20	10	5	15	Vegetables, woodlands, large ruminants
V	B	20	10	50	10	5	5	Woodlands, vegetables, small ruminants
VI	B	20	10	50	10	5	5	Woodlands, vegetables, small ruminants
VII	B	10	10	50	10	5	15	Forests, small ruminants
VIII	A	0	0	70	20	0	10	Biodiversity

Table 8 The ideal land use pattern of the dominant LQ/ALQ classes of Adana-southern, Karataş townships according to the Anthroscape approach

Stress Class (Beinroth et al. 1999)	1, 2, 3, 13, 21, 22-in DS and C
Land Quality Class/Anthroscape Land Quality class	III, VI, I/D, C, D
Agriculture (food and industrial crops) and animal husbandry	Intensive production of cotton, maize, cereals, soybeans, citrus and vegetables
Biodiversity zones	Forest Wilderness: Transition and coastal zone biodiversity. including land for halophyte and fisheries (coastal and lagoonal) man
Urban/industry/infrastructure	Expansion/settlement as satellite towns on the coast to the proximity of the "green belts" with special functions, e.g. Stone pine production/packing, based on the increasing population by migration and buffer zones of pre-settlement developed by the pioneering migrants
Major land stress-susceptibility factor/driving force	Soil sealing, excess use of water, pesticides and tillage due to population pressure. Impeded drainage, low organic matter and high shrink/swell potential. Eutrification of the coastal strip by chemicals and sediment accumulation degrading shallow coast marine environment of natural fish breeding
Criteria for assigning stress/pressure of driving force and state of the pressure	Increased ET and yield decline due to soil deg. via unsustainable/excess water and pesticide use. Improper use and tillage of calcretes-Cambisols Calcisols, increase of salinity at coastal margins. Declining income generation in fisheries. Abandonment of the land and the coast. Increased ET and yield decline due to soil degradation and pesticide use, and deep tillage of Vertisols/Fluvisols

DS downstream, *C* coastal

(b) The shallow to moderately deep soils-Leptosols, Cambisols, Calcisols and Luvisols-with slightly sloping topographies overlying limestones and flat-platforms/natural terraces of calcretes (sought to be allocated to vineyards, carobs and especially traditional olives and apiculture) (Akça et al. 2004), i.e., the transition zone.

(c) The shallow to moderately deep Leptosols, Cambisols, Luvisols and Calcisols of the karstic-limestone terrain of the upstream zone, with the naturally combined forest/biodiversity/olive and carob production, which bear minimal practices together with graze-lands for local species of goats allocated in maquis (Güney et al. 1999). Goat production in this area revealed a high rate in twinning with significance in income generation (Gürsoy et al. 2003) at areas adjacent to the traditional Ottoman grafted olive orchards (Kapur et al. 2002) i.e., the upstream-transition zone.

The LQCs/ALQCs in the Tables 5, 8–11, denote the level of the "soil class" i.e. the LCC based LQC bound to its inherent physical and chemical attributes, that is equivalent to the invert quality given to the agro-ecosystem/anthroscape i.e. the

Table 9 The ideal land use pattern of the dominant LQ/ALQ classes of the Adana-northern, Karaisalı, and Pozantı townships

Stress Class (Beinroth et al. 1999)	11, 13, 22, 23
Land Quality Class/Anthroscape Land Quality Class	IV, V, VI/A, A, A
Agriculture (food and industrial crops) and animal husbandry	Major production sites of apples, apricots, pears, and vineyards.
Biodiversity zones	Forest: (Forest)+ (grazelands) Comprising the poplar strips located along the courses of the river tributaries within the ancient deforested area converted moderately dense to dense rangelands FOR SMALL RUMINANTS on soils overlying Tertiary sediments facing minor erosion with a supply of app. 1,500 mm/annual rainfall to sustain its present condition. The trans-humance grazing system is in need of a sustainable grazing programme because of the increased population, since these two townships are the suppliers of meat to the SA Wilderness
Urban/industry/infrastructure	The decrease of the younger local population, is subsequently followed by an increase of population via migration to Adana from the poorer parts of the country, which is sought to be met by the expansion of satellite towns with special functionality in production/ manufacturing, e.g. olive/olive oil production satellites including the earlier/ancient wall terraced olive orchards dispersed throughout the karstic slopes of the Taurids; carob/carob powder production satellites; apiculture satellites for production and processing (the selected sites will be mapped upon particularly allocated functions at further stages of implementation by local administrations, which are at a stage of development)
Major land stress-susceptibility factor/driving force	Shallow soils of the indigenous olive/carob/fig agro-ecosystem, allocated together with patches of grazelands in between the Med. Macchia. Biodiv./macchia. Preservation for wilderness and apiculture. Calcareous conditions limiting factor for urban expansion on the karstic land for hydrologic significance.
Criteria for assigning stress/ pressure of driving force and state of the pressure	Degradation of the agro-ecosystem and biodiversity. As well as the grazelands

classes from A to E, based on the lands indigenous properties/qualities, i.e., the agro-ecosystem (rich in indigenous crops of olives and carobs, etc.) in the upstream as a prime anthroscape land quality class is equivalent to a constrained land or a lower quality LCC based LQC e.g. LQC-V to VIII in Beinroth et al. (1998) and Klingebiel and Montgomery (1961). However in the approach of the "Anthroscape," this class would be considered as an upper class LQC/AQC (A or B) for the upstream and transitional zones (Table 5).

Accordingly, Tables 8–11 given below illustrate the major land resource stresses or conditions – major actual/potential threats to land/soil degradation and resilience, listed in order of severity and the land quality/anthroscape land

Table 10 The ideal land use pattern of the dominant LQ/ALQ classes of the Saimbeyli, Tufanbeyli, Feke, Çamardı townships – Mediterranean highlands

Stress Class (Beinroth et al. 1999)	11, 13, 22, 23
Land Quality Class/Anthroscape Land Quality Class	VII–VIII/B, A
Agriculture (food and industrial crops) and animal husbandry	Major production sites of apples, apricots, pears, and vineyards. Intensive vegetable production along river banks lined traditionally with poplars
Biodiversity zones	Forest: Comprising apiculture and almond/carob agro-ecosystems and biodiversity Wilderness: Biodiv./macchia. Preservation for wilderness and apiculture. National Parks
Urban/industry/infrastructure	Expansion by satellite towns of special functionality in production/manufacturing of fruit trees and/or their products (selected sites are mapped upon particularly allocated functions)
Major land stress-susceptibility factor/driving force	Shallow soils of the indigenous olive/carob/fig agro-ecosystem, allocated together with patches of grazelands in between the Med. Macchia. Calcareous conditions limiting factor for urban expansion on the karstic land for hydrologic significance
Criteria for assigning stress/ pressure of driving force and state of the pressure	Degradation of the agro-ecosystem and biodiv. As well as the grazelands

Table 11 The ideal land use pattern of the dominant LQ/ALQ classes of the Ulukışla, Pınarbaşı, Sarız, Tomarza, Develi and Yahyalı-Margins of the Inner/Central Anatolian Plateau adjacent to the Mediterranean

Stress Class (Beinroth et al. 1999)	11, 13, 22, 23
Land Quality Class/Anthroscape Land Quality Class	VII, VIII, III/B, A, B
Agriculture (food and industrial crops) and animal husbandry	Major production sites of apples, apricots, pears, and vineyards. Intensive vegetable production along river banks lined traditionally with poplars
Biodiversity zones	Forest: (forest) + (grazelands). Larger grazeland areas than the other townships due to the countrywide increasing demand of meat for large and especially small ruminants Wilderness: National parks
Urban/industry/infrastructure	Decreasing population due to migration to city centers. Require subsidies in inducing development of grazeland satellites and production of small and large ruminant processing/packing of their by-products (meat and dairy)
Major land stress-susceptibility factor/driving force	The abandonment of the extensive grazelands
Criteria for assigning stress/ pressure of driving force and state of the pressure	Degradation-ion of the grazelands and increased erosion

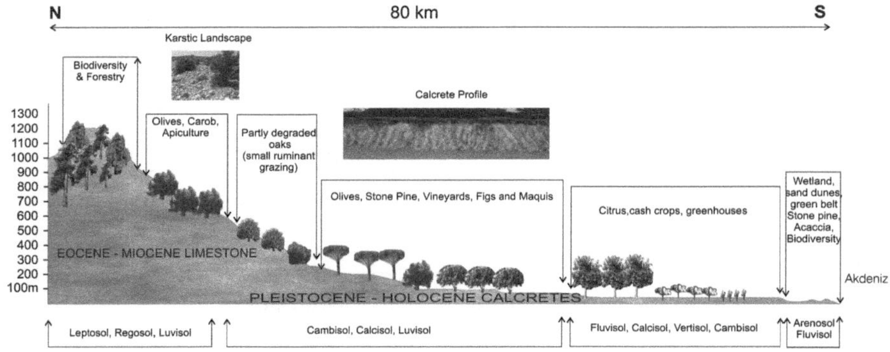

Fig. 15 The anthroscape section of the Seyhan Basin (Kapur et al. 2004)

quality classes for the Seyhan Basin town centers (modified from Beinroth et al. 1999). The Adana Center and the downstream part of the SA is the major part for the increasing land of cultivation as well as urbanization due to migration and population increase.

The unsustainable use of water by the furrow irrigation practiced in the basin for vegetables and the exotic crops has ample influence on changing the climate and the destruction of the soil structure and especially the soil horizons, which have developed since millennia by the constant and environmental friendly integrated cultivation habits of the earlier inhabitants.

The excess use of water in the downstream practiced at the irrigation network developed by the regional State Hydraulic Works at four different phases covering the delta from the north to the south is a partially managed system by the Water Use Associations. The system has caused an appreciable amount of decrease in the previous salinity determined by Dinç et al. (1991) in the upper three phases of the irrigation network and parallel to an increase in the fourth phase – the southernmost part of the Seyhan Basin – where the ground and table water depths increase to the south along with soils of specific sites around the lagoons and soils at low elevation topographies. A water management programme considering some halophyte cash crop management is sought to be developed for this area in order to produce alternative uses to prevent salinity built up which depends on the potential salinity of the area rich in swelling clay minerals.

2.9 Quantification of the Carbon by NPP at the ALQCs

Vegetation plays an important role in the energy, matter and momentum exchange between land surface and atmosphere. Through the process of photosynthesis, land plants assimilate carbon in the atmosphere and incorporate into dry matter while part of is emitted into the atmosphere through plant respiration. The remainder of photosynthesis and respiration is called the net primary productivity (NPP), which has significance in the global and regional carbon budgets (Berberoğlu et al. 2007). Thus, NPP is a function of standing biomass, an important component of the carbon cycle and a key indicator of ecosystem performance (Lobell et al. 2002). It is also a variable that will change differentially with latitude, altitude, land cover and management as being an indicator of the *Anthroscape* and the *Anthroscape Land Quality*. The remotely sensed data and the carbon models can be used to quantify the temporal behaviour of photosynthetically – active tissue and therefore provide information on the seasonal dynamics of biosphere production. For example, the Carnegie, Ames, and Stanford Approach (CASA) (Potter et al. 2003) terrestrial biogeochemical model, designed to simulate the terrestrial carbon cycle using satellite sensors and meteorological data, was used to estimate the annual regional NPP. This model is based on light-use efficiency (LUE), influenced by temperature, rainfall and solar radiation. NPP has been mapped on a global scale using the Normalised Differences Vegetation Index (NDVI) and models which calculate NPP as a function of irradiance, Absorbed Photosynthetically Active (400–700 nm) Radiation (APAR) and an average light use efficiency (LUE) (ε) (Prince and Goward 1995; Potter et al. 2004, 2007). Since, under ideal conditions, the rate of primary production is related linearly to APAR (Monteith 1977), the LUE can be regarded as a conservative quantity, which can be used to scale the integral Fraction of the Photosynthetically Active Radiation (FPAR)*PAR (over a growing season) to primary productivity.

The monthly NPP flux, defined as the net fixation of CO_2 by vegetation, is computed in CASA on the basis of LUE (Monteith 1972). The fundamental relation in the CASA model is $\text{NPP} = f(\text{NDVI}) \times \text{PAR} \times \varepsilon \times g(T) \times h(W)$ where APAR (in megajoules per square meter per month) is a function of NDVI and downwelling PAR and ε (in grams of C per megajoule) is a function of the maximum achievable LUE (ε) adjusted by functions that account for the effects of temperature $g(T)$ and water $h(W)$ stress (Potter et al. 2007). Whereas the previous versions of the CASA model (Potter et al. 1993) used NDVI to estimate FPAR and the current model relies upon canopy radiative transfer algorithms (Knyazikhin et al. 1998), which are designed to generate more accurate FPAR products as inputs to carbon flux calculations.

This study utilised the full spatial resolution Envisat MERIS data to drive an ecosystem productivity model for vegetation in the Seyhan River Basin. The current spatial distribution of climate data was generated with point data from 50 climate stations within the watershed using cokriging. Climate data outputs can be moderated using the four variables of percent tree cover, land cover, soil texture and NDVI. This study employed 47 MERIS images recorded between March 2003 and September 2005 to derive percent tree cover, NDVI and land cover. Soil texture was available from published sources.

The model was used to predict the annual regional fluxes in terrestrial NPP at various degrees of C (depending on the yearly conditions). The model outputs were monthly NPP maps at a 300 m grid cell size for each *Anthroscape Land Quality*

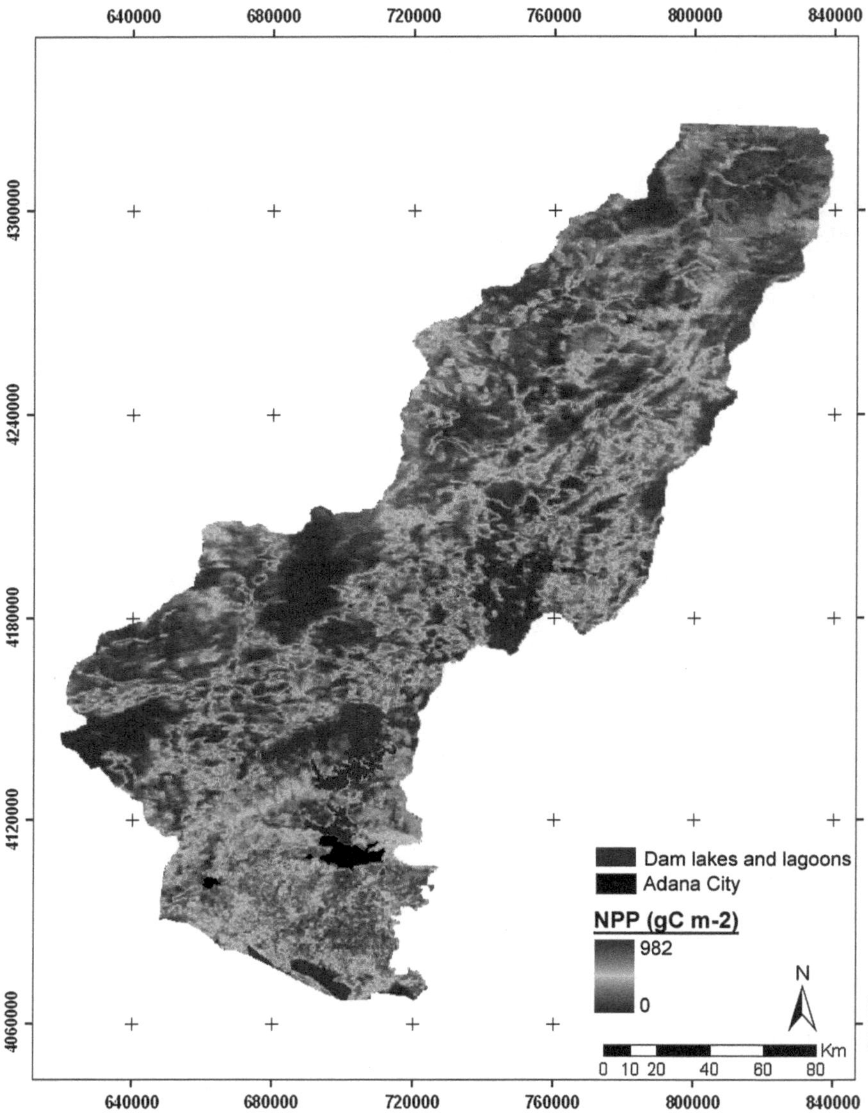

Fig. 16 Total NPP of the Seyhan Basin

Class (Fig. 16). In general, NPP increased from *Anthroscape Land Quality Class* A to D, as a result of intensive land use (Fig. 17).

The most obvious anthropogenic influence on NPP involves changes in land use and cover type caused by human activities, which change the natural environment and consequently influence, the ecosystem NPP. The influence of land use and

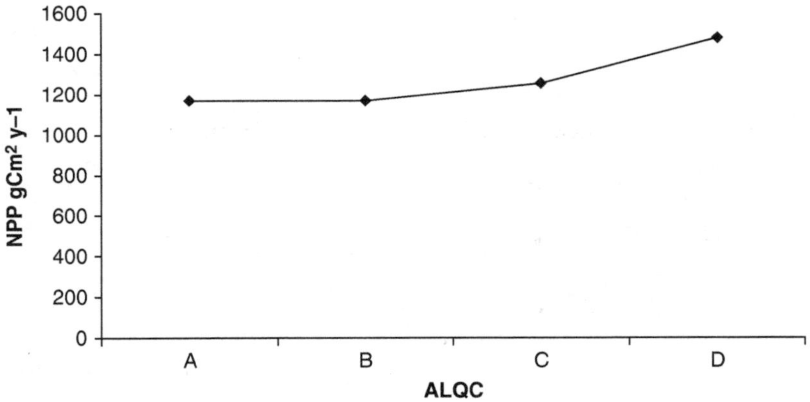

Fig. 17 Total NPP of Anthroscape Land Quality Classes in the Seyhan Basin

cover type changes on NPP includes both direct and indirect influences on the *Anthroscape Land Quality*. Direct influences involve a change in the type of *Anthroscape Land Quality Class*, whereas indirect influences involve changes in the natural environment that result from changes in land use and cover type. By studying NPPs response to changes in, land use, and cover type, we can better understand the *Anthroscape Land Quality* and its ability to respond to changes in the physical and social environment. The monthly spatial distributions of the NPP for zones in the Seyhan Basin are shown in Fig. 18.

3 Results

Regional and seasonal distributions of the NPP reflect the interface between physical (e.g., precipitation, PAR) and biological processes (e.g., species composition, microbial activity, and interactions among organisms) as the function of intensification of land use. The estimated average NPP was approximately 388.79 gC m^{-2} per year in the basin. The mean NPP of the *Anthroscape Land Quality Classes* for the basin, differed significantly among all months ranging from 6.67 to 225.73 gC/m^{-2} $year^{-1}$. There was virtually no change in the mean monthly NPP of ALQC for the upstream, where this zone largely consists of ALQC A and B (Fig. 19) and limited seasonal variation in NPP, due to the constant tree cover, i.e., minor variations occurring parallel to the tree growth. However, the high seasonal variation (despite the lower quality ALQC compared to the up spring) amongst the *Anthroscape Land Quality Classes and the higher npp-c were* distinct in the transition and downstream zones, where intensive land use took place (Fig. 20).

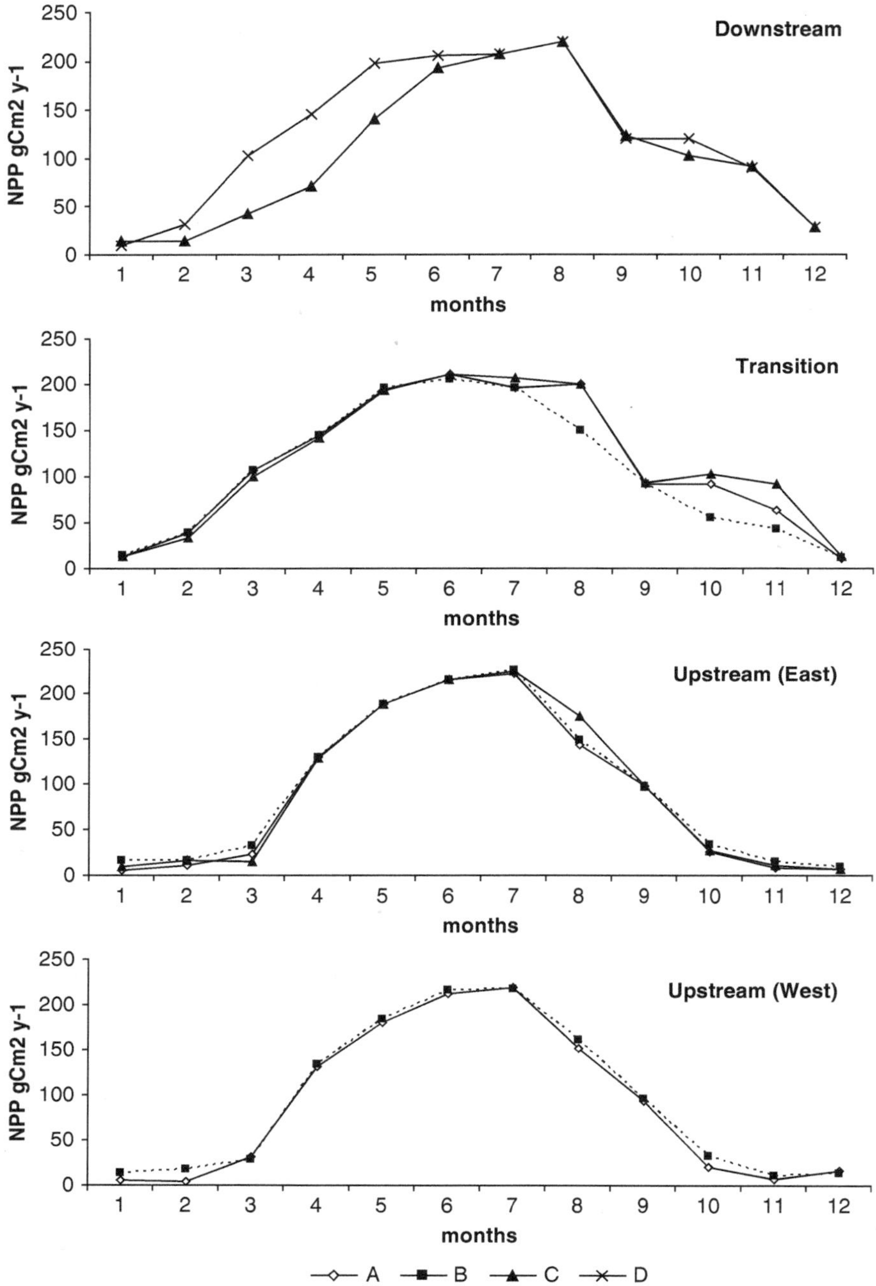

Fig. 18 Monthly NPP of Anthroscape Land Quality Classes for each zone in the Seyhan Basin

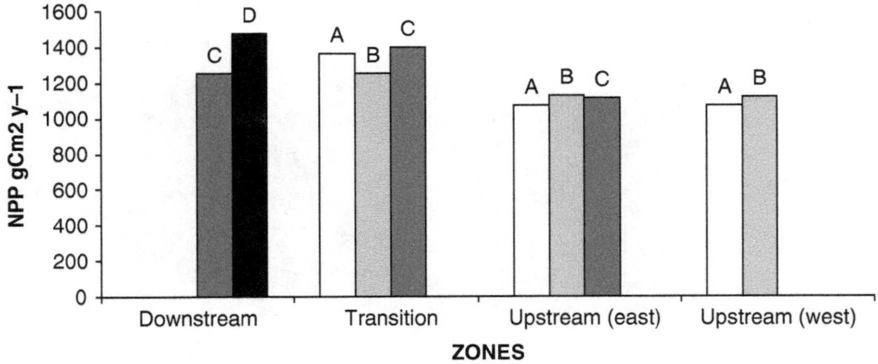

Fig. 19 Total NPP of Anthroscape Land Quality Classes for each zone in the Seyhan Basin

Figures 21 and 22 show the seasonality with high carbon rates in the winter months and low rates in summer in the Seyhan Basin for the year of 2005. The mean NPP for the basin differed significantly among all months and ranged from 15.42 to 481.80 gC/m^{-2} year^{-1} (Fig. 23). The highest mean NPP value for the Seyhan Basin was 481.80 gC/m^{-2} year^{-1} in June and the lowest NPP value was 15.42 gC/m^{-2} year^{-1} in January. The NPP in June–September, 2005 mainly ranged from 482 to 178 gC/m^{-2} year^{-1}, and there is an approximate increase of 130 gC/m^{-2} year^{-1} from September to October (Table 12).

4 Conclusion

The final outcomes of the approach of the anthroscape, seeking to develop a powerful tool for use by the local communities/administrations and relevant bodies for a bottom to top approach, as a quantified entity, meant to direct the future land and water use decisions to be taken at lower levels – as farm domains, etc. – has led to the development of an "Anthroscape Land Quality Class" map and the relevant "Ideal Land Use Patterns" of the Seyhan Anthroscape (Fig. 20 and Table 7). These two final products of quantification are capable in revealing the magnitude and the distribution of the degradation of the selected area, as well as allocating the ideal land use types given for the percentages of the distributions of land except their specific location. For example, the downstream part of the map shows the abundant degradation arising via the intensive cultivation practices where the class stated in the map reveals the urgent need of an integrated SLWM Programme to revert the lower class C and D ALQCs to higher ALQCs to meet the requirements of the sustainable use of the land. Whereas, the higher ALQC land in the transition or upstream zone stands for higher resilience and lower input requirements to meet the ideal use of the land.

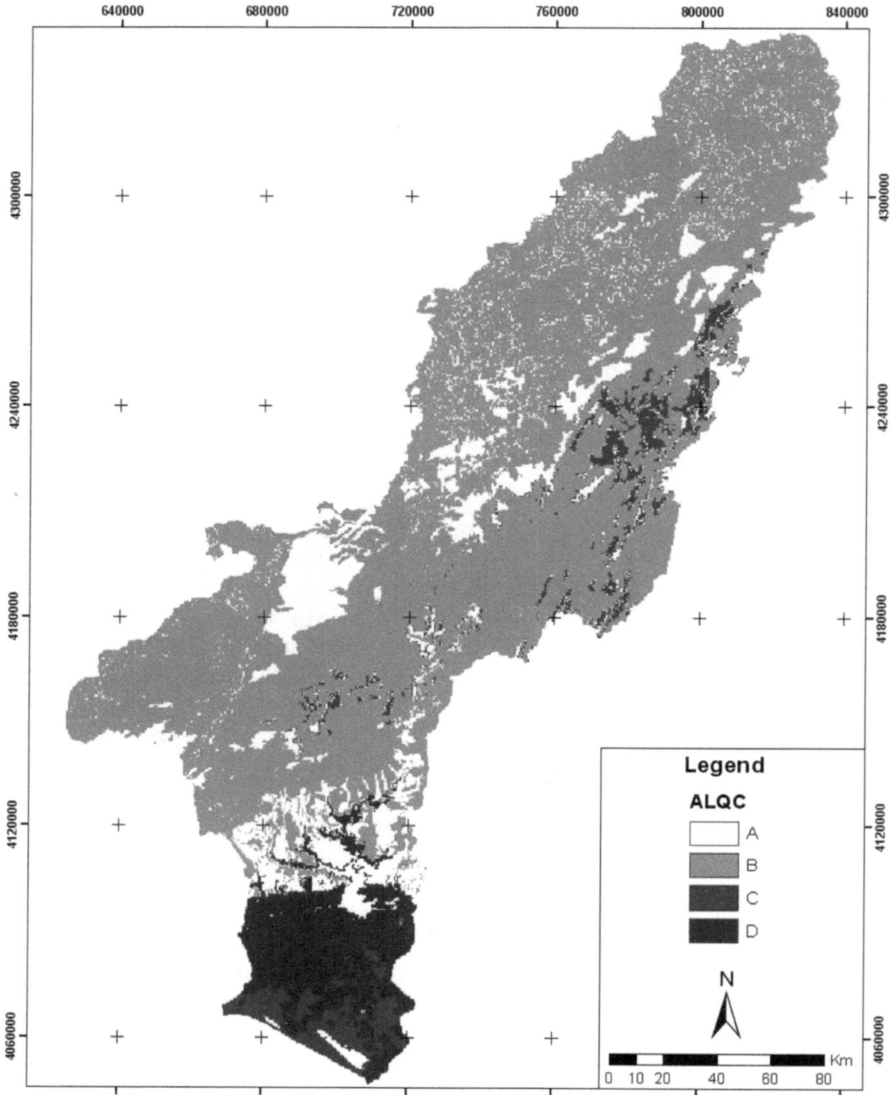

Fig. 20 The AQLC of the SB

The expectations related to the programming of the Sustainable Use of the Land and Water via the Anthroscape contextual approach mentioned above will most probably be dealt under the umbrella of the recently established "Regional Development Agencies" by the government and the "Çukurova Development Agency-ÇDA" (the Adana and Mersin provinces) selected as a pilot area in Turkey. The ÇDA is sought to construct an integrated approach to the rural/urban development

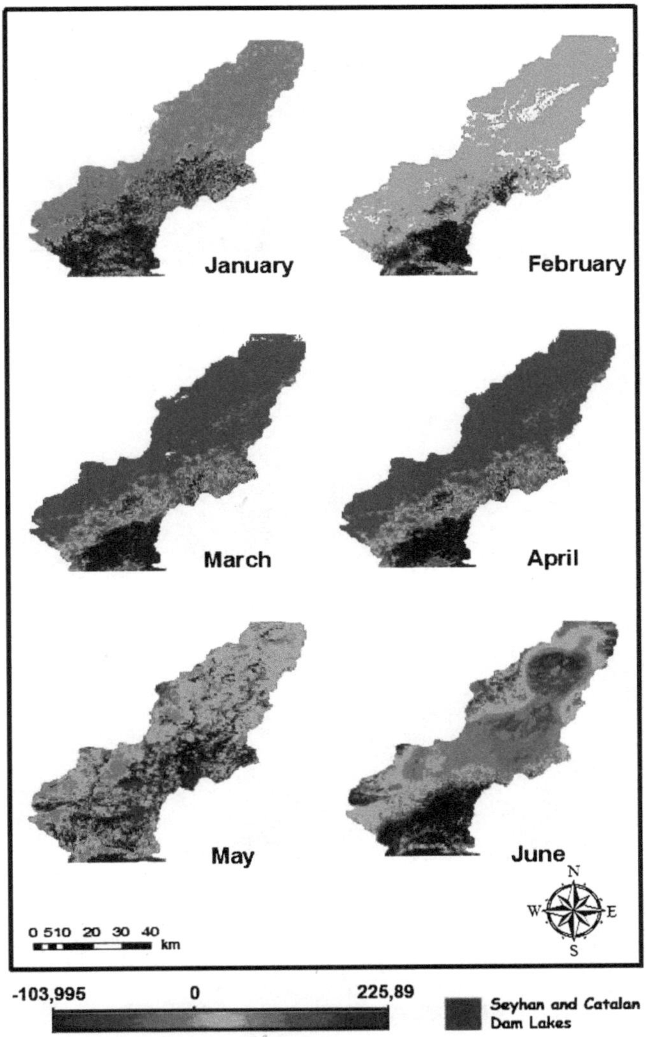

Fig. 21 The monthly spatial distribution and annual variations of the net primary production fluxes in the Seyhan Basin (January–June 2005)

of the country. This body will most probably attempt to amalgamate the long standing local trends, which are the indigenous technologies as well as knowledge on natural resource management in need of renovation with the cooperation of the public and the responsible formal bodies and the private enterprise. The increasing population via migration is also sought to be coupled to the renovated income generating tools and integrated to the local conditions. Ultimately the ÇDA and the related governmental bodies together with the NGOs are strongly required to look

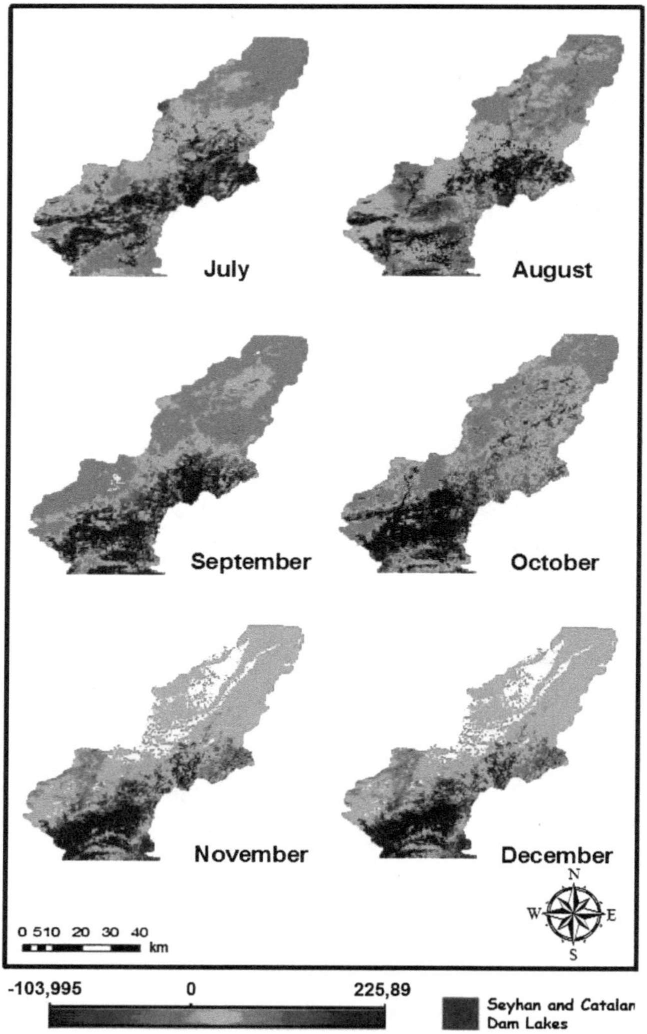

Fig. 22 The monthly spatial distribution and annual variations of the net primary production fluxes in the Seyhan Basin (July–December 2005).

ahead to create the opportunities of mitigating the present problems arising from the impacts illustrated in Table 1 and the figures of the chapter, as well as their future versions, which most likely will be of greater magnitude.

Table 1 and the attached figures/maps developed to address the magnitude of the driving forces of Land Degradation and Desertification along with their mitigations (mentioned earlier in this report) revealing the need for the urgent precautions to be taken

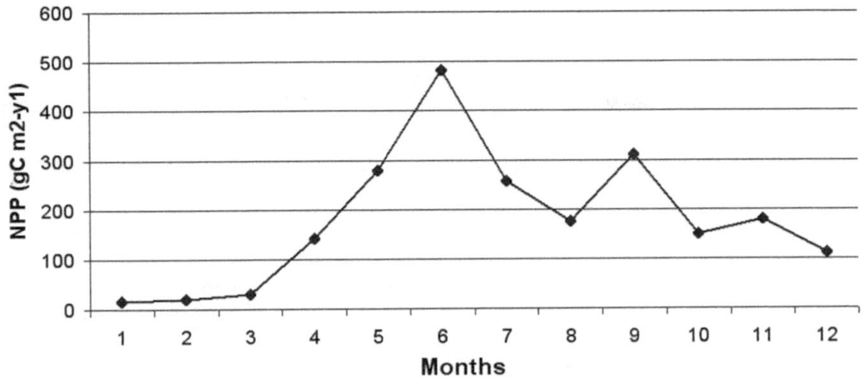

Fig. 23 Monthly mean NPP values for the Seyhan Basin (2005)

Table 12 Annual NPP of the major units of the Seyhan Basin (2005)

Classes	Mean total NPP
Masked area (water and urban)	
Broadleaf deciduous forest	588.71
Mixed broadleaf and needle leaf forest	469.11
Needle leaf evergreen forest	514.22
Grassland	233.09
Bare soil	185.17
Agriculture	342.44

(a) For the increasing socio-economic components, i.e., migration and the parallel increase in the lack of capital, education and low income causing problems related to the small and large ruminants, apiculture, and fisheries in the downstream. The problems on the ruminants are followed by the impacts of descending productions of apiculture, fisheries and vegetables along with the indigenous crops in the transition zone, whereas the upstream is mostly effected by the pressure of migration to the cities and increasing need of decreasing man-power expected to deal with the problems ruminants, cereal production, apiculture and especially indigenous crops.
(b) For the urgent needs of renovated measures to be accomplished for animal breeding with special care on forming the stud flocks as materials for optimal breeding, artificial insemination, the construction and the improvement of animal shelters complemented by the low production trends especially in the upstream are followed by the transition and the downstream zones.
(c) For the urgent needs in optimal crop production the low productivity coupled with the low inputs and proper mechanization along with land fragmentation seem to be the main drivers responsible for the pressures developing especially in the upstream followed by the transition and downstream zones. The main problems faced in the upstream on the loss of production by the indigenous crops and cereals is most probably due to the lack of proper mechanization

suitable to agro-ecosystem management and environmental friendly technology, land fragmentation inducing abandonment of valuable and unique indigenous knowledge, and increasing losses of natural fodder production sites – the grazelands. The low inputs and problems of lacking mechanization that might seem to be one of the major drawbacks for income generation in the upstream, is actually the renovation of the traditional technologies that are not as costly as the imported exotic production needs.

(d) The sustainable use of water (the lack of regulation of water use and the inadequate use of the water resources) seems to be the major drawback for all the sectors of production in the downstream and transition zones, but most of all in the upstream zone, which is a major dependant of indigenous production technologies in need of renovation due to the increasing population pressure demanding the so-called income generating exotic know-how and misuse of the natural resources of the agro-ecosystem.

(e) The lack of the extension services and of the information on the organization for optimal land use seem to be a major handicap for the will to integrate the different land and water management programmes developed by the different levels of the communities at all sectors of production. The problem of extension concerning women is due to the incomparable traditional contribution of women to animal husbandry especially in the highlands/upstream.

(f) The lack of a well established marketing infrastructure demanding quality standards for products and processing units functioning over the national and international trade trends of the goods produced is the major legislative/economic shortcoming for the integrated development of the region. Moreover, the high credit rates and lack of subcidies are the supplementary obstacles for an integrated economic/legislative development. Ultimately, the integration of the farmers and the consumers together with the development of awareness between these two parties is a must to optimize the use of the natural resources.

(g) The ultimate outcomes of the Anthroscape should be studied in comparison with relevant contexts developed elsewhere in the world. These comparable contextual approaches may be the "Plaggen Böden" of Northern Europe; the long standing cultivated land enriched by organic matter, the "Satoyama" landscapes of Japan: the long cultivated paddy fields on sloping land integrating the highlands with the low via the trade of rice as a major commodity and the rice fields of the far east.

References

Akça E, Pekel M, Kapur B, Zdruli P, Eswaran H, Kapur S (2004) The Upstream and Downstream Integrated Resource Management of the Adana Anthroscape. In (P. Zdruli and G.T. Liuzzi eds) Workshop proceedings of promoting participatory management of the land system to enhance soil conservation – MEDCOASTLAND Project. Alexandria, Egypt, 9–13 October 2004, 243–250

Alphan H (2003) Land-Use Change and Urbanization of Adana, Turkey. Land Degradation and Development. 14: 575–586

Atalay İ (2002) Ecoregions of Turkey. Publication of the Ministry of Environment and Forestry. Pub. No. 163. 2nd Edition. ISBN: 975-8273-41-8, Meta Press, 266 P. Bornova, İzmir

Beinroth FH, Eswaran H, Reich PF (1999) Global Assessment of Land Quality. Sustainable Global Farm. In (D.E. Stott, R.H. Mohtar and G.C. Steinhardt eds) 10th international soil conservation organization meeting, Purdue University, 24–29 May 1999

Berberoğlu S, Dönmez C, Curran PJ (2007) Quantifying Forest Ecosystem Productivity of Eastern Mediterranean Using Envisat MERIS Data. European Space Agency Environment Satellite Symposium, Montreux/Switzerland

Cangir C, Kapur S, Boyraz D, Akça E, Eswaran H (2000) An Assessment of Land Resource Consumption in Relation to Land Degradation Turkey. Journal of Soil and Water Conservation. 55: 253–259

Dinç U, Akça E, Dinç O, Özden DM, Tekinsoy P, Alagöz U, Kızılarslanoglu HA, Köroğlu B, Serdem M, Gültekin E, Zoroğlu L, Fisunoglu M, Eswaran H, Kapur S (2001) Soil Sealing: The Permanent Loss of Soil and Its Impacts on Land Use. In First MEDRAP workshop on sustainable management of soil and water resources – Greece/European Union concerted action to support the Northern Mediterranean RAP, 18–19 December 2001, Athens, Greece

Dinç U, Şenol S, Kapur S, Sari M, Derici R, Sayin M (1991) Formation, Distribution and Chemical Properties of the Saline and Alkaline Soils of the Çukurova Region. CATENA, 18 (2),173–183

Eswaran H, Kapur S, Akça E, Reich P, Mahmoodi S, Vearasilp T (2005) Anthroscapes: A Landscape Unit for Assessment of Human Impact on Land Systems. In (J.E. Yang, T.M. Sa and J.J. Kim eds) Application of the emerging soil research to the conservation of agricultural ecosystems. The Korean Society of Soil Science and Fertilizers, Seoul, Korea, 175–192

European Union (EU)-European Environmental Agency (EEA) (2002) http://ec.europa.eu/comm/

Güney O, Öztürk D, Biçer O (1999) The Present Situation of Extension Service Organization in Turkey. In (R. Rubino and P. Morand-Fehr eds) With: systems of sheep and goat production: organization of husbandry and role of extension services. Options Méditerranéennes, Série A, 38, 63–67

Gürsoy O, Acuz S, Kırk K, Şentut T, Bilgiç I (2003) Preliminary Evaluation of the Project "Use of Kermes Oak" (Quercus Coccifera L.) Acorn for Increasing Reproductive Performance of Goats on the Highlands of Anti-Taurus Mountains. In (P. Zdruli, P. Steduto, S. Kapur and E. Akça eds) Proceedings of the MEDCOASTLAND project. Ecosystem based assessment of soil degradation to facilitate land users and land owners prompt actions, Adana 2–7 June 2003

Kapur S, Akça, E (2002) Global assessment of degradation. Encyclopedia of soil science, Marcel and Dekker, Inc. 296–306

Kapur S, Eswaran H, Akça E, Dinç O, Kaya Z, Ulusoy R, Bal Y, Yilmaz T, Çelik I, Özcan H (1999) Agro-ecological management of degrading coastal dunes in southern Anatolia. In (E. Özhan ed) The fourth international conference on the Mediterranean coastal environment, Volume 1, Medcoast 99, Antalya, Turkey, 347–360

Kapur S, Eswaran H, Akça E, Dingil M (2002) Developing a sustainable land management research strategy fort he southeast Anatolian Irrigation Project. Option Mediterranean, Serie A n. 50, 325–334

Kapur S, Zdruli P, Akça E, Arnoldussen AH, Kapur B (2004) Anthroscapes of Turkey: Sites of Historic Sustainable Land Management (SLM). In (S. van Asselen, C. Boix-Fayos and A. Imeson eds) "Briefing papers" of the second SCAPE workshop in Cinque Terre (Italy), 13–15 April 2004

Karlen DL, Stott DE (1994) A framework for evaluating physical and chemical indicators of soil quality. In (J.W. Doran, et al. eds) Defining soil quality for a sustainable environment. Proceedings, symposium of division S-3, S-6,S-2, Soil Science Society of America, Division A-5 of the American Society of Agronomy, and the North Central region Committee on Soil Organic Matter (NCR-59), 4–5 November 1992, Minneapolis MN. SSSA Special Publication No. 35

Klingebiel AA, Montgomery PH (1961) Land-capability classification. USDA-SCS agricultural handbook. U.S. Government Printing Office, Washington

Knyazikhin Y, Martonchik JV, Myneni RB, Diner DJ, Running SW (1998) Synergistic algorithm for estimating vegetation canopy leaf area index and fraction of absorbed photosynthetically active radiation from MODIS and MISR data. Journal of Geophysical Research 103:32257– 32276

Lobell DB, JA Hicke GP, Asner CB, Field CJ, Tucker SO Los (2002) Satellite estimates of productivity and light use efficiency in nited States agriculture 1982–1998. Glob. Chem. Biol. 8:722–735

MEF (2005) The National Action Program on Desertification. In (M. Düzgün, K. Selim, C. Cangir, E. Akça, D. Boyraz and N. Gülşen eds). Publication of the Ministry of Environment and Forestry, No. 250. ISBN: 975-7347-51-5. 110 P, Ankara

Monteith JL (1972) Solar radiation and productivity in tropical ecosystems. Journal of Applied Ecology, 9, 747–766

Monteith JL (1977) Climate and the efficiency of crop production in Britain. Philosophical Transactions of the Royal Society of London, Series B 281: 277–294

Potter CS, Davidson A, Klooster S, Nepstad D, Gustavo H, Denegreiros H, Brooks V (1993) Regional Application of an Ecosystem Production Model for Studies of Biogeochemistry in Brazilian Amazonia, Global Change Biology pp. 315–333

Potter C, Klooster S, Myneni R, Genovese V, Tan P, Kumar V (2003) Continental scale comparisons of terrestrial carbon sinks estimated from satellite data and ecosystem modeling 1982–98. Global and Planetary Change, 39:201–213

Potter C, Klooster S, Steınbach M, Tan P, Kumar V, Shekhar S, Carvalho C (2004) Understanding global Teleconnections of Climate To Regional Model Estimates of Amazon Ecosystem Carbon Fluxes, Global Change Biology, 10:693–703

Potter C, Klooster S, Huete A, Genovese V (2007) Terrestrial carbon sinks for the United States predicted from MODIS satellite data and ecosystem modeling. Earth Interactions 11:1–21

Prince SD, Goward SN (1995) Global primary production: a remote sensing approach. Journal of Biogeography, 22:815–835

Toprak Su (1974) Soils of the Seyhan Basin. The General Directorate of Soils and Water, The Ministry of Agriculture and Rural Affairs. No. 286, Ankara

Zucca C, Bautista S, Previtali F (2006) Desertification prevention and reversal through mitigation and restoration programs. Encyclopedia of Soil Science. Taylor & Francis Group. New York. http://www.informaworld.com/smpp/title~db=all~content=g740169536~tab=toc

Anthroscapes in the Light of the EU Soil Thematic Strategy

Luca Montanarella

Abstract The EU Soil Thematic Strategy is the new framework of the EU for achieving sustainable soil use in Europe. It is based upon the concept of soil multi-functionality and recognizes the various soil functions relevant for European economy and the environment. These functions are under threat by various soil degradation processes that are severely limiting soil functionality and therefore economic growth. Soils in Europe are recognized as being the result of millennia of human interaction with the landscape. Therefore the new paradigm of "anthroscape" is highly relevant to Europe and to effective soil protection in Europe. The anthroscape of the Cinque Terre National park is used as an example to demonstrate the potential of this concept as an effective tool for implementing soil protection strategies.

Keywords Soil protection · Legislation · Landscape · Soilscape · Anthroscape

1 Introduction

During the past decade there has been an intensive scientific and political debate in Europe around the necessity of a comprehensive approach to soil protection at European Union level. The debate was essentially rotating around the basic questions: "Do we have a soil degradation problem in Europe?" and "Do we need a coherent approach the soil protection at EU level?".

The adoption of the EU Thematic Strategy for Soil Protection by the European Commission on 22 September 2006 has provided a reply to the above questions by giving formal recognition to the severity of the soil and land degradation processes

L. Montanarella (✉)
European Commission, Joint Research Centre, Institute for Environment and Sustainability, 21020 Ispra, Italy
e-mail: luca.montanarella@jrc.it

S. Kapur et al. (eds.), *Sustainable Land Management*,
DOI 10.1007/978-3-642-14782-1_2, © Springer-Verlag Berlin Heidelberg 2011

within the European Union and its bordering countries. The Strategy includes a communication (European Commission, COM (2006a) 231) outlining the strategy, a proposal for framework directive for soil protection (European Commission, COM (2006a) 232) as a legally binding instrument and an extended impact assessment (European Commission, SEC (2006b) 620) that has quantified soil degradation in Europe, both in environmental and economic terms.

This impact assessment is based mainly, but not exclusively, on reports (Van-Camp et al. 2004a, b, c, d, e, f) by the Joint Research Centre (JRC) of the Commission and the Working Groups set up to assist the Commission, and reports carried out for the Commission in assessing the economic impacts of soil degradation and economic, environmental and social impacts of different measures to prevent soil degradation.

Available information suggests that, over recent decades, there has been a significant increase in soil degradation processes, and there is evidence that these processes will further increase if no action is taken. Soil degradation processes are driven or exacerbated by human activity. Climate change, together with individual extreme weather events, which are becoming more frequent, will also have negative effects on soil.

Soil degradation processes occurring in the European Union include erosion, organic matter decline, compaction, salinization, landslides, contamination, sealing and biodiversity decline.

The strategy proposed by the European Commission is based on four pillars: A binding legislative instrument (the proposed soil framework directive), integration of soil protection into existing legal instruments at European level, new and enhanced research activities related to soil protection and a renewed effort in awareness raising initiatives.

2 The Importance of Anthroscapes Within the Soil Thematic Strategy

The strategy recognizes the need for soil protection in Europe and puts the accent on the need to protect the soil's multifunctional role. Soils have not only to be protected for the sake of assuring sufficient food, fibre and biomass production, but also as a fundamental ecological compartment assuring clean water and air, high biodiversity, raw materials, etc... One of the main functions of soils, as identified within the thematic strategy, is to provide the substrate for our cultural development by supporting housing and infrastructure and by acting as a cultural archive of our history (archaeological and cultural heritage function). Recognizing the fundamental role of human activities in shaping the landscape of Europe is crucial for any integrated assessment of soil degradation. In Europe there are probably no soils that have not yet experienced some form of impact of the human activities. We therefore live in Europe in a cultural landscape shaped by the hard work of the Europeans over millennia. Particularly the role of the early agricultural development, starting from the first agricultural revolution in the Neolithic have substantially shaped the landscapes surrounding us in Europe today. More then in landscapes (Fig. 1) we are living in Anthroscapes that are the result of the interaction of human activities and the biophysical landscapes of Europe. Understanding the processes that have led to the definition of these Anthroscapes of Europe requires an in depth analysis of the historical development of the various socio-economic factors that have been driving the human activities shaping our landscapes. A full historical analysis is beyond the scope of this chapter. Here we would like to simply relate the soil protection policy development to the recognition of the anthroscape paradigm as described elsewhere in this book.

Developing policies implies the recognition that the main drivers are determined by human activities and therefore regulating instruments, like specific legislation addressing soil protection, can be effective in steering the development of Anthroscapes and their protection. Effective legislation can indeed shape Anthroscapes dramatically, determining in many cases actual land use extensively. Certainly, one of the most recent examples in Europe is the Common Agricultural Policy (CAP), which has substantially oriented agricultural production and practices in most parts of Europe.

There is therefore the need to develop for Europe a novel approach integrating human activities and landscape features in a more holistic and coherent manner. Anthroscapes are therefore the necessary operational unit for future soil protection policy implementation frameworks.

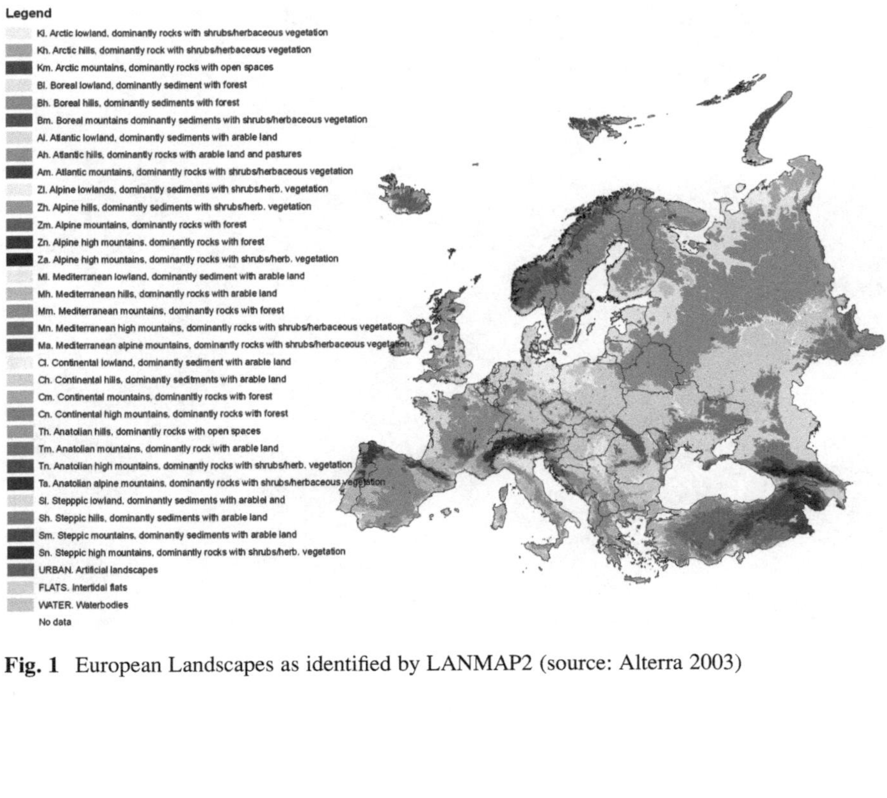

Fig. 1 European Landscapes as identified by LANMAP2 (source: Alterra 2003)

Unfortunately we are still lacking a full inventory of European Anthroscapes allowing for a complete geographic coverage of Europe of the location and boundaries of these fundamental operational units. We have a full coverage of landscape units (Mücher et al. 2003) and of soilscapes (Jones et al. 2005) but we are still lacking a full inventory of Anthroscapes.

3 Some Relevant Anthroscapes in Europe

Some European Anthroscapes are well known and extensively documented, often also because they are recognized as a major European cultural heritage, or sometimes even a world heritage, like the Cinque Terre National Park in Italy, which is classified as a UNESCO world heritage site. A more detailed analysis of this anthroscape may allow gaining further insight into the relations between biophysical conditions and socio-economic drivers that are at the origin of most European Anthroscapes.

The area of the Cinque Terre (five lands) includes the five communes of Levanto, Monterosso Vernazza, Riomaggiore e La Spezia (Fig. 2). Since 1997 it is classified as a UNESCO world heritage site. The Committee decided to inscribe this site considering that the eastern Ligurian Riviera between Cinque Terre and Portovenere is a cultural site of outstanding value, representing the harmonious

Fig. 2 The area of the Cinque Terre National Park

interaction between people and nature to produce a landscape of exceptional scenic quality that illustrates a traditional way of life that has existed for a 1,000 years and continues to play an important socio-economic role in the life of the community. A unique characteristic of the area are the terraced vineyards sustained by dry walling, a perfect example of landscape architecture created by man in inaccessible surroundings.

The recently created National park has as one of its main tasks to develop strategies for the protection of this cultural heritage. Land abandonment is threatening the stability of the terraced areas, leading to extensive soil erosion and landslides. A shifting economy from agriculture to mass tourism is at the origin of these negative trends. The development of options for a sustainable tourism in the area requires also a constant monitoring of this very fragile landscape.

The full analysis of the Driving Forces-Pressures-State-Impact-Responses (DPSIR) framework, as proposed by the European Environment Agency (EEA) has allowed to fully understand the socio-economic drivers that are at the origin of this anthroscape as well as their role in determining the current degradation processes the area is experiencing.

The Cinque Terre is a typical Anthroscape of the Mediterranean area, resulting from centuries of human interaction with harsh biophysical environments. The result is a perfectly balanced and sustainable land management system that has lasted for centuries. Sudden modification of external socio-economic conditions has triggered the collapse of this delicate system, with the consequence of massive land abandonment, collapse of terraces and ultimately land degradation leading to total soil loss by erosion and landslides.

Reversing these negative trends is possible, as demonstrated by the successful measures implemented by the National Park authorities. Sustainable tourism, “adoption” of terraced systems by foreign visitors committing themselves to minimal terrace maintaining work, revenue generating land restoration activities, like new terracing systems with vineyards and olive tree plantations, recognition of typical local food products with higher price tags, etc.... are some of the many measures successfully implemented in this area that could serve as a model for other, similar areas around the Mediterranean basin.

A major challenge ahead is therefore to complete a full inventory of high value Anthroscapes in Europe in order to establish an effective network of such sites and initiate an effective experience sharing and exchange programme that could be the first step towards effective soil protection in Europe.

4 Conclusions

The work of the European Commission in developing a coherent approach to soil protection in Europe has paved the way to a radical re-thinking of the existing scientific paradigms applied so far to the understanding of soils and their functioning within the landscape. The recognition that the landscapes of Europe, and therefore most of their soils, are extensively the result of human activities has allowed identifying a new scientific object: The "Anthroscape".

Transferring the Anthroscape concept into operational policy decisions and land restoration measures has so far only being possible in some limited local cases, like in the area of the Cinque Terre in Italy. A strong local regulatory authority, like the National Park of the Cinque Terre, could effectively implement policy measures reversing the degradation trend and generating substantial socio-economic benefits to the local population. Full inventory of high value Anthroscapes in Europe could allow the transfer of successful experiences across the various Regions of Europe, initiating a network of successful land restoration experiences as a first step towards soil protection in Europe.

References

European Commission (2006a) Thematic Strategy for Soil Protection, COM (2006) 231.

European Commission (2006b) Impact Assessment of the Thematic Strategy ON Soil Protection, SEC (2006) 1165.

Jones A, Montanarella L, Jones RJA (eds.) (2005), "Soil Atlas of Europe", European Soil Bureau Network, European Commission. Office for Official Publications of the European Communities, Luxembourg, 128 pp.

Mücher CA, Bunce RGH, Jongman RHG, Klijn JA, Koomen A, Metzger MJ, Wascher DM (2003) Identification and Characterisation of Environments and Landscapes in Europe. Alterra rapport 832, Alterra, Wageningen.

Van-Camp L, Bujarrabal B, Gentile AR, Jones RJA, Montanarella L, Olazabal C, Selvaradjou S-K (2004a) Reports of the Technical Working Groups established under the Thematic Strategy for Soil Protection. Volume I Introduction and Executive Summary. EUR 21319 EN/1, Office for Official Publications of the European Communities, Luxembourg, 126 pp.

Van-Camp L, Bujarrabal B, Gentile AR, Jones RJA, Montanarella L, Olazabal C, Selvaradjou S-K (2004b) Reports of the Technical Working Groups established under the Thematic Strategy for Soil Protection. Volume II Erosion. EUR 21319 EN/2, Office for Official Publications of the European Communities, Luxembourg, p. 127–309.

Van-Camp L, Bujarrabal B, Gentile AR, Jones RJA, Montanarella L, Olazabal C, Selvaradjou S-K (2004c) Reports of the Technical Working Groups established under the Thematic Strategy for Soil Protection. Volume III Organic Matter. EUR 21319 EN/3, Office for Official Publications of the European Communities, Luxembourg, p. 311–496.

Van-Camp L, Bujarrabal B, Gentile AR, Jones RJA, Montanarella L, Olazabal C, Selvaradjou S-K (2004d) Reports of the Technical Working Groups established under the Thematic Strategy for Soil Protection. Volume IV Contamination and Land Management. EUR 21319 EN/4, Office for Official Publications of the European Communities, Luxembourg, p. 497–621.

Van-Camp L, Bujarrabal B, Gentile AR, Jones RJA, Montanarella L, Olazabal C, Selvaradjou S-K (2004e) Reports of the Technical Working Groups established under the Thematic Strategy for Soil Protection. Volume V Monitoring. EUR 21319 EN/5, Office for Official Publications of the European Communities, Luxembourg, p. 653–718.

Van-Camp L, Bujarrabal B, Gentile AR, Jones RJA, Montanarella L, Olazabal C, Selvaradjou S-K (2004f) Reports of the Technical Working Groups established under the Thematic Strategy for Soil Protection. Volume VI Research, Sealing and Cross-cutting issues. EUR 21319 EN/6, Office for Official Publications of the European Communities, Luxembourg, p. 719–872.

Soil Erosion-Desertification and the Middle Eastern Anthroscapes

Uriel N. Safriel, Pedro Berliner, Ariel Novoplansky, Jonathan B. Laronne, Arnon Karnieli, Itzhak Moshe, A. Kharabsheh, A. Ghaleb Mohammad, and G. Kusek

Abstract A 5-year (2001–2005) joint study of research teams from the Palestinian National Authority, Jordan, Israel and Turkey explored the responses to land management of dryland watersheds in each of the respective countries. The study watersheds differed in their land uses and applied management but were unified by apparent land degradation expressed in water-driven soil erosion due to removal of vegetation cover, and causing on-site loss of land productivity and off-site clogging of water reservoirs.

The study describes and quantifies the rainfall-vegetation-runoff-erosion and the rainfall-soil moisture-soil organic matter-vegetation chain links associated with specific site and management attributes of the study watersheds, spanning from semiarid to arid drylands and from dry woodland to dry rangeland ecosystems. The effectiveness of trees in controlling soil erosion was reaffirmed while discovering that in-spite of their transpiration trees in drylands need not necessarily reduce soil water storage more than the herbaceous vegetation. It was also found that transforming rangelands to planted forest does not necessarily reduce overall plant biodiversity but does change its species composition; that runoff-harvesting practices become less effective as the inherent site's aridity increases; that traditional runoff-harvesting

U.N. Safriel (✉), P. Berliner, A. Novoplansky, and A. Karnieli
Blaustein Institutes for Desert Research, Ben-Gurion University, Negev, Israel
e-mail: uriel36@gmail.com

J.B. Laronne
Department of Geography, Ben-Gurion University, Negev, Israel

I. Moshe
Keren Kayemet Le'Israel, Jerusalem, Israel

A. Kharabsheh
Al-Balqa'Applied University, Al-Salt, Jordan

A.G. Mohammad
Hebron University, Hebron, Palestinian National Authority

G. Kusek
Agricultural Rural Services, Malatya, Turkey

S. Kapur et al. (eds.), *Sustainable Land Management*,
DOI 10.1007/978-3-642-14782-1_3,

practices (e.g. terraces) are effective but non-traditional ones (i.e., large-scale furrowing) are both simple and effective too.

Keywords Soil erosion · Desertification · Middle Eastern Anthroscapes

1 Introduction

Soil constitutes the infrastructure for manufacturing, through the biological process of primary production, food, forage, fiber, timber, and biochemical compounds of medicinal, herbal, cosmetic and other industrial uses. Yet, soil is a fragile fabric, sensitive to the elements and to human impact. These jointly can reduce the soil's capacity to provide for primary productivity. This reduction, which intensifies with the mounting pressure on soils to provide for the needs of the burgeoning global human population, is termed "land degradation", and its salient expression is soil erosion.

Land degradation through soil erosion is a worldwide phenomenon occurring in most types of terrestrial ecosystems (human shaped/reshaped landscapes: Anthroscapes), but most prevalent in dryland ecosystems, which include dry woodlands, shrublands, grasslands, savannahs and deserts, and cover some 40% of the global land. When land degradation occurs in drylands, it is termed "desertification" and defined (by the Millennium Ecosystems Assessment) as "persistent reduction of productivity in drylands". Desertification, i.e., land degradation in drylands has drawn the attention of the international community ever since the Rio Summit, which highlighted all expressions of environmental degradation, land degradation included, as impediments to sustainable development and as drivers of rural poverty. This awareness resulted in the establishment of the United Nations Convention to Combat Desertification (UNCCD) that entered into force in 1997, a couple of years before the US-AID project "Monitoring and Evaluation of Watersheds in the Middle East Region" (or the "MEW-MERC") was conceived. This heightened awareness to land degradation in the drylands emanated from both the high fragility of the thin topsoil of drylands, and from the rate of population increase in the drylands, higher than in any other global ecosystem (Adeel and Safriel 2005).

Compared to all other internationally recognized regions, the Middle East is unique in that more than 90% of its area is drylands and is cultivated for millennia. This, together with the overall high annual population growth rate of 1.7% (in 2004), exposes most of the Middle East region to the risk of desertification, and its consequent rural poverty (in 2001) 23% of the Middle East's population lived on an income of under 2 US$ a day, leading to economic, social and political instabilities. These led the Working Group on Environment of the Multilateral Middle East Peace Talks to define Middle East desertification as an impediment to the Middle East Peace Process, and to commission a joint project entitled

"Initiative for Collaboration to Control Natural Resource Degradation (Desertification) of Arid Lands in the Middle East" (or the "Desertification Initiative", Safriel and Abu-Sliman 2007). This project started in1996, thus preceding the MEW-MERC that was launched in year 2000; both projects, however, terminated in 2005.

Like the Desertification Initiative, the MEW-MERC was born out of an expectation that regional technical cooperation could be an instrument for peace and stability in the Middle East, and between Israel and her Arab neighbors in particular. It was believed that such cooperation on desertification, which is of a common concern, could potentially establish channels of technical dialogue and exchange, thus supporting the peace process. In addition, the Israeli-Arab cooperation in this MERC project was enhanced by including a fourth partner-non Arab and non-conflict partner, but a dryland country in the eastern Mediterranean basin, experiencing severe soil erosion and struggling with technical problems identical with those confronted by the Middle Eastern partners.

Due to the geographical scale and severity of desertification, it attracted much research that generated knowledge regarding the causes and spatial dimensions of soil erosion especially in the drylands. Similarly, many practices and methods to restore degraded lands and arrest soil erosion have been devised. The prevailing problem, though, is that the causes of desertification and the appropriate means to avoid it greatly vary depending on local biophysical, ecological, social and political circumstance. Hence, it is often not known which land management practice is best to apply to a given area/Anthroscape in order to minimize risks of land degradation. Detecting and quantifying soil erosion is elusive, due to the diversity of both its drivers and expressions (wind or water erosion, gully or sheet erosion, respectively). Hence monitoring erosion is not straightforward, and it is often preferable to measure erosion proxies. For example, instead of measuring how much soil was lost, it is often preferred to measure how much soil was deposited at the receiving end of the erosion chain. Furthermore, soil deposition (or sedimentation) due to soil erosion may be harmful just as, or even more, than the productivity lost due to the erosion: it can block water reservoir or damage roads. Furthermore, soil erosion can be inferred by measuring the suspended matter in floodwater. Similarly, it is often easier and more reliable to evaluate methods of arresting soil erosion by measuring primary productivity, assuming direct relations between soil erosion and reduced productivity.

In whichever way erosion is measured, the manager and land user need to attribute the obtained value of eroded soil to its driver, or cause. However, the amount of soil eroded from a given land unit depends not only on that unit's properties, but mostly on the nature and human use of adjacent land units, especially those above it within a watershed system. Thus, for both measuring erosion and understanding its causes it is necessary to move from the local to the watershed scale. Furthermore, especially in drylands, where productivity is constrained by water, the most effective means of using land is through the integrated "Anthroscape approach" at the watershed scale, whereby people manage dryland watersheds for an efficient use of the water these geographical units collect, store and release. This is because, especially in drylands, soil productivity is constrained by water, which depending on management practices can either increase productivity when penetrating and being stored in the soil, or decrease it by eroding the soil through intense surface runoff. In this chapter we report on observations and measurements in selected arid and semiarid watersheds in the Eastern Mediterranean (Turkey) and the Middle East (Jordan, Palestinian National Authority (PNA) and Israel), and discuss gained insights useful for dryland watershed management.

2 Materials and Methods

2.1 Materials

2.1.1 The Study Watersheds-Features, Land Use, Management and Monitoring Objectives

Though each of the national teams had a previous experience and knowledge of its project's watershed prior to the project's initiation, once the project took off they collected more information on the study watersheds, what assisted in further implementation of the project. The following section describes the study watersheds by countries, whereas the rest of this chapter reports on project implementation and its results by themes.

2.1.2 The Yatir Forest (Israel): Afforestation of a Semiarid Rangeland

The Yatir Forest extends over 30 km^2 of hills (elevation ~600–650 m above sea level), at 31°21′N and 35°02′ E, within the sharp transition between the semiarid to the arid drylands of the Negev Desert (Fig. 1). The mean annual rainfall (in 1964–2000) is 255 mm, usually during ~10 days between November and March, and an average maximum and minimum temperatures of 32.3°C and 6.9°C, respectively. Yatir lies mostly on rendzina soil and lithosols on a hard chalk of the Upper Cretaceous Cenoman-Turon origin, and is covered with steppe vegetation, mainly *Sarcopoterium spinosum*, *Phlomis brachyodon* and *Artemisia sieberi*. But in much of the area the natural rangeland vegetation was replaced by Aleppo Pine (*Pinus halepensis*) as well as a fewer stands of – *Atlanta terebrinths*, cedars, oaks, Judas trees, cypresses and others. This afforestation was initiated by an afforestation agency (the Jewish National Fund, JNF) in 1964 that by the time the project took place planted there more than 4 million trees. The older part of the forest is located at the higher reaches of the Yatir watersheds where runoff harvesting was not implemented, but younger plantations ("newly-afforested"), mainly in the lower reaches are runoff-supported by extensive landscape manipulations that include digging run-off harvesting contour trenches. The land use is pastoral, by

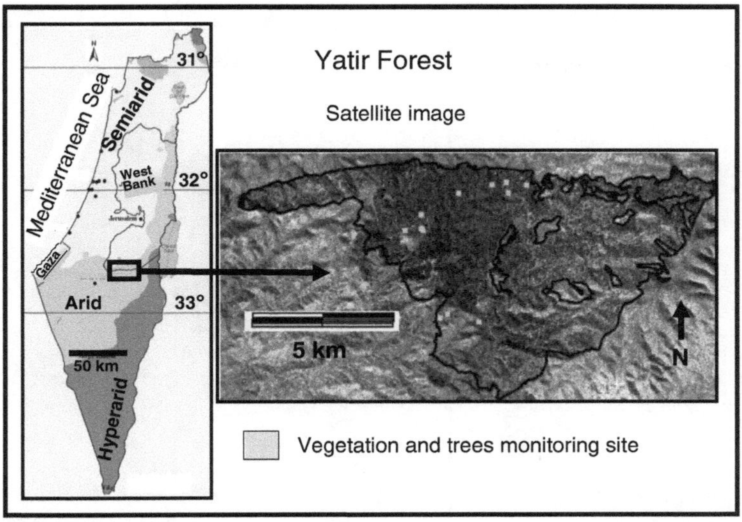

Fig. 1 Location of the Yatir Forest, Israel. *Left* – dryland zones of Israel and the Palestinian National Authority; *Center* – Afforestation projects in Israel; *Right* – Satellite image of the Yatir Forest, numbered points are the vegetation research sites

local settled Bedouins that use the forest for foraging and firewood, and by visitors from outside the region, for recreation.

The project objective was to monitor the effect of the old afforestation and the newly afforested sections of this watershed on runoff generation, soil erosion and sedimentation, as well as on plant cover and biodiversity. This (a) by equipping three sub-watersheds within the major Yatir areas, the Bikhra watershed (of 18.5 km^2) with monitoring devices – an afforested sub-watershed (1.6 km^2) mostly covered by pine, a non-afforested sub-watershed (0.6 km^2) and a newly afforested sub-watershed (2.1 km^2) for establishing rainfall-runoff-erosion relations at the whole watershed scale; and (b) by establishing experimental plots in which the indigenous vegetation cover within and off the afforested area was studied at the local scale, for evaluating the independent effect of the vegetation and the effect of the trees on runoff, erosion, soil properties and plant biodiversity features.

2.1.3 Humret Es-Sahin Watershed (Jordan): Rangelands Rehabilitation over an Aridity Gradient

Humret Es-Sahin is a watershed starting in the Al-Salt city region (32°05′N, 35°45′E) and draining into the Jordan Rift Valley, from an altitude of 437 m above sea level to 200 m below sea level (Fig. 2). The watershed constitutes an aridity gradient; from roughly 500 mm mean annual rainfall at its highest, semiarid dryland section, to some 50 mm at its lowest, arid dryland section. This aridity

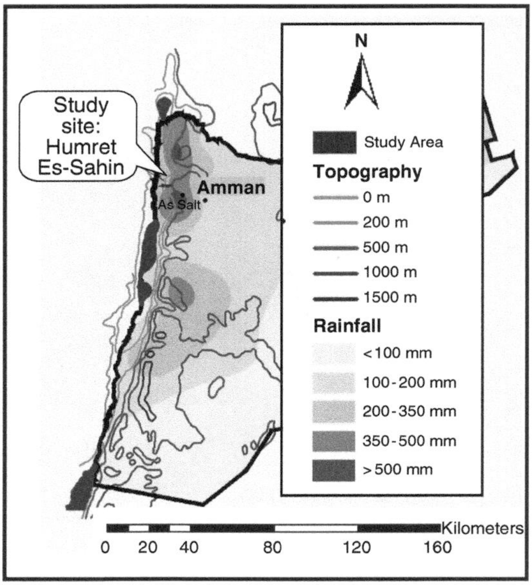

Fig. 2 Map of Jordan, with elevation and rainfall contour lines. Inset on *top left* – the study watershed of Humret Es-Sahin

gradient is expressed not only in precipitation but also in evaporation, which was measured by the Jordanian team (in 2004, Wadi Dafali catchment) – 2,535 mm in the mesic section of the watershed and 2,716 mm in its drier section. The rainy period is between October and March. The watershed is of about 10 km^2, and within it several sub-watersheds were studied-Wadi Um Al-Shananir (398–100 m above sea level), Wadi El-Mintar (0.42 km^2, 184–148 m above sea level), and Wadi Ed-Dafali (1.6 km^2, 435–50 m above sea level). Mean monthly temperature ranges from 11.6°C in February to 29.1°C in July at Wadi Dafali, and 12.8°C in January to 30.5°C in July for Wadi Mintar. Due to being a part of the Rift Valley deep depression, the study area has steep slopes, of 8–20%.

Soils are aridisols of high carbonates and low organic matter contents and with no distinct horizons, especially at steep slopes. The underlying rocks are lower cretaceous sandstone and are covered with a mix of Mediterranean and Asian vegetation shrubs and annual plants, as well as scattered acacia and *Ziziphus* trees, and exotic eucalypts. There are many contact-fault springs that discharge water from the underlying sandstone, and these water sources are augmented with many collecting wells. The study site is virtually non-inhabited, and most of it is protected from grazing, except one non-protected site. The management is geared towards range rehabilitation, more for demonstration and training in conservation by the local Al Balqa Agricultural University, than for rehabilitating its pastoral function for the benefit of a local pastoralist community.

The objective was to use this watershed for studying rainfall patterns and their relation to generated runoff and sedimentation, and for monitoring the effects of soil surface management practices, mainly furrowing, on erosion rates, soil moisture dynamics, surface runoff and sediment transport (in one experimental site within Humret Es-Sahin, watershed the University of Balqa Research Station, 20 km west of Al-Salt city, 197–397 mm of rainfall, minimum and maximum temperatures of 9°C and 37°C, respectively). Four locations within the watershed (Kena, Arsad, Mintar and Ghor, as well as one non-protected site) were used for studies of the local plant biodiversity.

2.1.4 Southern Hebron Highlands (Palestinian National Authority): Promotion of Agro-Pastoral Systems Along an Aridity Gradient

The Palestinian team established three research sites, about 15 km apart, at the southern part of the Southern Hebron Highlands, which are at the southern section of the West Bank. The three sites represent an aridity gradient: Sorif with 350–400 mm rainfall and potential evapotranspiration (PET) of 1,400–1,600 mm being thus a semiarid site; and Dura with 300–350 mm rainfall and 1,600–1,800 mm PET, thus of higher aridity than Surif but still defined as a semiarid site (Fig. 3). Whereas these two sites are on the Mediterranean side of the Hebron Highlands' water divide (mean temperatures of the hottest month are ~28°C), on Upper Cretaceous Cenomanian and Turonian limestones, the third site, Bane Noem is on the Dead Sea side of the water divide, on Upper Cretaceous Senonian chalky rocks. It receives 250–300 mm rainfall and its PET is within the higher part of the 1,600–1,800 mm range (mean temperatures of the hottest month is 32°C), thus this site is in the arid zone. The three sites are close to villages, and are under intensive agro-pastoral use. The objective was to monitor the effects of soil surface and landscape management practices, such as terracing, furrowing and several types of water-harvesting structures, on erosion rates, surface runoff, soil

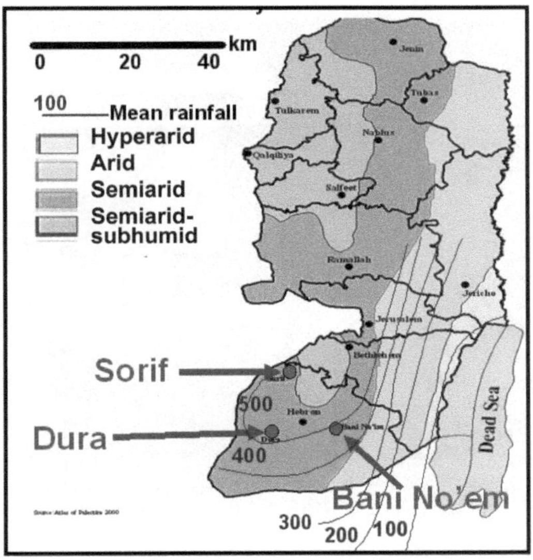

Fig. 3 The Palestinian project sites (*red dots*) within the West Bank, PNA, with rainfall contours and the extent of aridity zones

moisture, and the response of crop plants (mainly olive trees) and plant biodiversity to these management practices. Each of the three study sites was divided to a representative control and "treated" area with "treatments", or management practices, already in place or constructed by the research team.

2.1.5 Kızılöz Microcatchment (Turkey): Rehabilitation of Sylvi-Agro-Pastoral Watershed

The Kızılöz Microcatchment is a 64.5 km^2 watershed in a semiarid–dry subhumid highland region (watershed elevation ranges 1,250–2,050 m above sea level) of Eastern Anatolia in the Malatya region (ranging between 38°54′N–38°49′N and 37°41–37°47′E), some 100 km off the City of Malatya (Fig. 4). Mean annual precipitation of 387 mm including 170 mm of snow in winter. Precipitation occurs the year round but there is three relatively dry months (July, August and September). Mean annual temperature (in Malatya) is 13°C, and 25°C and 42°C are minimum and maximum temperatures, respectively. The Kızılöz watershed is one of 17 watersheds under the Eastern Anatolia Watershed Rehabilitation Project carried out by the Government of Turkey. Some of the largest dams of Turkey are at

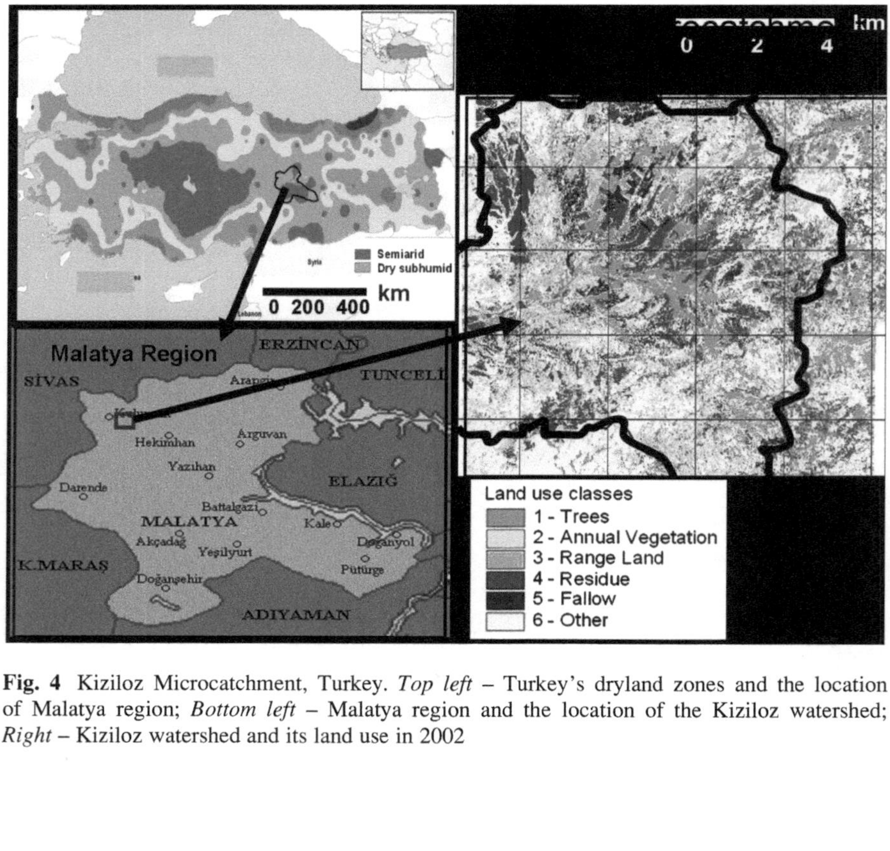

Fig. 4 Kiziloz Microcatchment, Turkey. *Top left* – Turkey's dryland zones and the location of Malatya region; *Bottom left* – Malatya region and the location of the Kiziloz watershed; *Right* – Kiziloz watershed and its land use in 2002

the lower reaches of this region, suffering from sedimentation of soil eroded in the upper reaches' watersheds, including the Kızılöz Microcatchment. The Eastern Anatolia Rehabilitation Project's objective was to control soil erosion using means that would also benefit land productivity, thus both reducing sedimentation damage in water reservoirs and improving livelihoods of the rural population. The Turkish team of the MEW-MERC project was engaged in both implementing the Government's project, and in monitoring its effectiveness in controlling soil erosion and improving land productivity. The Kızılöz Microcatchment includes forests, orchards, croplands and rangelands, used by local rural populations. The management included terracing and contour furrows for soil conservation and water harvesting, and planting forest and fruit trees for both expanding vegetation cover and improving economic productivity.

Three sub-catchments within the microcatchment, at elevations of 1,300–1,355 m above sea level and sized 25.7, 20 and 66.6 km^2, were fitted with monitoring devices, and six rain gauges were positioned within the microcatchment, as well as a flume at the major outlet of the whole microcatchment. This equipment was used for monitoring precipitation and runoff, as well as sediment discharge and deposition. The response of trees in growth parameters to the management practices was monitored on the ground and satellite imagery of the watershed was analyzed for detecting overall vegetation response.

2.1.6 Summary of Watersheds' Features, Land Use, Management and Research Objectives

With the common denominator of addressing soil conservation/erosion and monitoring both directly and indirectly, the watersheds selected by the four partners created a diversity of environmental conditions, land uses and management practices, thus providing a potential for an extensive coverage of the project's theme.

The watersheds of the non-Middle East country are at much higher altitude and geographical latitude and enjoy somewhat higher precipitation than the three Middle Eastern watersheds; the latter are in the arid/semiarid zones, whereas the watershed of the Turkish partner is more mesic, and have snowfall in the winter. The studied watershed in the PNA and Turkey are inhabited by local rural communities who depend on their productivity, which supports sylvi-agro-pastoral livelihoods. The Israeli watershed is used too in supporting pastoral activities, but these constitute only a small part of the income of the local communities. The Jordanian watershed is managed for educational and demonstration purposes and is not used by local communities. The Palestinian and Israeli teams studied the effects of ongoing management, terraces in the PNA and afforestation and furrows in Israel. But on top of it, both teams as well as the Jordanians experimented with various water-harvesting practices. The Israeli team studied an afforested and a forest watershed, respectively, managed by a national authority. The Palestinians and the Jordanians explored watersheds that are grassland/shrubland ecosystems. Both the Palestinian and the Turkish teams explored watersheds that supported orchards. Finally, whereas three of the teams selected watersheds sections with rather identical climate within it, the Palestinians and in some way the Jordanians, established several research sites along an aridity gradient. In spite of this diversity, all the national teams studied the results of management aimed at conserving soil, reducing erosion and deposition, reducing runoff, and increasing land productivity.

2.2 *Methods*

2.2.1 Precipitation

The number and density of rain gauges installed in the watersheds depended on the size of the study plots and the variability in environmental features within the watershed (e.g. altitudes, slopes, aspects, vegetation cover, etc.) and in the experimental or monitoring objectives and designs. The number of rain gauges ranged from total of 6–10 for each watershed, except for Yatir watershed where a total of 97 gauges were installed. Most of them were simple "mini" gauges (a glass tube within a hollow metal pole) and makeshift leveled food tin cans (of 8.4 cm diameter) installed on 0.2 m poles, to reduce vandalism and theft. The Israeli team found a strong correlation between data obtained by the "mini" gauges and data of adjacent standard met station ($r^2 = 0.92$, $P < 0.001$), and between the tin cans and the "mini" gauges' ($r^2 = 0.98$, $P < 0.001$). In each watershed, several of the gauges were of the recording type which enabled the computation of rain intensity (one of them with means to measure melt snow, in Kızılöz).

2.2.2 Runoff and Sediment

Runoff as well as its associated sediment concentration was measured in several manners. All partner teams monitored runoff generated from bounded plots (either treated experimentally or representing landscape features) by funneling it through a plastic pipe into a collecting container protected from rain. The sediment was allowed to settle and the amount of water and of sediment following a storm event was measured separately. Alternatively, the collected runoff was automatically recorded (by the Israeli team), using a specially built "tipping bucket" mechanism. For measuring the discharge of sub-watersheds and entire watersheds more elaborate weirs were constructed by damming the watershed exit and forming a small pond where maximal water depth was measured, or the instantaneous discharge was determined by automatic level recorders or by a depth sensor, their signal being transmitted to a data logger. Another variant was a limnograph installed at the outlet of the Kızılöz watershed, next to an old bridge whose cross section was used for the calculation of the total stream flowing through the bridge's cross-section. In some cases, not only water volume was measured but also chemical features of flow samples.

In all these weirs the water-transported sediment was measured in several manners: Automatic 24-bottle pump samplers were installed in Parshall flumes; next to the end pipes that siphon water into the samplers a turbidity sensor was installed to continuously determine the turbidity of the water.

Sedimentation plates were placed within the small pond of the weir or a concrete layer covered the pond surface and the accumulated sediment was collected after each rainstorm. Next to the limnograph in Kızılöz streamflow sediment was sampled by hand during flow events, for sediment analysis and the overall amount of transported sediment, which was calculated using the recorded stream flow. Sediment production from road construction and use, from thinning, as well as from "non-treated" areas were estimated independent of runoff, using replicated silt fences. All the national teams constructed a relatively large number (~10–20) of the simple devices to collect and monitor runoff and its suspended sediment from relatively small plots, and all (except the PNA team) installed a much smaller number (1–5) of the more elaborated weirs; the Israeli team collected sediments from sediment traps.

2.2.3 Soil-Erosion, Properties and Moisture

Soil erosion was measured directly by the Palestinian and the Jordanian teams by monitoring the degree of exposure of permanently located pins in experimental plots, and the Israeli team monitored gully head retreat in Yatir.

The Palestinian, Israeli and Jordanian teams evaluated soil properties and responses to different treatments by collecting soil samples of the top 2 cm (for soil seed bank survey, Israel), 10 cm, 15 cm, and/or at different layers down to 90 cm depth. These were collected, dried, crushed, sieved and subjected to analyses using conventional methods specific to each of the parameters – pH, electrical conductivity, soil organic matter, $CaCO_3$ content, nitrogen, phosphorous and potassium concentrations. The Palestinian team also implemented a soil bioassay to evaluate soil productivity by germinating corn seeds in soil samples kept in a greenhouse for 35 days, and measuring stem diameter, height, dry weight of shoot and root and surface root area.

Soil moisture was measured using two methods: In the fields – by the gravimetric method (PNA), applied to samples taken from several depths, from surface layer to a depth of 70 cm; in the field-by neutron probe (TROXLER 43001), using access tubes to a depth of 90 cm (Jordan) and 5 m (Israel). In Yatir an in-situ calibration of the neutron probe was carried out for each of the 0.15 m. upper layers (0–1.5 m) and for the other depths. Also in Yatir CS620 HydroSense® Water Content Sensor (Campbell Scientific Inc., Logan, Utah, USA) was used to measure soil moisture at 12 cm depth. The Jordanian team measured field capacity and permanent wilting point by using ceramic plate pressure apparatus at 0.33 bar (33 kpa) and 15 bars (1,500 kpa), respectively and calculated soil water storage using the difference between the volumetric soil water content (%) after each rainfall storm and before it, multiplied by soil depth.

2.2.4 Vegetation

Two types of vegetation were separately studied: trees in forest stands (Israel, Turkey) or in orchards (PNA, Turkey) – parameters that relate to their growth, surface cover and density; and non-woody indigenous natural vegetation, in rangelands or afforested areas – its cover, biomass and biodiversity components.

2.2.5 Trees

For monitoring tree growth and cover plots within stands that represent different habitats or treatments were selected. Within each of them, a set number of individual trees were randomly selected for repeated measurement. Diameter at Breast Height (DBH) – (Turkey, Israel) or trunk width (PNA), canopy diameter (using tape), and tree height (using clinometer) were measured. These measurements were either used to evaluate tree growth (PNA). Or, data were used for establishing "training" sites (areas representing known land cover that appears homogenous on the image) for ground-truthing that enables using satellite imagery for monitoring at much larger spatial scale (100 plots with a 5.6 m radius in Turkey, 72 circular 200 m^2 plots positioned within 2 km^2, Israel).

Several IKONOS images of the studied watersheds were used for this training, which enabled estimating tree density. The images covered 10 × 10 km at 1 m (Turkey) or 2 m (Israel) spatial and panchromatic resolutions, and 4 m multispectral (4 bands, in the blue, green, red, and near-infrared regions). They were used for calculating Normalized Difference Vegetation Index (NDVI) that estimates primary productivity, to be used for estimating tree density and canopy cover (Israel) and to determine land use changes (Turkey).

Regarding tree density, even the high-resolution space-born sensors that produced these images cannot detect a single tree in the stand, but can provide overall canopy cover. Correlations between stands' density and overall canopy cover measured in Yatir enabled a large-scale estimation of tree density using the images. To control for background reflectance minimum and maximum NDVI were calculated, the minimum denotes the background reflectance of areas covered with litter only. In-situ measurements were done using LICOR LI-1800 high spectral resolution field spectroradiometer in the range 400–1,100 nm with spectral resolution of 2 nm, attached to a telescope positioned 80 cm above the surface. Seven plots of different tree densities were used for comparison with imagery's NDVI results.

Regarding trunk diameter two spectral vegetation indices were calculated by the Israeli team and regressed on trunk diameter in four vegetation-signature reflecting sampling sites. Leaf Area Index was measured in the field to compare with its calculation from images. ΔT-SanScan Canopy Analysis System for measuring photosynthetically active radiation beneath the canopy was used for defining transmittance at different sun angles during the day. In all uses of the satellite imagery, the field sites were detected by Global Positioning System (GPS) and were subseted from the image using ERDAS Imagine image processing software.

The Israeli team measured light penetration that is indicative of the Leaf Area Index on the ground also by using hemispherical photography by Nikon Coolpix 950 camera and a Nikon FC-E8 Fisheye converter mounted 1 m above the ground with the lens facing skyward and the camera body leveled and aligned with the magnetic North. The photographs were taken under clear sky conditions shortly after sunset, in order to avoid large variation in brightness across the images, and obtained midway along each sampling transect. The Fisheye photographs were analyzed using the XLSCANOPY program (Regent Instruments Inc) in order to compute and monitor canopy cover in stands of planted trees.

To evaluate the impact of planted tree on soil water, transpiration of selected trees was measured by the Israeli team using the heat-pulse technique, whereby two temperature sensors (3 mm diameter and 60 mm long rods), one with six embedded thermistors placed at 8 mm intervals and the other with only one thermistor served as a reference. The six-thermistor probe was inserted radially into the trunk in a hole drilled 15 mm above the heater, while the reference probe was inserted into a hole drilled at least 100 mm below the heater. The heating element was a 0.8-mm diameter, 80-mm long Ni-Chrome wire, enclosed in a stainless-steel tube with an outer diameter of 1.8 mm. A battery-operated data logger connected to a junction box enabled measurement on seven trees at hourly intervals. These trees were selected such that their diameters at breast height matched the diameter distribution of the population. Transpiration fluxes were computed from the measured data using the algorithm of Schiller and Cohen (1998). Stemflow, a potentially important pathway of intercepted rainfall that could affect soil moisture and runoff generation was measured by installing collectors on the tree trunks, one meter above ground. The collectors were made from plastic rings sealed with silicone rubber and the intercepted water conveyed to a container that was emptied after each rainfall event (Israel).

2.2.6 Non-woody (Herbaceous) Natural Vegetation

The vegetation was sampled using several methods, as follows:

- Transects: All plants are clipped along 10 cm wide and 3–21 m long transects, length depending on habitats' spatial distribution, or the length of all transects maintained constant, at 10 m (Israel).
- Step point transects: Samples are taken or estimates are made only on what occurs below the tip of a boot along one hundred steps transect started from randomly determined point (PNA).
- Permanent line transects: Samples are taken or estimates are made across permanent 10 m long transect line, established within an experimental plot (PNA).

- Quadrates: 100 × 100 and 300 × 300 cm (Jordan), 20 × 30 or 20 × 200 cm (Israel), 25 × 25 and 100 × 100 cm (PNA), randomly set up within each habitat and/or land use and/or experiment types.
- Permanent plots: were monitored throughout the project period (Turkey).

Number of transects and quadrates was set to enable full coverage of habitat types and experimental set up, and number of transects and quadrates amounted to scores and hundreds, respectively.

Within such sampling and monitoring units total percent cover (also bare soil or rock, PNA), or percent cover of each species (Turkey) and vegetation height (Jordan) were estimated. When plants were collected, the collected plants from each sampling unit were separately bagged and then frozen until further processing, then identified and weighted after drying for 48 h at 70°C in a ventilated oven (Israel) or at 105°C or 65°C and until weight loss ceased (Jordan).

Individual plants were identified to species level, number of individuals of each species was counted, and species were categorized by families, functional groups (shrubs, grasses, forbs, legumes) and biogeographical distribution (chorotypes – Israel, PNA, Jordan). Several plant community indices – richness, diversity, evenness, density, relative abundance and relative frequency were calculated (different assortments of these variables calculated by Jordan, Israel and PNA) as well as community similarity indices – Jackard (PNA) or Sorensen (Israel) indices – for quantifying similarity of plant communities of different habitats or "treatments" were computed. Following their identification, all plants were dried for 48 h at 70°C and weighed, for obtaining overall dry biomass (all partners).

Though no study of belowground plant biomass was carried out, the soil seed bank was assessed by collecting soil samples by the Israeli team (uppermost 2 cm of soil, with a metallic cylinder of 8.4 cm diameter). The samples were sieved and placed over sterile soil in a greenhouse and watered to field capacity every second day as the subsequent growing season started. The seedlings were grown until their identification was feasible. For comparing the germination achieved in the greenhouse and the germination in the field a circular subplot of size as that of the cylinder with which the soil samples were collected, was marked close to each plot in which a soil sample had been taken, and all the plants that germinated in this plot were collected, pressed, identified and counted. Species richness, diversity and evenness of the seed bank that germinated in the field and in the laboratory, were then calculated.

The IKONOS imagery was used for both trees and non-woody vegetation areas in the studied watershed for between-year (Turkey, Israel), or overall primary productivity expressed by NDVI (Turkey, Israel), and for assessing changes in land uses (Turkey).

3 Results

In the following, the salient results and their interpretation are presented, dealt with by themes rather than by countries.

3.1 Precipitation Features

Rainfall was monitored in all sites and analyzed in different ways by each team. Comparing with long-term means the average rainfall for the 5-years of the project was slightly higher than the long term mean in Yatir but within one standard deviation about the long-term average. The variability between years was high in all project sites. For example, the lowest annual rainfall was 61% and 73% of the highest annual rainfall, in Kızılöz and Yatir, respectively, a trend that is compatible with the position of these two sites along the aridity gradient. Other analyses were on rainfall properties, which corroborated what has been already known of rainfall regimes in dryland areas. In Yatir, for example, the rainfall events of up to 10 mm depth (such events were the most numerous) did not contribute by more than 20% to the total annual rainfall (Schachnowitz et al. 2008). Where explored (in Yatir), no significant differences (ANOVA, $P > 0.05$) between afforested and non-afforested sites, nor between slope aspects or between positions along the sites slopes, as well as their interactions, were found for total seasonal rainfall.

In the afforested sites no significant differences were observed between gauges placed under high and low canopy cover (ANOVA, $P > 0.05$, Ariza 2004). Thus, precipitation variability (snow included, in the Turkish site) within the project sites of each partner was temporally high but spatially low. Since rainfall is the driver of runoff, erosion, soil moisture and land productivity, the rainfall data were used by all the teams in conjunction with these variables.

3.2 Rainfall, Throughfall and Stemflow

Whereas in all the study watersheds rainfall was measured and then related to other variables monitored in their project sites, in the Yatir forest (which is a result of afforestation with a locally exotic species) in addition to rainfall above the forest canopy, also the rainwater that reach the forest's soil surface was measured. In Yatir (Israel) a large spatial variability in the rainfall reaching the forest soil surface ("throughfall"), which was captured far better by a dense array of rain gauges than by a thin alignment, and is likely to be associated with the structural diversity provided by the trees' architectures. Nevertheless, when all mean throughfall and rainfall data were used in a regression analysis the team found a very strong ($r = 0.997$) and highly significant linear regression ($b = 0.93$, $P < 0.01$) of rainfall reaching the soil surface in the forest plots on rainfall monitored in adjacent treeless plots (and assumed to be identical with the one reaching the forest canopy). Using the obtained regression equation (Throughfall $= 0.93 \times$ Rainfall $= 0.76$), 77% of the smallest rain depth (5 mm) and 93% of the highest one (88 mm) reached the forest soil surface. The high-depth rainfall storms are those that contribute most to the total seasonal depth of the throughfall and only less than 5% of the total seasonal above canopy rainfall does not reach the soil surface. Such a high throughfall in a forest ecosystem is due to its low planting density, and hence canopy cover (Schachnowitz et al. 2008) .

Stemflow was very small – only 1–2% of rainfall. However, this flux could be more important than its low fraction would suggest, as the volume of water is concentrated on a very small perimeter surrounding the trunk that would penetrate deeply into the soil through the gap between the trunk and the soil. The depth of penetration minimizes the evaporative losses and thus most of it would be available to the tree (Schachnowitz et al. 2008).

3.3 Runoff and Sediment Discharge at the Watershed Scale

Runoff and sediment discharge from entire watersheds were monitored at watershed outlets. The results that follow represent an integration of processes over the whole watershed area, and thus provide a useful insight to regional planning and policy-making. Soil erosion at the watershed scale is driven by surface runoff which, in turn, is generated by rainfall storms, and the features of these relationships in the studied watersheds are described in the following sub-sections.

3.4 Relations Between Runoff and Rainfall

Measurements of rainfall and runoff generated by it were taken during individual rainstorms in the Israeli and the Jordanian sub-watersheds. The total runoff generated measured in the Jordanian sub-watersheds in 1 year amounted to 20% of the total annual rainfall for that year. Using a modeling approach, the measurements enabled reconstructing the runoff coefficient of that watershed for the years 1975–2003, the mean of which was 10%. The Jordanian values are much lower than those for the Israeli sub-watersheds of the Yatir area, which for individual rainstorms amounted to 50–60%. This difference is mostly due to the Jordanian sites being grazing-protected watersheds, whereas the Israeli one was an overgrazed watershed that has not been afforested. However, afforestation of the Israeli previously-overgrazed rangeland reduced runoff significantly.

Depths of the 45 rainfall storms that occurred in Yatir during the study period ranged between 1 and 85 mm (Cohen et al. 2008). Only one of these storms (of 61 mm depth) generated runoff in the afforested sub-watershed. Unfortunately, the amount of generated runoff was not measured, but the maximal discharge was less that 0.1 m^3/s (22 times lower than the highest maximal discharge in a non-afforested sub-watershed). In contrast, the newly (and more sparsely) afforested sub-watershed generated runoff from storms of 16–25 mm and 53–63 mm (Fig. 5). Yet some storms within these ranges or of higher rainfall did not generate runoff, both in the Israeli and the Jordanian sites (Fig. 6).

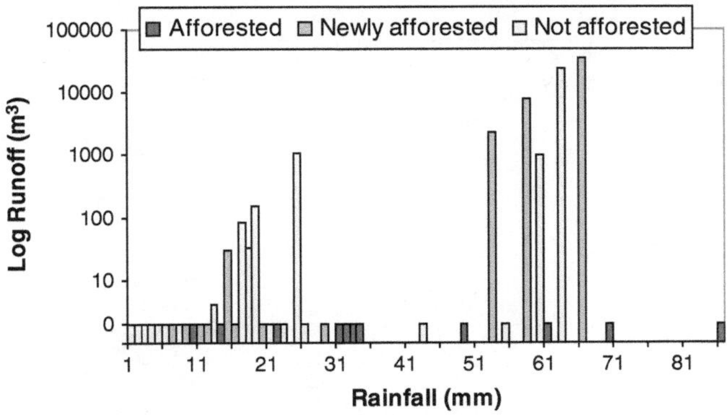

Fig. 5 Runoff discharge from Yatir sub-watersheds, Israel during four rainy seasons, 2001–2004, as a function of storm rainfall depth (discharge is log-scaled, but the filled-in columns in this level signify zero runoff generated by the various rainfall depths. Where these zero columns are absent, no storms of such depths occurred). Note that the afforested area generated no runoff

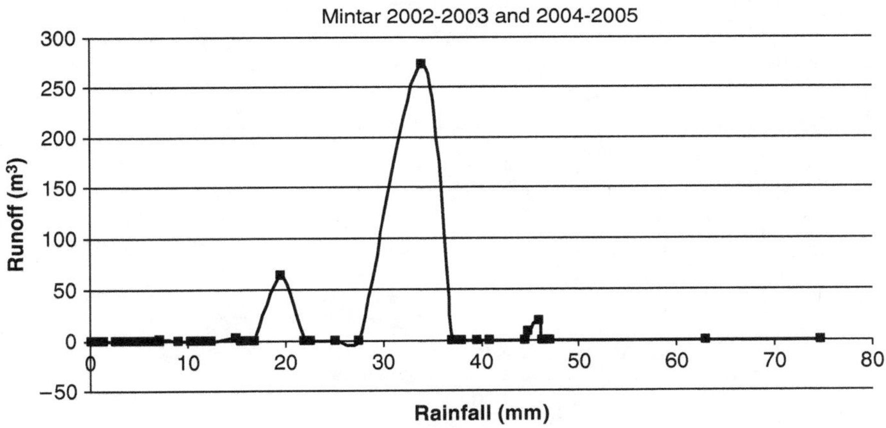

Fig. 6 Runoff discharge from Mintar sub-watershed, Humret Es-Sahin, Jordan for two rainy seasons, as a function of rainfall depth

Once runoff was generated, its overall discharge was strongly correlated with the rainfall depth in both the Israeli and the Jordanian sites (Fig. 7). Runoff coefficients in Yatir ranged 0.01–0.59 for individual storm events, with highest values for the non-afforested watershed, and lowest values for the young afforestation watersheds, but each also varied depending on storm rainfall (Frank-Wener 2006). Non-afforested overgrazed watersheds in Yatir also had much higher peak discharges than "treated" watersheds, when "treated" means sparse afforestation along contour lines, behind contour dykes constructed to reduce flow and store runoff (Fig. 8). The relations between runoff and precipitation in Turkey are different, since runoff is snow melt-dependent, such that runoff simply occurs only in spring, with no direct relations to rainstorm incidence and magnitude (Fig. 9).

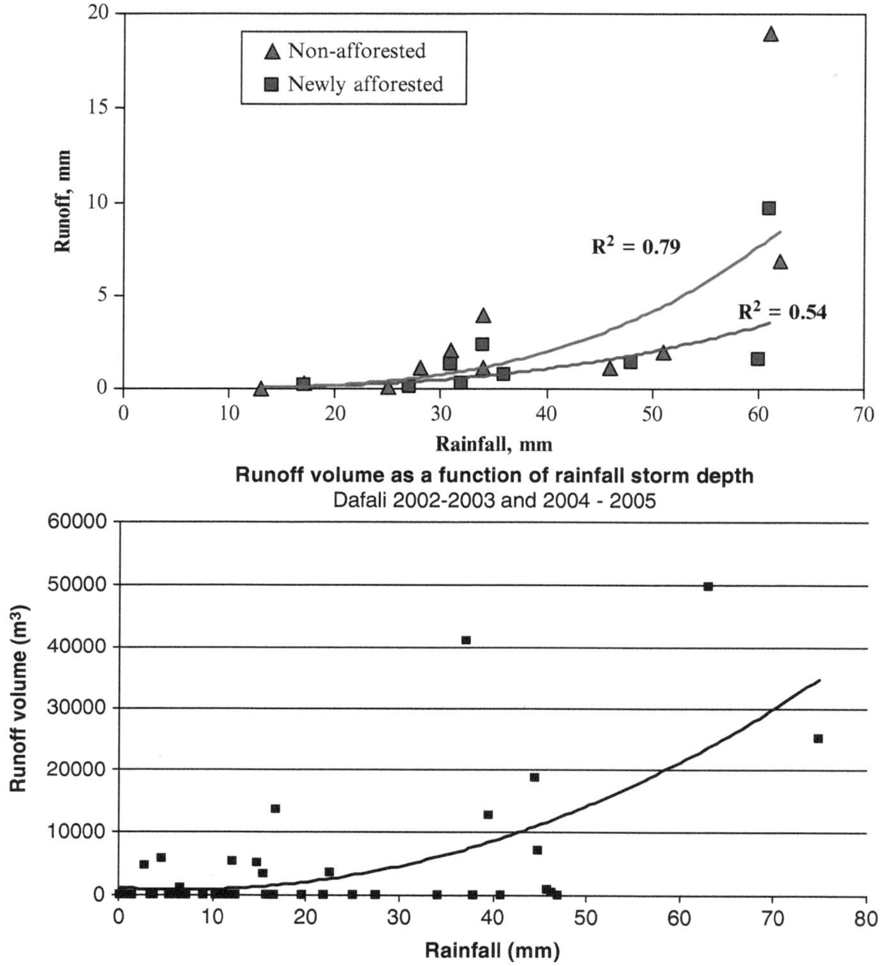

Fig. 7 Regressions of runoff on storm depth. *Top* – Yatir, Israel (runoff is expressed in mm, i.e., volume of water monitored at the weir divided by watershed area). Note that runoff increases with the storm rainfall depth more rapidly and more consistently (as expressed by the higher coefficient of determination) in the non-afforested than in the afforested sub-watershed; *Bottom* – Wadi Dafali in Humret Es-Sahin, Jordan (runoff is expressed in m^3). The Israeli "blue" curve is for a newly and sparsely afforested rangeland, and the Jordanian curve is for a protected rangeland (runoff is expressed in watershed water volume)

3.5 Relations Between Suspended Sediment and Runoff

Surface runoff has the potential to remove topsoil, and in drylands, more than 90% of the eroded soil carried by the runoff flow is transported in suspension (Powell et al. 1996). Therefore, when both water discharge and suspended sediment

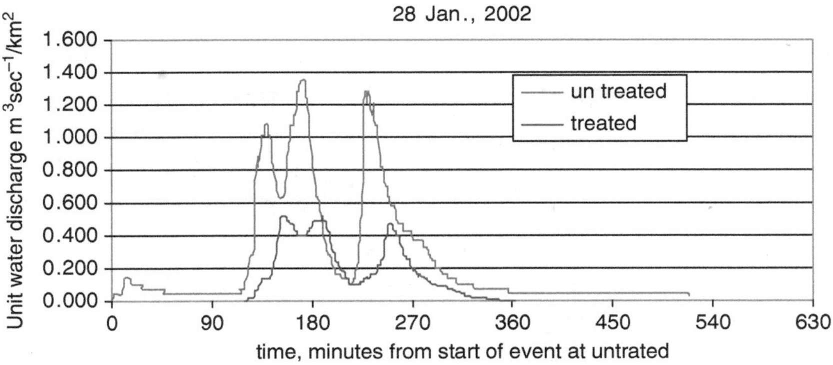

Fig. 8 Peak discharges of a rainstorm in Yatir, Israel, calculated as $m^2/s\ km^{-2}$ of non-afforested ("untreated") and newly-afforested ("treated") watersheds

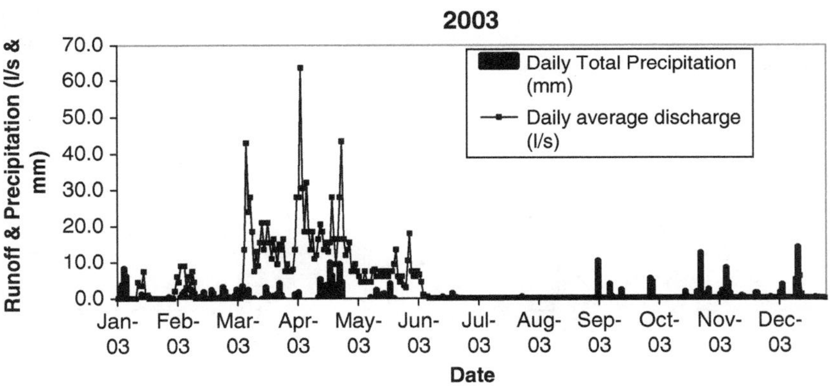

Fig. 9 Precipitation and runoff in the Kızılöz Microcatchment, Turkey

concentration are monitored the sum of their product represents the total sediment yield transported in suspension, and can serve as a reliable surrogate of the amount of soil eroded from a watershed during a storm. The study in the Turkish site demonstrated that the mean daily runoff discharge measured in liter of surface runoff per second is a good predictor of an overall suspended sediment discharge, expressed in ton per day (Fig. 10).

The relations between sediment concentration (expressed in g sediment per liter, Fig. 11) and runoff discharge (m^3/s), as well as the relations between the sediment concentration and turbidity were studied in Yatir (Fig. 12). Since calibration of turbidity depends on soil properties too, calibration was undertaken on sediment that had been transported in suspension. Using a dynamic calibration, a power relationship was found, which is sufficiently strong to recommend this measure as a substitute for (or as a predictor of) suspended sediment concentration.

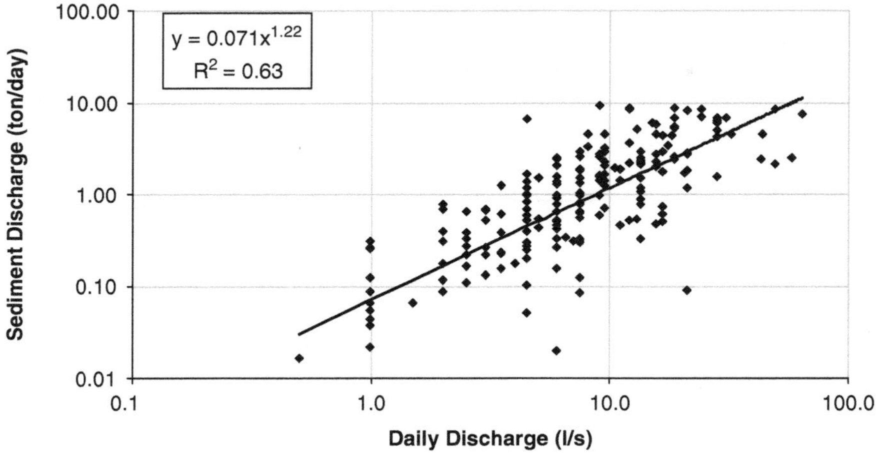

Fig. 10 The overall daily sediment yield as a function of runoff discharge in three experimental sub-watersheds in the Kiziloz Microcatchment, Turkey, during the only rainstorm that generated runoff from these sub-watersheds during the 5 years of the project

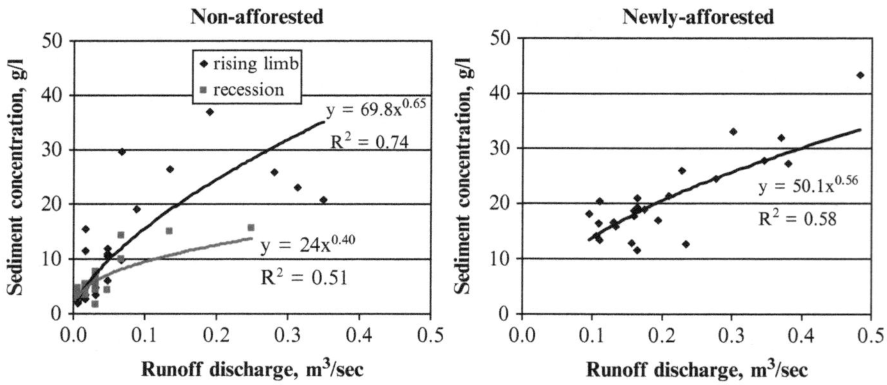

Fig. 11 Regressions of sediment concentration on runoff discharge for all the monitored runoff events during the rainy season of 2002–2003; 5 events in the newly-afforested sub-watershed and 6 events in the non-afforested sub-watersheds. *Left*: Sediment concentration increases with water discharge, more so for the "rising limb" (the initial runoff surge) of the hydrographs but less on the "recession" (the slow wane of the runoff flow), implying more sediment availability at the onset of a runoff event in the non-afforested sub-watershed, typical of dryland watersheds; *Right*: In the newly-afforested sub-watershed sediment concentration also increases with increase in water discharge, but less so than in the non-afforested basin Thus, although concentrations of sediment in the discharged runoff are comparable, they grow with increase in storm depth slower than in the non-afforested sub-watershed, indicating less soil eroded in the afforested than in the non-afforested sub-watershed

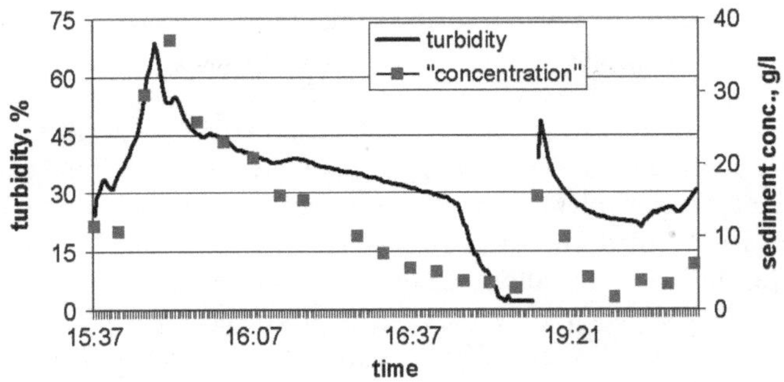

Fig. 12 Turbidity and sediment concentration sedigraphs – non-afforested watershed, 20–21 December, 2002 rainstorm, Yatir, Israel

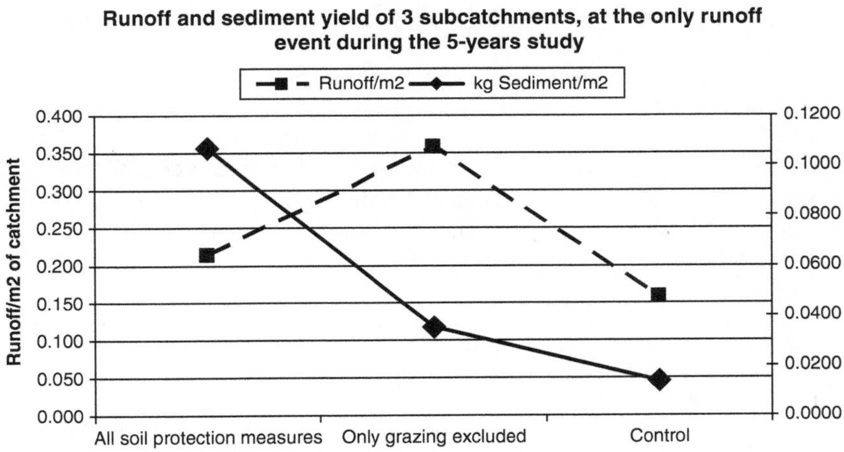

Fig. 13 Runoff and sediment yield of sub-catchments differing in their soil conservation measures, Kızılöz microcathcment, Turkey

3.6 Sediment Discharge in Relation to Watershed Management

Sediment discharge was explored in Israel and Turkey through surface runoff from sub-catchments representing different water management practices. The need to reduce soil erosion at the regional scale motivated exploring the effect of different measures for reducing runoff-driven soil erosion (carried out at the local scale, within the Kızılöz microcatchment) on the amounts of sediment suspended in the runoff generated from their experimental plots (Fig. 13). Contrary to expectation, runoff was negatively related to soil conservation measures such as terracing, planting trees and grazing exclusion; and the amount of sediment carried by the runoff was not directly linked to the magnitude of the runoff discharge. These results, however, derive from the sole runoff event that the studied sub-watersheds experienced throughout the whole period of the project, such that this observation does not have temporal (and also spatial) replication.

In Yatir measurements were made of runoff and its suspended sediment discharged by three different watershed management – afforested watershed (35 years, mono-specific, with no runoff harvesting structures), a newly-afforested watershed (4 years, mixed-species, planted in contour trenches behind contour dykes), and a non-afforested watershed grazed or over-grazed by local Bedouin herds (for the latter see Alexandrov et al. 2008). Whereas the afforested sub-watershed did not generate runoff in any of the 5 study years and hence no sediment, the newly afforested sub-watershed released both runoff and sediments. The regression of sediment concentration on water discharge for the non-afforested watershed is steeper than that for the newly afforested one (Fig. 11); the non-afforested watershed lost sediment with low runoff discharge, whereas the newly afforested started losing sediment only in higher discharges. It is likely, though, that the afforestation activities cause destabilization of the surface, which is thereby eroded by water during runoff-generating storms.

All the 12 monitored rainstorms combined discharged 18 and 38 mm of runoff, and 200 and 240 ton/km^2 of sediment from non-afforested and newly-afforested sub-watersheds, respectively (Fig. 14). However, whereas the difference in area-specific runoff yield between newly-afforested and non-afforested watersheds increased with decreasing storm depth, the sediment yield in the non-afforested watershed was higher than in the afforested basins for events of all magnitudes (Chocron 2009).

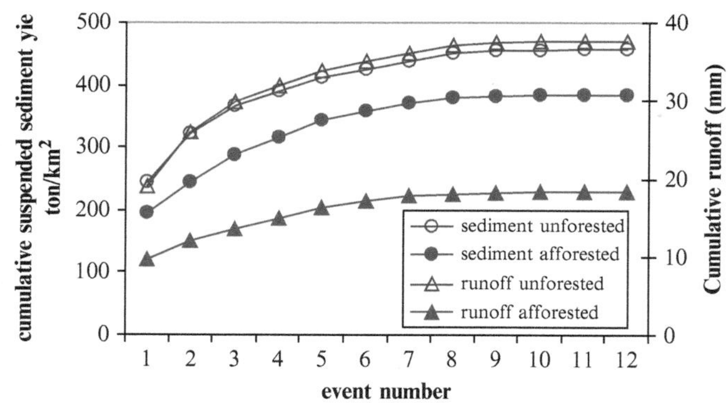

Fig. 14 Cumulative amounts of sediment and runoff during the rainy seasons of 2001/2002 and 2002/2003, Yatir, Israel. *Top*: Runoff generation and sediment production from the non-afforested and newly-afforested watersheds, Yatir, Israel. For the 12 monitored flow events (of the two rainy seasons) – from largest (#1) to smallest – the specific runoff (the volume of water per drainage area) of the newly-afforested basin is smaller than that of the non-afforested, the difference markedly large for small (right side of diagram) events when hardly any runoff was generated in the newly-afforested basin. The sediment yield in the non-afforested basin is higher than in the newly-afforested basins for events of all magnitudes; *Bottom*: Total yields for each of the 2 years, in each of the two sub-watersheds, newly-afforested and non-afforested

The effect of surface runoff on soil salinity in Yatir was indirectly assessed through measuring solute concentration in runoff during storm events. As expected, runoff washes away not only soil, but also surface solutes.

3.7 *Soil Erosion Through Gullying in Relations to Watershed Management*

A visible expression of runoff-driven soil erosion is the phenomenon of gullying and the retreat of headward gullies. This retreat was monitored in Yatir (in 9 gully heads during 2003 and 2004). A few of these gullies did not retreat, but for those that did maximal retreats were 45 and 88 cm for 2003 and 2004, respectively and the mean retreats were of 45 and 48 cm for the 2 years, respectively, whereas the means for all the 9 gullies were 12 and 38 cm for the 2 years, respectively. Since each of these gullies received runoff from catchments of different sizes, the ratio of retreat to catchment area was calculated, and the differences between the 2 "treatments" and the 2 years were related to their runoff discharge as measured in the weirs (Table 1). Indeed, the difference between the 2 years in gully headward retreat can be attributed to the higher intensities of runoff in 2004 than in 2003, as measured in the weirs that collected runoff from each of the watersheds, newly-afforested and non-afforested, in which these monitored gully heads occurred. This analysis corroborates that (a) gully erosion, a phenomenon at the local scale, is controlled by the magnitude of flow events detected at the whole-watershed scale, and (b) that the measures to increase runoff harvesting for the afforestation effort also reduce further headward retreat of already existing gullying activity, an effect that is more pronounced in years of high rainfall intensity.

3.8 *Watershed Processes at the Local Scale*

Though the runoff and sediment generation measured at the watershed scale provide for an integrative overview, studies at the local scale provide an important

Table 1 Gully retreat statistics (in erosion monitoring sites) related to runoff flow measured in sub-watersheds' weirs, Yatir, Israel

	2003		2004	
	Non-afforested	Newly-afforested	Non-afforested	Newly-afforested
Range of retreat in m	0–0.5	0–0.4	0–0.88	0–0.5
Range of retreat in m/dunam watershed area	0–0.11	0–0.11	0–1.87	0–0.12
Mean of retreat (in m/dunam area)	0.061	0.028	0.455	0.051
Annual flow volume (m^3) of whole	310	0	2.042	12.443
Mean rainstorm flow of whole watershed (m^3)	52	0	681	3.111

insights into local processes that affect integration at the watershed scale. For example, rainfall may be temporally and spatially variable over the whole area of the watershed, a feature that is often not captured by conventional rainfall monitoring networks. Also, the rainfall depth generating soil moisture at the different parts of the watershed cannot be directly derived from the measured volume of runoff discharge at the outlet. The MEW-MERC therefore paid attention to processes related to water, soil and productivity at the local scale.

3.9 Rainfall, Runoff and Sediment Deposition

In contrast with an annual runoff coefficient of 20% at the watershed scale, runoff coefficients ranging 14–83%, range depending on rainstorm depths and intensities, at small plots within a protected watershed originally used as rangeland (20 m × 6 m, slopes of 3% and 7%, Table 2) were found in the Jordanian sites. The regression of the runoff coefficient on rainfall depth is logarithmic (Fig. 15, red curve) – with much rainfall the soil becomes saturated and higher percentage of the rainfall runs-off; with even greater storm depths further increase in the percentage that runs-off is probably constrained by evaporation. The relations between the runoff coefficient and storm depth at the local scale of the Yatir Israeli site (measured within the pine forest) is logarithmic too but of lower correlation and much smaller runoff coefficients compared to the Jordanian ones, measured in a rangeland (Fig. 16).

The runoff generated at the local scale of the Jordanian within-watershed plots also generated sediment, yet even in this scale the relations between runoff and sediment discharge are not linear, as evidenced by comparing the rainfall depths and intensities that generated the minima and maxima of runoff (Table 2) and sediment discharge (Table 3). This is also supported by the Palestinian finding (Table 4) that the within-site differences between the 2 years (one more rainy than the other) in the amount of sediment carried by the generated runoff, did not correlate with the difference in rain amount between these year; less sediment was generated in the more rainy year, except in the driest site. At the same time, though, the between-site comparison is compatible with the annual rainfall difference between the three sites – more sediment with higher rainfall.

Table 2 Minima and maxima of runoff coefficients measured during 23 rainstorms in 2001/2002 and 2002/2003 rainy seasons, Humert Es-Sahin, Jordan

	Lowest runoff			Highest runoff		
	Storm			Storm		
	Depth (mm)	Intensity (mm/h)	% Runoff	Depth (mm)	Intensity (mm/h)	% Runoff
2001/2002	17	1.5	53	14	1.2	67
2002/2003	3.8	1	14	106	1.4	83

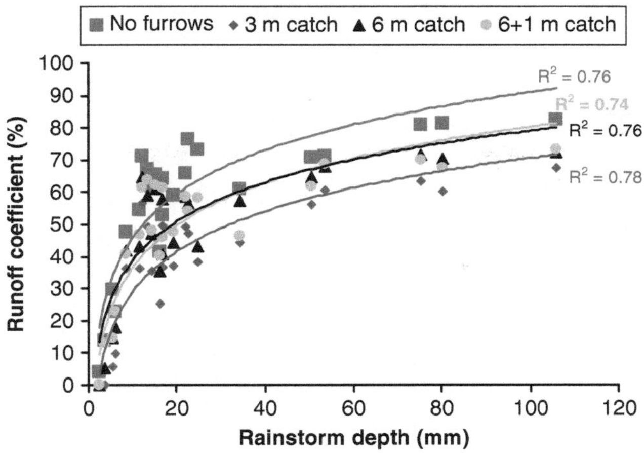

Fig. 15 Logarithmic regressions of the runoff coefficient of four 6 m × 20 m plots each, on rainstorm depth of 4 (2002/2002) and 19 (2002/2003) rainstorms at Humret Es-Sahin, Jordan "No furrows" is the non-treated control, "3 m catch" is the runoff produced from a catchment of 6 m width between two successive furrows, and "6 + 1 m catch" is that one produced by a catchment of 6 m, draining into two furrows separated by a 1 m catchment

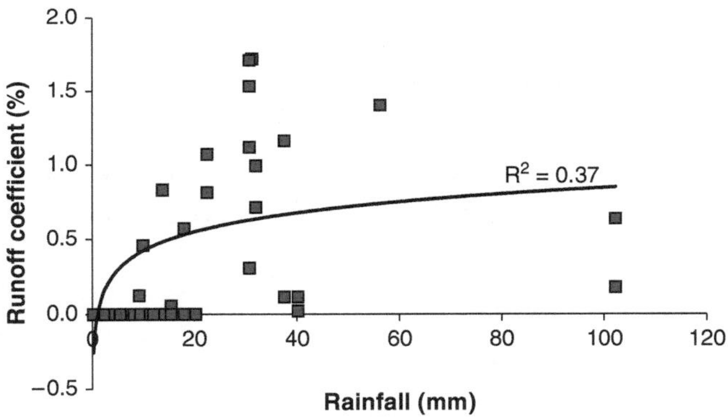

Fig. 16 Regression (logarithmic) of runoff coefficient on rainfall storm depth in Yatir, in three 70 m^2 plots having 3–5 pine trees each, within the Yatir forest, Israel (correlation not improved when runoff is regressed on storm intensity)

3.10 Soil Cover, Runoff and Erosion

The effect of non-woody vegetation on runoff generation was studied in non-afforested patches within the Yatir forest, Israel. In plots within this forest that did not have trees but just typical herbaceous rangeland vegetation (Fig. 17a, b), the generated runoff decreased as the rainy season progressed, with minimal runoff coefficient towards the end of the rainy season, when the standing crop of the

Table 3 Minima and maxima of thickness of sediment discharged by the runoff measured during 23 rainstorms in 2001/2002 and 2002/2003 rainy seasons and deposited in the runoff measuring container, Humret Es-Sahin, Jordan

	Lowest sediment accumulation (mm)		Highest sediment accumulation (mm)	
	Storm depth	Sediment depth	Storm depth	Sediment depth
2001/2002	14	12	22	23
2002/2003	2.4	0	80	44

Table 4 Sediment accumulated during 2003/2004 and 2004/2005 (g/m^2) in each of the three Palestinian sites

	Sorif		Dura		Bani Noem	
	2003/2004	2004/2005	2003/2004	2004/2005	2003/2004	2004/2005
Total annual rainfull (mm)	317	400	250	385	225	334
No runoff-harvesting	48 a	24 a	24 b	11 a	8 b	14 a
Terracing	23 b	–	36 b	12 a	26 a	–
Runoff harvisting for olives	2 d	6 b	65 a	17 a		
14 a Other runoff harv-esting structures[a]	15 c	23 a			13 ab	16 a

Means followed by same letter in the same column – not significantly different ($P \leq 0.05$). "No runoff-harvesting" are plots under traditional rangelands use

[a]Semi-circle bunds (Sorif), contour ridges (Bane Noem)

herbaceous vegetation reached its maximal size. As to the nearby woody, afforested plots, the generated runoff was negligible and did not change throughout the rainy, growing season (Fig. 17c).

3.11 Soil Moisture Response to Soil Conservation Measures

When measures that conserve soil and prevent runoff and its associated erosion are successful, the result is increasing soil moisture contents. Therefore, soil moisture is an indirect indication of soil conservation and runoff prevention, and especially in drylands it is also an indicator of a potential for primary productivity and maintenance of biodiversity. The MEW-MERC project also explored the effect of soil conservation and runoff management on soil moisture, at the local scale.

The effect of several soil surface furrowing treatments applied at two sites differing in slope (7% and 3%) on soil moisture storage, as well as on related variables such as soil erosion and rainstorm-driven runoff generation was studied in the Humret Es-Sahin watershed (Jordan). The experiment included four replicates of 20 m × 6 m plots used for testing three 30 cm width × 30 cm depth furrow alignments: 6 m apart, 3 m apart between successive furrows, and 6 m apart

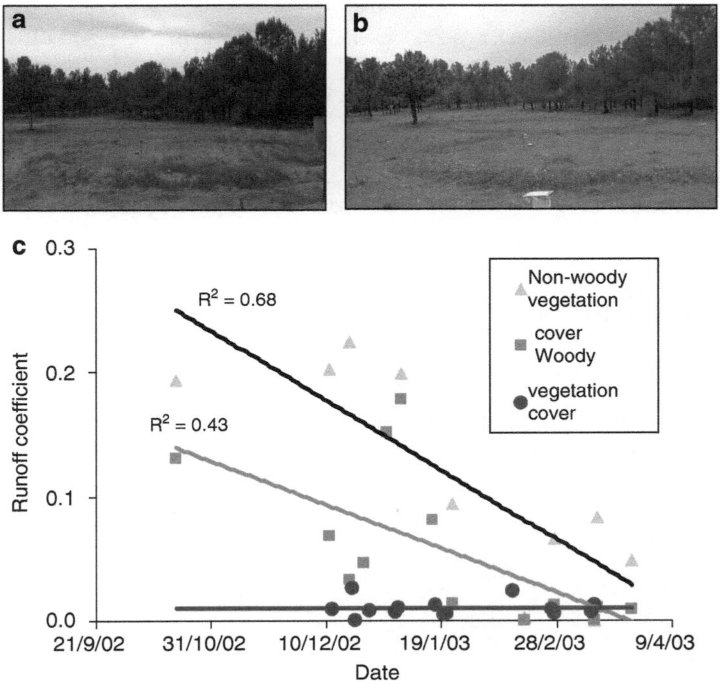

Fig. 17 The effect of seasonal vegetation growth on runoff generation, Yatir, Israel. (**a**) Photo of one of the plots located in the clearing, taken at the beginning of rainy season; (**b**) Same plots towards the end of the rainy season – seasonal growth of the non-woody vegetation; (**c**) Seasonal changes in the runoff coefficient for two plots of non-woody vegetation within the forest clearing, compared with a plot within the forest (*blue, horizontal regression line*), i.e. having woody vegetation. Values on the left side of (**c**) stand for runoff coefficients during the period (**a**) was photographed, *Right side* – for runoff coefficients during the period (**b**) was photographed. The low vegetation cover in the non-afforested plots just prior to the onset of the rainy season may be a combination of preceding two drought years and heavy grazing

between groups of 2 furrows each with a 1 m within-group distance. These three treatments were compared with a non-furrowed control, whereby all treatments were assigned to plots through a randomized complete block design.

All the furrowed plots generated less runoff than the non-furrowed plots, with runoff coefficient logarithmically increasing with rainstorm depth and with catchment's width (Fig. 15), such that the areas of only 3 m width (between furrows) generated less runoff than the wider catchments. The generated runoff caused topsoil erosion, and the sediment was deposited in the furrows in amounts positively correlated with the amount of runoff, with highest amount of sediment measured from catchments not protected by furrows and hence generating the highest amount of runoff (Table 5). Finally, with reduced runoff generation due to furrowing, and especially where the areas for runoff generation were small (i.e. only 3 m between successive furrows), as expected, soil moisture increased, and

Table 5 Increasing soil moisture by runoff control measures – furrowing experiment at Humret Es-Sahin, Jordan. Blank rows – 2001/2002 season with 69 mm rainfall; Grey rows – 2002–2003 season with 563 mm rainfall

Distance (m) between successive furrows	None (no furrows)	3 m	6 m	6 m, between groups of two 1 m-separated furrows
Variables	Measured values	Percentages relative to no furrowing[a]		
Runoff (%)	63	74	87	92
	74	73	84	83
Depth of accumulated sediment (mm)	69	49	51	48
	423	37	62	65
Soil water storage	20	156	129	118
	123	193	153	157

[a]All values of the 3 m treatment are based on measured values which are significantly different (Duncan test, 5% level) from those of all other treatments, except sediment depth in 2001/2002 which is significantly higher than "no furrows" only

Table 6 Differences in water storage between furrowing treatments and sections of the soil profile, furrowing experiment, Humret Es-Sahin, Jordan

Soil water storage at different soil depths (2001–2002)	% Relative to no furrowing		
	3 m	6 m	6 m, between groups of two 1 m-separated furrows
30–60 cm profile	300	243	200
60–90 cm profile	300	175	186

was nearly twice higher than in the non-treated plot (Table 5). Furthermore, furrowing increased water storage throughout the different layers of the soil profile (Table 6), and the differences between no-furrowing and furrowing was maintained from the measured 10–90 cm depths, but was much more pronounced in the rainy year as compared to the low-rainfall year (Fig. 18).

These results confirm earlier studies, suggesting that furrowing increases infiltration (vertical as well as lateral infiltration within furrows, thus the furrows improve storage across the between-furrows areas). This increased infiltration reduces surface runoff, and the reduced surface runoff reduces soil erosion and hence sediment transport and accumulation.

The experiment also demonstrates the need to further testing with different furrowing densities in order to obtain the best alignment. The performed experiment demonstrated that there must have been a threshold value, since the intermediate (3 m) distance between furrows performed significantly better in increasing soil moisture and reducing erosion than the 1 m and the 6 m alignments. Namely, there must be an optimal distance between furrows, which would vary depending on rainfall regime, soil type and profile, and surface aspect and slope, for maximizing soil moisture storage and minimizing soil erosion.

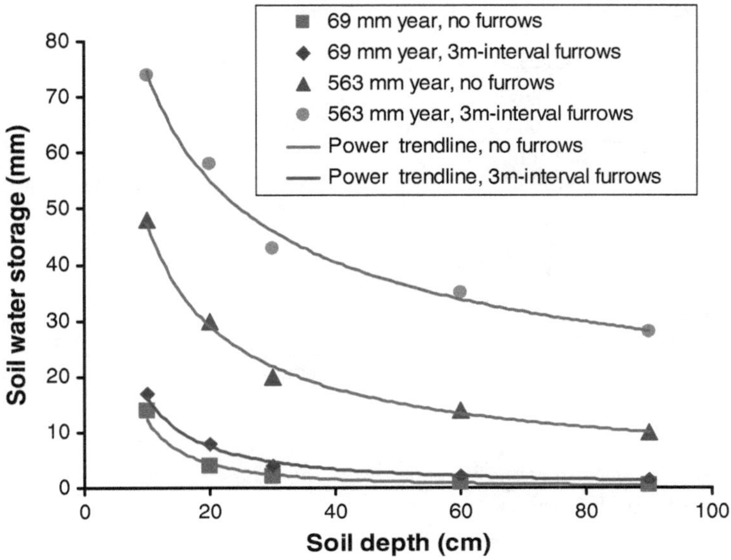

Fig. 18 Soil moisture profile (cumulative of all rainstorms of each rainy season), from 10 cm to 90 cm depth, furrowing experiment in Humret Es-Sahin, Jordan. Two *lower curves* are for 2001/2002 with low annual precipitation; two *upper curves* are for 2002/2003 with high annual precipitation. *Red curves* are for the "control" no-furrowing plots, *blue curves* are for plots with 3 m interval furrowing, which reduce runoff and erosion and increase soil moisture

Unlike the Jordanian sites that have been under a very little if any current human pressure, all the Palestinian sites have been rather intensively used, and in each of them, the team either installed or used existing soil conservation and/or runoff harvesting measures for comparing their effectiveness in promoting soil moisture. These included the following: already existing "forests" or "plantations" of pines planted in 1960; plantation of bushes as a restoration measure; stone terraces (constructed by hand to reduce machinery-caused disturbance); contour ridges for rangeland improvement constructed as earth ridges along contours; semi-circulars bunds of earth embankments with their tips on the contour and runoff producing catchments between them, a techniques used for supporting olive trees and for rangeland rehabilitation; and similar but v-shaped bunds. Each replicated set of these runoff-harvesting structures was matched by plots with no "treatment", i.e. of "natural vegetation".

Except for plots of natural vegetation and terraced slopes in the driest site (Bane Noem), the cumulative amount of generated runoff was positively related to the annual rainfall amount in the controls ("natural vegetation") and in all the other "treated" plots subjected to the different types of "treatment" (Table 7). In Dura, though, the area that generated most runoff was a terraced olive plantation (36,000 cm^3/m^2 in 2004–2005 season, compared to the smallest volume generated by non-cultivated terraced plots, in Bane Noem – 1,200 cm^3/km^2 in the 2004–2005 season).

Table 7 Total surface (cm^3/m^2) runoff in different "treatments", in each of the three Palestinian sites

	Sorif		Dura		Bane Noem	
	2003/2004	2004/2005	2003/2004	2004/2005	2003/2004	2004/2005
Total annual rainfall (mm)	317	400	250	385	225	334
Natural vegetation	13.483 a	26.584 a	7.332 b	19.634 b	13.761 a	6.047 a
Terraces	7.089 c	9.278 b	5.803 b	14.456 b	5.143 b	1.195 a
Olive Trees (Terracing)	10.390 b	7.437 bc	17.827 a	36.071 a		
Semi-circle bunds	3.718 d	4.448 c				
Reforestation (shrub)	5.213 cd	4.448 c				
Trees (Acacia)					4.205 b	4.462 a
Contour Ridges					2.314 b	1.619 a

Means followed by the same letter in the same column are not significantly different at the $P \leq 0.05$ level

The effect of reducing runoff by the tested measures is also evident when runoff is related to individual rainstorms – the effectiveness of these measures increased with increasing rainfall storms' depth (Fig. 19). This analysis also reveals the exception of the olive plots in Dura. It can be due to the intensive olive cultivation in this site, in which the herbaceous vegetation cover of well-attended olive groves is much reduced and the soil may be compacted by the farmer's trampling.

It can be concluded that in each of the sites each of the treatments (except the terraced olives in Dura) reduced both runoff and the resulting sediment removal (Table 4). However, no significant regression is apparent when values of sediment discharge for all treatments and sites are lumped and plotted against runoff discharge (Fig. 20). This finding highlights the difference between local scale processes, which must depend on the high spatio-temporal variability at that scale, and the integrative whole-watershed scale measurements, which produced high correlations between whole-watershed sediment discharge and runoff generation (e.g. Fig. 11).

Following the within- and between-treatment and site differences in runoff response to rainfall and in erosion response to runoff, the response in soil moisture was also variable in the Palestinian sites, with only one unifying feature – higher soil moisture in all sites and treatments in the rainy year, as compared with the low-rain year (Fig. 21). But in sharp contrast to the Jordanian furrowing experiment whereby the (single) treatment, furrowing, unequivocally increased soil moisture, in Sorif, a site of the highest rainfall (compared to the other two Palestinian sites and the Jordanian site) soil moisture in natural, non-treated area was highest at the end of the rainy season as compared to all other "treatments" in that site (Fig. 22, top). This suggests that where rainfall is relatively high hence agriculture is intensive, the pressure on soil moisture is high even though the runoff-harvesting measures are in place; with no runoff harvesting but also no agricultural use as in the "natural" areas, soil moisture regime is the best. As to the comparison between the different treatments, soil moisture at the end of the rainy season was lowest in olive orchards terraces, as compared to the semi-circle water harvesting structures, and areas planted with shrubs.

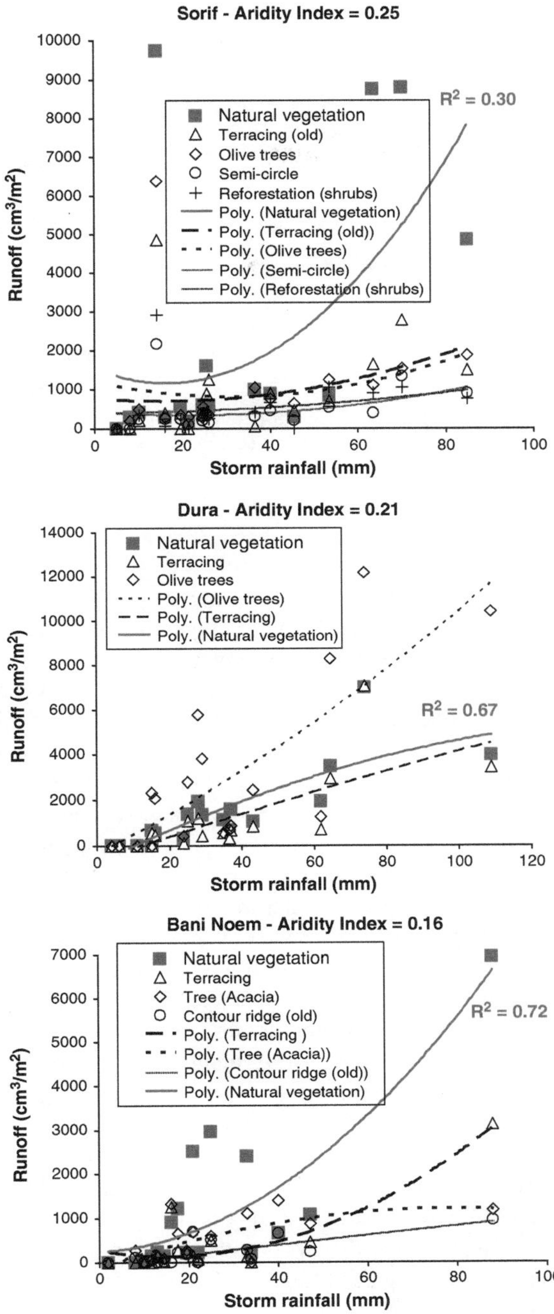

Fig. 19 Runoff volume (per unit area) generated by the 18 rainstorms of 2003/2004 and 2004/2005 in the 3 Palestinian sites. "Natural vegetation" (*red points and line*) are "non-treated" plots, used as rangelands under traditional livestock grazing, and other data are for "treated" plots. The aridity index (ratio of precipitation to potential evapotranspiration) values are derived from rainfall map

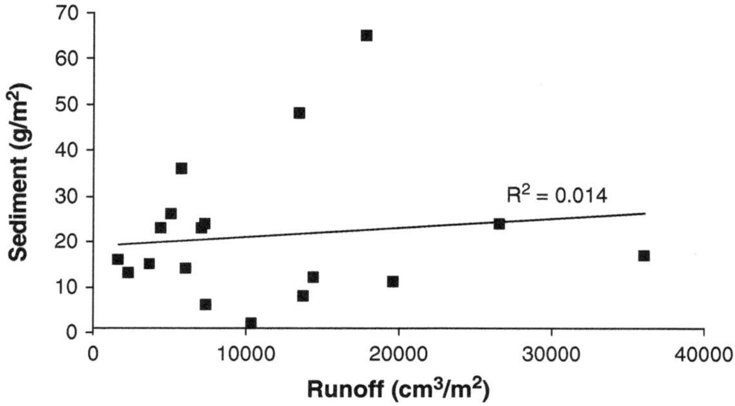

Fig. 20 Relations of sediment discharge and runoff for all Palestinian sites and seasons combined

More important is how much of this moisture "survived" to the end of the rainy season. In the low-rain year, 2004 terraces lost up to more than half of their moisture, whereas all other treatments and "control" lost far less. In the rainy year, however, all areas lost by the end of the dry season more of what they initially stored, as compared to the wet year. This is probably due to transpiration loss associated with the higher productivity of that year, driven by the higher rainfall. In Dura, with lower rainfall than Surif, all water-harvesting structure had more soil moisture than the "natural vegetation", and the semi-circle and V-shaped structure were doing better than the terracing, in both years (Fig. 22, center). Furthermore, the two former structures also retained more moisture by the end of the dry season, as compared to all other "treatments" and "control". Comparing with Surif, though, all areas lost more moisture, probably due to their relatively higher summer temperature and hence evaporation. In the driest site, Bane Noem, terracing was more efficient than all other measures (also than "natural" areas) in storing water at the end of the rainy season (Fig. 22); and though the difference in rainfall between the 2 years in this site was not high, the response of the "treatments" to the slightly improved rainfall was very intense. As to losses by the end of the dry season, again, there were relatively more losses in the rainy year as compared to the less rainy one, and the terraces and forest plantation lost less than other "treatments". These measurements demonstrate the significance of measures such as water harvesting structures and planting vegetation in dry sites.

Finally, the effectiveness of the different water-harvesting/soil-conservation measures in improving soil moisture along the soil profile was assessed at one of the PNA sites (Dura) and in one rainy season only (Fig. 22). It was found, for example, that the V-shaped bunds were more effective than the semi-circular ones but at the deep part of the profile the semicircular band was more effective than its control, as compared with the V-shaped bunds at this depth. Furthermore, at the end of the dry season the semicircular bunds and their control have only 3% soil moisture content at 15 cm depth (compared to 18% and 16%, respectively, at the

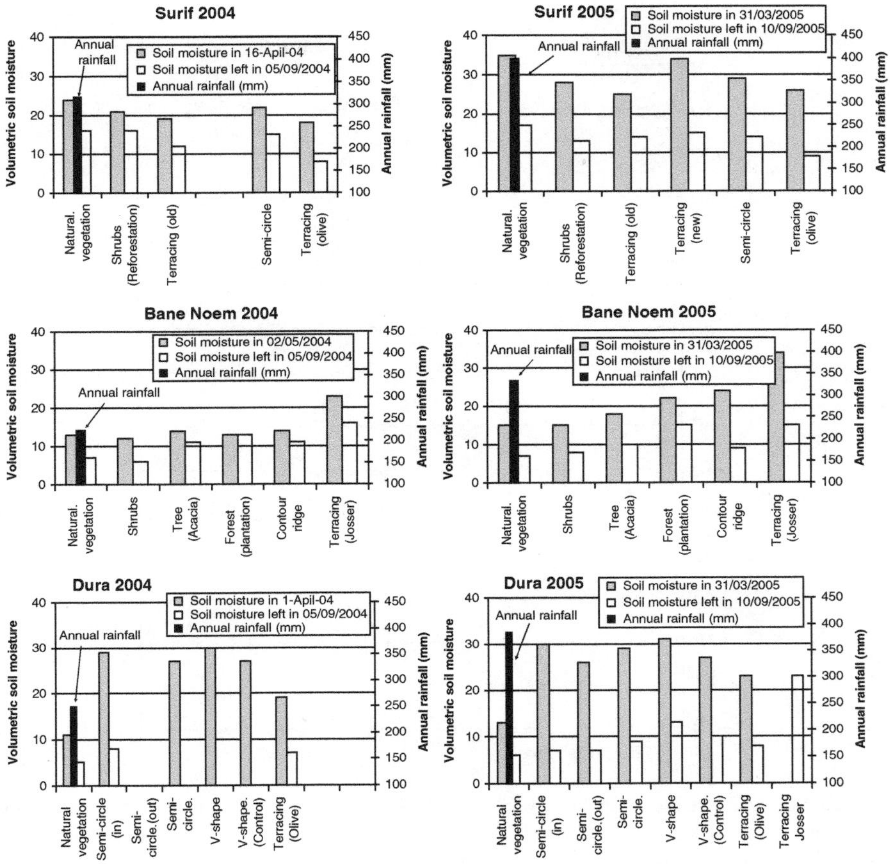

Fig. 21 Volumetric soil moisture (wet weight of soil sample – dry weight of soil sample/dry weight of soil sample × Bulk density × 10) at 30 cm depth in three sites, the Southern Hebron Mountains, PNA under various runoff harvesting and soil conservation measures at the start and end of the dry season in each of the 2 years (2004 and 2005)

beginning of the dry season); but at 65 cm depth the circular bunds soil has more stored water than its control. The distribution of soil moisture along the soil profile in these two runoff harvesting measures (Fig. 22), as well as the others suggest that by the end of the dry season there is a threshold at the 30–40 cm depth, below which more moisture is stored than above it.

Several vegetation cover intervention practices used to reduce surface runoff and topsoil erosion at Sorif, their least dry site were explored. The site's indigenous vegetation is mostly Mediterranean, dominated by the thorny shrub *Sarcopoterium spinosum*. The cover of this shrub is reduced by two land uses, cultivation and afforestation, and by experimental removal during this study, in 2004/2005. The effects of these three "treatments" were assessed by measuring soil chemistry and moisture, and runoff and sedimentation. Surprisingly, runoff and sedimentation

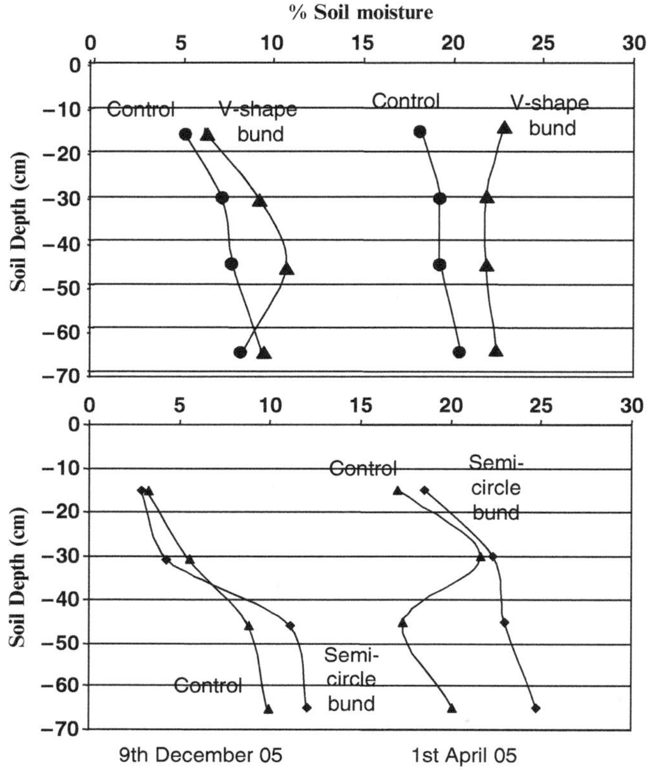

Fig. 22 Soil moisture profile in Dura, PNA in 2005 in two of the runoff harvesting measures and their specific controls. The *right hand side curves* are for the beginning of the dry season (1st April) and the *left hand side curves* are for its end (9th December)

were positively associated with vegetation ground cover. Interestingly, runoff and sedimentation peaked at medium vegetation biomass and were low where the biomass and ground cover of wild vegetation was lowest (Fig. 23). It is expected, though, that runoff and erosion would be affected by the trees and the cultivated crop more than by the *Sarcopterium* under-story. Indeed, peak runoff and sedimentation were at the forest edges and where the *Sarcopoterium* was experimentally removed, whereas the lowest rates were within the forest, in cultivation, and in the natural community ("*Sarcopoterium* dominant"). Thus, though *Sarcopoterium* is not favored by livestock and hence dominates under overgrazing, it reduces soil erosion and runoff.

Contour trenches (constructed by bulldozing top soil from stripes of contributing slope areas and piling them into elongated soil mounds along contour lines), widely used in a newly-afforested areas were explored in the Yatir site (Fig. 24). Water was unevenly (and significantly so) redistributed within these structures. Soil moisture was 20% higher in the swale than in the "control" (non-"treated" site), whereas soil moisture of the intact section, which is the runoff generating area, was similar to

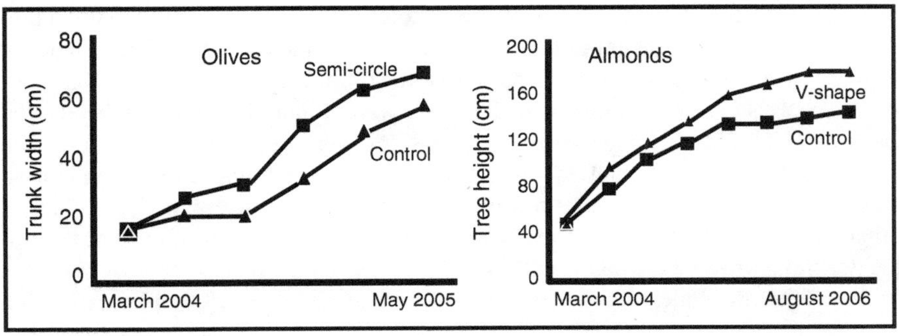

Fig. 23 Soil cover "treatment" and their effect on runoff, sedimentation and soil organic matter (*top*), and the effect of the soil organic matter on soil moisture (at two depths, at the start and end of the dry season), in Sorif, PNA

Fig. 24 Contour trenches inYatir, Isarel, the "newly afforested" watershed. (**a**) View of a whole watershed slope. *Bottom right* of photo is a close up of two consecutive contour trenches, with a contributing area ("source") inhabited by grassland vegetation, and the trench (or "swale") by planted trees; (**b**) Transect of a contour trench. Words in quotes are names of sampling sites

that of the control. Other sections of the structure transect were drier compared to the control, probably meaning that the water gain of the sink (the swale) is at the expense of the water that did not infiltrate into soil of the runoff source (Fig. 25a). As to chemistry, no significant differences were found for nitrogen along the transect (Fig. 25b); phosphorus and potassium were significantly lower in the disturbed microhabitat than in other microhabitats, and phosphorus content in swale and potassium in swale and the control were significantly higher than in other sections of the transect. The high salinity found in the swale can be a result of the leaching from the other sections (Fig. 25c). These results corroborates the finding of in the Palestinian sites – that olives grew better when planted within

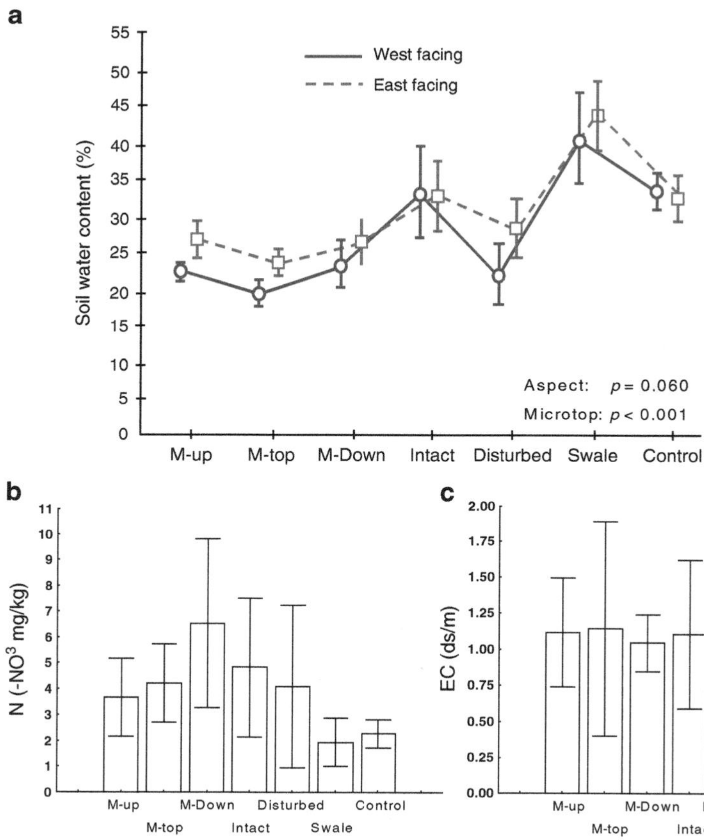

Fig. 25 Water and chemistry measurements in contour trenches, Yatir, Israel, in the six microhabitats along transect of a contour trench, also compared with a control area with no water harvesting manipulation (right hand item). (**a**) Soil water contents at 12 cm depth, measured 1 week following a 15 mm rainstorm, in west facing and east facing replicates. Vertical bars are 0.95 confidence intervals. Analysis of variance across the transect – $F(6.210) = 33.7$, $P < 0.001$, and of the aspects (east and west) – $F(1.210) = 3.6$, $P = 0.06$; (**b**) and (**c**) – nitrogen contents and salinity, expressed as nitrate ion concentration and electrical conductivity, respectively

the "sink" of the semi-circle bunds than when growing in the contributing area of these bunds, suggesting that this difference was due to the effect of the uneven water redistribution within this water harvesting structure.

3.12 The Effect of Afforestation on Soil Moisture

The tradeoff posed by rangeland afforestation, between soil conservation due to runoff reduction, and water loss due to transpiration was explored in Yatir. In both cases the latter option was studied at the local scale, but the motivation to the study was an observation at the whole watershed scale that pointed at the former benefit (i.e., the watershed scale observation that the forested watershed did not generate any runoff in Yatir). Plots of 70 m^2 – three were established in the forest and three in non-forested areas near the forest. Since none of the rain falling on the forested plots leaves the watershed as runoff, it was explored how the presence of trees affects the ratio of transpiration to evapotranspiration and soil water storing patterns (depth and time) at the local plot scale.

Since on the average about 1% of rainfall in the forest generated local surface runoff, stemflow was very low and through fall was high, more than 90% of the rainfall was expected to penetrate the soil. Using neutron probes inserted through permanent access tubes in one of the Yatir tree covered plots enabled to reconstruct by interpolation the fate of water infiltrating from the soil surface into the soil profile in both the horizontal and the vertical dimension, thus forming a 3-dimensional image of moisture distribution (Fig. 26). This approach demonstrated that much lateral movement takes place within the soil profile, since an increase in a deep layer relative to a decrease in an upper layer (marked by the arrows in Fig. 26) is likely to be due to such movement rather than to a vertical infiltration. Nevertheless, in-spite of the spatial variation within each depth the spatially averaged soil water content enabled an insight into the soil moisture content dynamics throughout the whole season in the forest (Fig. 27, left, Schachnowitz et al. 2008).

The soil moisture profile studied in Yatir is the deepest compared to such profiles studied by the other teams, but in the Jordanian profile (which is only 90 cm deep) soil moisture decreased with depth whereas in the Israeli one it increased (in the first 90 cm of the Yatir profile, Fig. 28). This may demonstrate the combined effect of decreasing evaporation and water consumption due to afforestation of a semiarid rangeland.

The monitoring in Yatir also extended to a depth of 5 m, and the salient feature is the relative stability in the (low) water content below the depth of 1.5 m. The Israeli team, however, wished to construct a full water budget at the forest's local scale; given that very little if any is lost as runoff, evapotranspiration and soil water storage should be balanced by throughfall. It was, therefore, necessary to assess whether the storage measured in the 4 m profile represents most of the overall soil storage. To elucidate this, it was assumed that the depth at which moisture seasonally fluctuates at the lower limit of the instrument ability to detect this fluctuation, no more measurable water is found. Calculating the coefficient of variation of all access tubes throughout the year and comparing it with that of the used instrument (Fig. 27, right), showed that all tubes

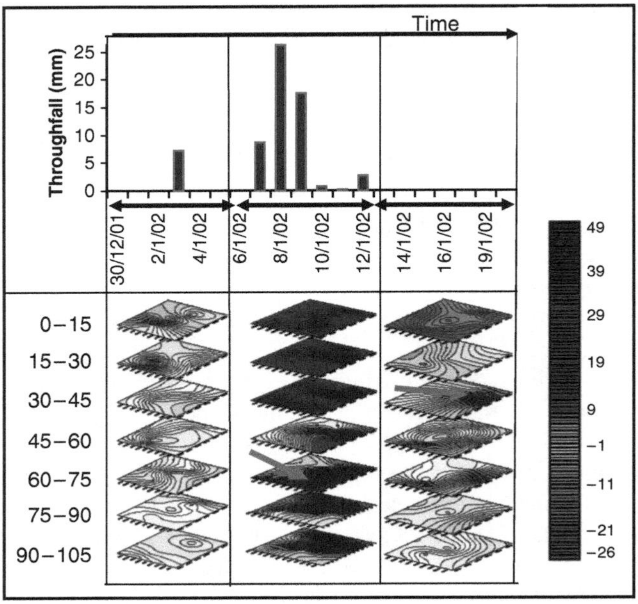

Fig. 26 Three profiles of soil moisture change, Yatir forest, Israel. Each column represents the changes in water content (expressed in mm [=change in water content* depth interval]) measured between the dates marked by the *horizontal arrows* at the top (e.g., first column measuremets were carried out on the 30/12/01 and on the 6/1/02), and affected by the amount of throughfall during each of the periods (column diagram on the top). Soil depth intervals appear on the *left side* of the figure and the color key for water contents (mm) on the *right*. The *red arrows* indicate an increase in moisture relative to the layer above it

reached the critical value of the instrument's sensitivity at a depth of 1.65 m, which also corresponds to the first reading taken within the rocky underlying layer. This assertion led to calculating the 7-access tubes' overall seasonal changes in total soil water content to a depth of 1.5 m, of the forest plots during the three seasons of study (Fig. 29). It is evident that while peaks of total soil water content vary, the lower amount of water content shows temporal stability (Schachnowitz et al. 2008).

To evaluate the effect of this afforested rangeland on soil water storage the same procedure carried out in the afforested plots was also undertaken on forest clearings, which represent other parts of the Yatir rangeland before it has been afforested, 35 years earlier. The depth of highest temporal stability in water content was very similar to that within the forest, 1.5 m, a determination which enabled to compute the water storage dynamics of non-afforested areas too. No differences were found between afforested and non-afforested plots (Fig. 30). Moreover, total water contents in the soil prior to the onset of the rainy season were similar for the 3 years, with no significant differences between the two sites. It thus seems that afforestation does not compromise the soil water storage, as compared to non-afforested areas (Schachnowitz 2006).

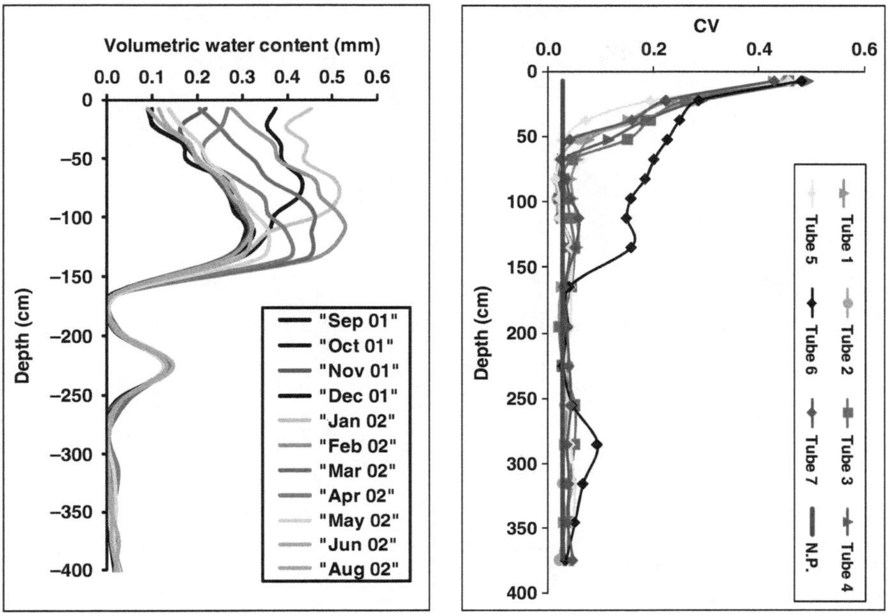

Fig. 27 Soil moisture profile, Yatir, Israel. *Left*: Volumetric water content distribution with depth throughout one season for one representative access tube (the other five access tubes generated similar results); *Right*: Coefficient of variation (CV) for the various tubes within the afforested area as a function of depth (NP corresponds to the CV of the standard of the instrument used for measuring soil moisture)

Finding out that both the afforested and the non-afforested plots have similar water regime and hence the forest does not seem to impact water resources compared to the rangeland vegetation, the ratio of transpiration to evapotranspiration in the forested area was assessed. Transpiration of several trees was measured using the heat pulse method, and this method was evaluated by comparing totals of transpiration against totals of evapotranspiration derived from changes in the soil moisture profiles during periods in which it could be safely assumed that evaporation was negligible. The results indicated that the two measuring systems produced similar results (Schachnowitz 2006). It was also attempted to discern transpiration from evaporation by measuring transpiration at events of climatic conditions during which transpiration was expected to be high while evaporation negligible (such conditions occurred during days following rainfall but having high cloud cover, or in times when the upper soil layer was dry hence evaporation was minimal, but the deeper layer had water to be used by vegetation). The finding suggested that on the average (of the 3 years) about 30% of the throughfall in the forest was lost by evaporation (Schachnowitz 2006).

The overall transpiration of the forest, calculated at the end of the dry season, would therefore be the difference between the rainfall of the previous rainy season

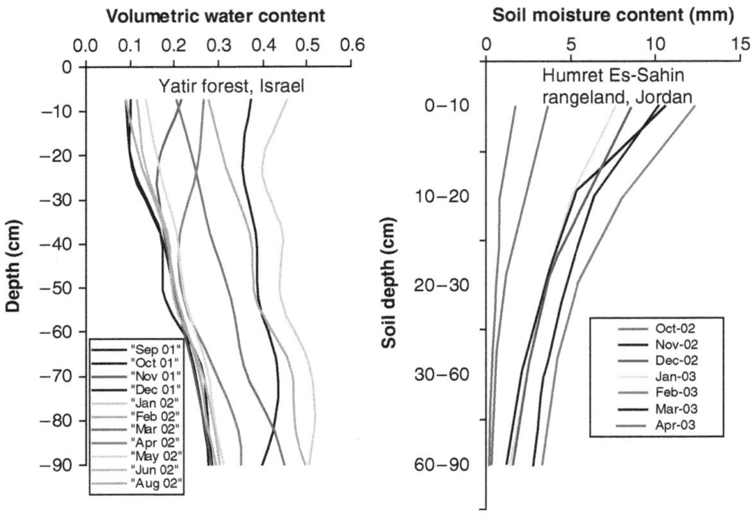

Fig. 28 Comparison of soil moisture profiles (first 90 cm) of the Israeli afforested rangeland (Yatir forest, Israel) and the Jordanian protected rangeland (Humret Es-Sahin, Jordan). *Left*: Yatir – measurements throughout a whole year (September 2001 to August 2002); *Right*: Humret Es-Sahin – measurements from October 2002 to April 2003, i.e., dry season excluded; These are the "control" plots of the furrowing experiment

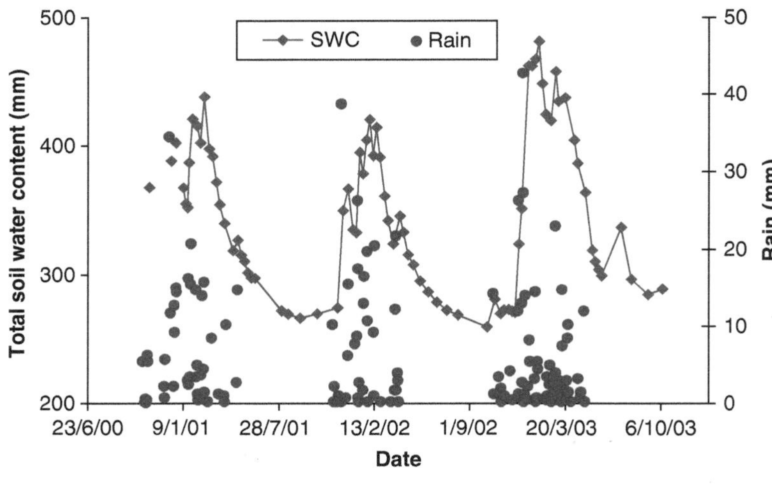

Fig. 29 Averaged total soil water (TWC) content, to a depth of 1.5 m in one of the afforested plots during three consecutive seasons, Yatir, Israel

(plus the soil water storage remaining before the onset of that rainy season) and the soil water storage left at that end of the dry season. These were, indeed, calculated for each of the three seasonal cycles, both for the forest and for the non-afforested plots. These calculations revealed small and statistically non-significant differences

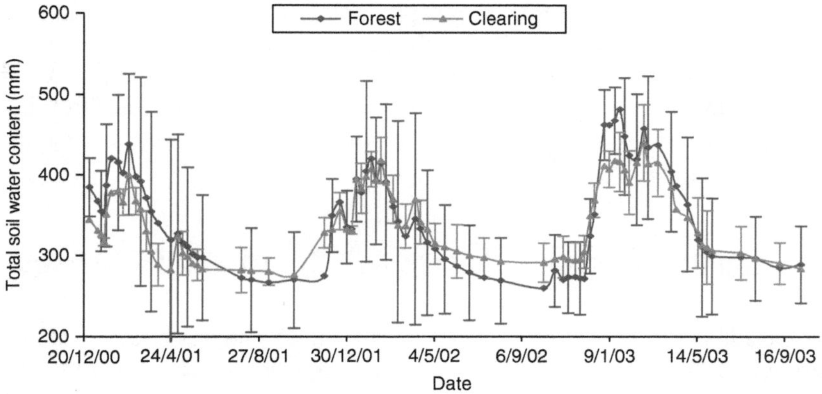

Fig. 30 Total soil water content for afforested and nearby clearing representing non-afforested areas in Yatir, Israel, during three consecutive seasons. Bars denote one standard deviation

Table 8 Major components of the local water balance at the two adjacent studies sites (afforested and non-afforested clearing), Yatir Forest, Israel, during three consecutive seasons

Site	Afforestation			Non-afforested clearing		
Season	2001–2002	2002–2003	2003–2004	2001–2002	2002–2003	2003–2004
Water source	Throughfall + Stemflow-Runoff			Rain-Runoff		
Net water input	276.3	307.4	195.3	298.5	328.0	209.8
Storage (mm)	4	19	−22	18	14	−33
Water use (mm)	272.8	298.9	223.0	280.4	314.1	243.2
Water use as % of water input	99	97	114	94	96	116

"Storage" was computed as the change in the total soil water content between measurements carried out before the onset of two consecutive rainy seasons. Since runoff was found to be negligible the net between the input and soil storage change represents evaporation and transpiration. In the afforested area actual transpiration was measured and evaporation was computed from the equation Throughfall + Transpiration + Evaporation + Storage = 0, whereas in the clearing only overall evapotranspiration was estimated from rainfall and storage change. However, none of the differences between afforested and non-afforested plots is statistically significant

in water use (evaporation plus transpiration) between the afforested and the non-afforested plots (Table 11), whereby a few percent of water input is left in the soil storage by the end of the dry season. Furthermore, it is conceivable that the percentage use higher than 100% in the last year is due to this year's rainfall being considerably low, thus the vegetation of both the non-afforested and the afforested one (including the trees) used more than that year's rainfall, i.e., most of the water storage remaining after the 2 previous years of high rainfall.

This is evidenced by the negative values of storage of the last season that equal the sum storages of the 2 previous years, (Table 8, Schachnowitz 2006). To conclude, the afforestation project found that soil and water conservation at the whole watershed scale in Yatir does not affect negatively the soil water storage at the local scale. These findings call for exploring means for rangeland management in the Yatir region, given the projection that the rangeland is as good as the forest in locally conserving soil water, provided it is not overgrazed. The question remains, though, whether a well-managed rangeland may also be as good as the forest in soil conservation and the resulting water conservation at the whole watershed scale.

3.13 Soil Cover, Runoff, Soil Organic Matter and Soil Moisture

No correlation between vegetation biomass and soil organic matter was found in the Palestinian project sites (Fig. 23, top), but the expected association between soil organic matter and soil moisture was stronger at 30 cm than at 15 cm depth (Fig. 23, bottom). This finding suggests that soil organic matter has a positive effect on deep soil water storage. In all "treatments" applied in the Palestinian sites soil organic matter declined with the site's aridity (i.e. highest in Sorif and lowest in Bane Noem, Fig. 31). Within sites, it was lowest (for example 2.7% in Sorif, the least dry site) in olive groves on constructed terraces as compared to most other "treatments", whereby plots reforested by the team (shrubs and tree saplings) thus not grazed and not cultivated, had the highest amount of soil organic matter (e.g., 6.4% in Sorif). Nitrate concentration followed similar trend to that of soil organic matter – reduced with aridity and lowest in terraces with olives, as compared to "natural vegetation" and to grazed plots

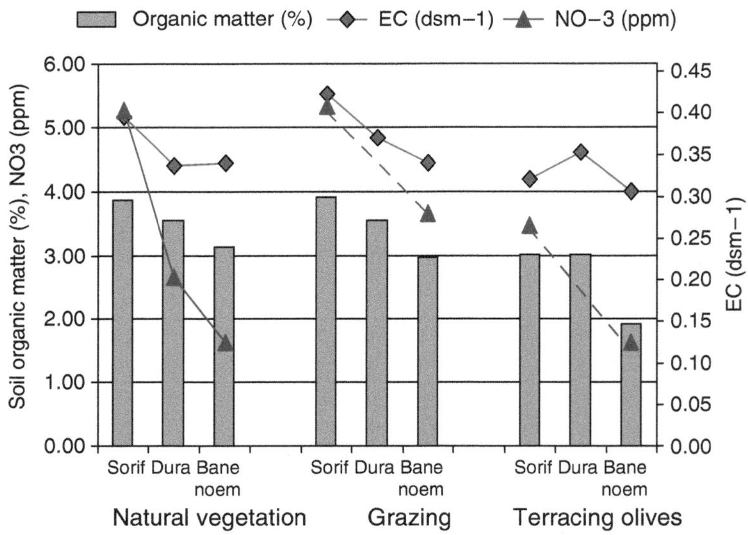

Fig. 31 Means (over the 3 study years, 2003–2005) of all sites within each category, of soil organic matter, salinity (measured by electric conductivity) and nutrients (represented by nitrate ion). Values of nitrate for Dura sites, subjected to fertilizer application, are not provided (signified by broken *blue line*)

(Dura site not included, since in much of it fertilizers have been used, Fig. 31). On the other hand, salinity values (measured by EC, electrical conductivity) were rather low and similar (ranging 0.31–0.59 dsm) in most sites and treatments, but within this range salinity was highest in grazed plots (in Sorif) and lowest in terraced olives, especially in Bane Noem, the driest site, for unknown reasons (Fig. 31).

A comparison between the soil organic matter in the Israeli and the Palestinian sites (Fig. 32) demonstrates the potential of afforestation in drylands to increase soil organic matter, as shown in Sorif (PNA site) and Yatir (Israeli site), whereas in Bane Noem the difference was insignificant. Despite the 35 years during which Yatir forest more than doubled the local soil carbon contents, its current value is still half the value of that of Sorif "control", whereas Sorif's rainfall was only about 30% higher than that of Yatir (Fig. 32). Finally, the effect of olive groves, on the other hand, seemed to be negative, with no apparent explanation.

The hypothesis that across a watershed slope affected by soil erosion, the lower sections will be more productive than the upper ones, due to the downslope movement of eroded soil, was explored. This hypothesis was tested by conducting a bioassay in which several attributes of corn seedlings grown in soil samples collected from various habitats of these two slope sections were measured at their 35th day of age. Regarding stem height, for example, the seedling did significantly better on soils from lower slope sections than those from upper sections, but only when soils were from grazing-protected sites (Table 9). No significant differences between high and low slope section were found in grazed sites, what can be attributed to soil fertilization by livestock droppings masking the erosion effect. Thus, the difference

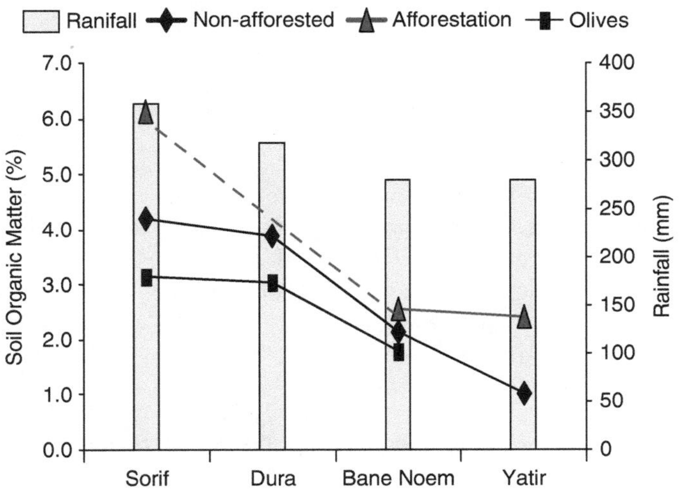

Fig. 32 Soil organic matter in "control" and plots covered with woody vegetation – old afforestation in Yatir (1 year), Israel and Dura, PNA, recent afforestation in Bane Noem, PNA, and olives on terraces in the three PNA sites (2-year means). Rainfall values are for the years of soil organic matter determination (means of 2003/2004 and 2004/2005 for the Palestinian sites and value of 2001/2002 for the Israeli site)

Table 9 Bioassay of soils from upper section and lower section of slopes, in grazed and grazing-protected plots, expressed by stems of *Zea mays* seedling grown in the laboratory, at their 35th day, South Hebron Mountains, PNA

	Bane Noem		Sorif	
	Grazed	Not grazed	Grazed	Not grazed
Downslope	39.9 ab	37.8 b	42.6 a	35.8 c
Upslope	38.7 a	30.7 c	43.8 a	30.0 b

Means followed by the same letter in the same column not significantly different (at the $P \leq 0.05$)

Table 10 Vegetation changes between 2002 and 2004 of the whole Kiziloz microcatchment, detected from space and expressed by NDVI derived from IKONOS images

% of watershed that was not vegetated		100%
and became vegetated	7.1	
and remained not vegetated	77.7	
% of watershed that was vegetated		
and remained vegetated	6.8	
and lost its vegetation	2.2	
% of vegetated watershed that remained vegetated but vegetation increased or decreased	6.2	
% watershed that did not change	84.5	100%
% watershed in which vegetation increased	12.7	
% watershed in which vegetation decreased	3.3	

between slope sections is likely resulting from long-term processes, suggesting that indeed eroded soil accumulating in the downslope increases fertility there.

3.14 Vegetation

Vegetation productivity is the ultimate target of promoting soil water content by reducing erosion by preventing surface runoff. At the same time it is the soil vegetation cover that reduces surface runoff and, hence, soil erosion. It was attempted to detect changes in vegetation cover and productivity at the Turkish watershed scale, using remotely-sensed IKONOS-generated images from which NDVI was calculated. The researchers determined the change (pixel by pixel of the watershed's NDVI presented as "vegetation") between the 2002 and 2004 images, taken in late July/beginning of August. This analysis demonstrates (Table 10) more increases than decreases in vegetation during this period, though the rainfall was decreasing from 2002 to 2004 (from 276 mm in 2002 to 272 and 259 in 2003 and 2004, respectively). At least part of this apparent anomaly can be explained by agricultural and forestry activities in the watershed during the study periods. Whereas this analysis encompasses both woody and non-woody vegetation, all other teams assessed these two very distinct types of vegetation life forms separately, as presented in the following sections.

3.15 Trees

The research in the Turkish sites established that planting indigenous forest trees – Almond, *Cedrus* and *Pinus nigra* was much more successful than planting *Quercus*. It was therefore concluded that only a longer monitoring period and more extensive replication could reliably assess tree performance under the relatively extreme local conditions in their sites. It was found that the Palestinian water harvesting structures positively affected growth of olives and young almond trees, expressed in trunk diameter and height – by the end of 1-year almond heights and olives trunk width grew ca 22% and 17% more than those without harvested runoff supply (Fig. 33). It was also found that trunk diameter of the pine trees in the Yatir forest growing on east facing slopes were 17% greater than of those growing on west facing slopes (20 trees in each of 3 plots for each slope, $P < 0.01$).

Detecting changes in tree density of the Yatir forest using satellite imagery, thus addressing the whole-watershed scale would enable correlating tree properties such as their cover of soil surface, with the degree of reducing runoff at the watershed scale. This approach would also facilitate elucidating whether afforestation compromises or promotes soil moisture storage, as addressed by the researchers at the local scale. Data of a ground survey of 646 pine trees demonstrated a significant positive linear dependence of trees'height ($R^2 = 0.96$; $p < 0.01$) and a positive curvilinear dependence ($R^2 = 0.95$; $p < 0.01$) on crown width (expected to be detected from space), on trunk diameter at breast height (DBH, which is habitually measured in ground surveys). It was then found, as expected, that tree density is negatively correlated to both DBH (measured on the ground) and canopy cover (expected to be detected from space) – R^2 is 0.64 and 0.69 respectively, $p < 0.01$. These relationships are critical for deriving tree density in the Yatir forest from canopy cover detected by satellites, matched by ground sampling of DBH (Sprintsin et al. 2007, 2009).

Canopy cover of the "training"plot in Yatir forest was derived from a high resolution multispectral remote sensing IKONOS imagery. It enabled discerning the total specific forest surface reflectance measured by the satellite and expressed

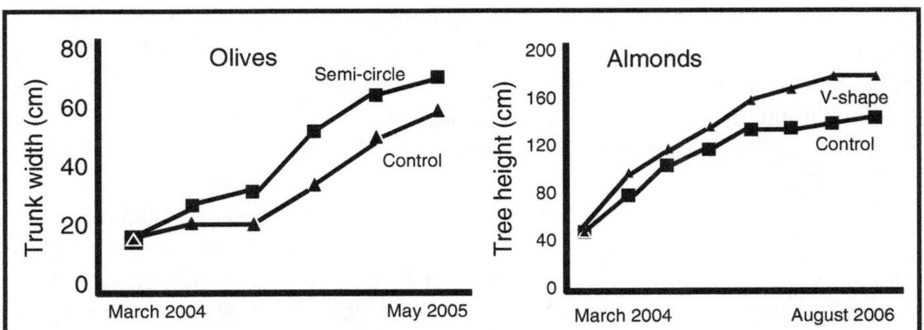

Fig. 33 Tree growth (trunk and height) under two runoff-harvesting/soil protection measures during 1 year, Dura site, South Hebron Mountains, PNA

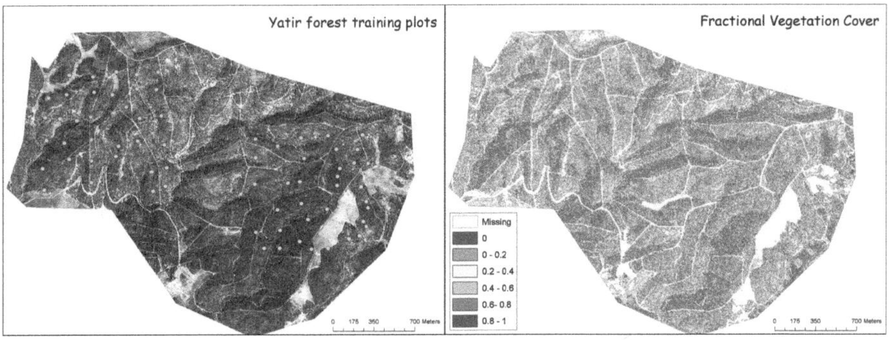

Fig. 34 The Yatir forest remote-sensing "training" plots (*yellow dots*, *left*) and its satellite image featuring the "Fractional Vegetation Cover" – an NDVI-based estimate of crown cover classification (*right*). The range of FVC of this area is 0.4–0.7 of a pixel area, and the overall forest canopy cover for this area is 55% (*right*), compared to ground truthing in the plots (*left*) which was 52%

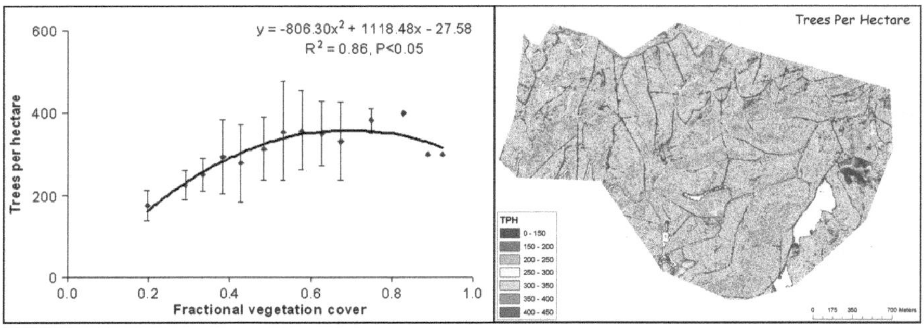

Fig. 35 Satellite imagery and a remote-sensing training area in Yatir, Israel, utilized for generation of tree density as a function of fractional vegetation cover derived from satellite imagery (*left*); also, use of this function for producing tree density map (*right*)

by maximum NDVI, from the non-canopy background reflectance expressed by minimum NDVI (pine needles' litter, responsible for most of this background reflectance was measured in situ using a high spectral resolution field spectroradiometer). These measures were used for calculating a weighted sum of canopy and background reflectance, from which the Fractional Vegetation Cover (FVC), a summation of the crown areas as seen from above, was derived.

The resulting FVC map showed a 55% canopy cover detected from space, whereas the calculated canopy widths derived from the ground measurements and calculated correlations resulted yielded an estimate of 52% canopy cover (Fig. 34). Once FVC map was available, it was used to correlate its values to actual tree densities measured in the training plots. A hump-shaped curve was obtained, with low density at low canopy cover but also at very high canopy cover, where large trees overshadow and outcompete small ones (Fig. 35, left). The resulting regression function was used to derive a tree density map for the training area (Fig. 35, right), showing that the dominant density in the training plot was 200–400 trees per hectare, a value within the range found in the ground survey (Sprintsin et al. 2007, 2009).

To test this methodology of combining remote sensing with ground sampling for assessing tree density on a large scale, the satellite imagery was used for estimating tree density at the whole watershed scale. The results were then tested using information from a tree census conducted in 2004 over a large area within the forest, what added another set of 50 randomly distributed plots covering a diversity of stand features within the Yatir forest. It was found that the satellite imagery estimate had an error not exceeding 10% for 61% of the ground-censused stands and only for 17% of them the error exceeded 20% (Sprintsin et al. 2007, 2009).

3.16 Non-woody Vegetation

Non-woody vegetation responses to environmental variables were expressed by aggregate measures such as vegetation cover, biomass and density that are independent of which and how many species are involved, and plant community measures-species richness, diversity, composition, functional groups and chorotypes.

3.17 Aggregate Measures of Vegetation Cover

Overall rangeland biomass in the Palestinian and the Turkish sites, as well as overall density in the Palestinian sites were not affected much by the inter-annual rainfall variability, it hardly changed within a range of 170–270 mm in Turkey (Fig. 36). Density, on the other hand, might have been negatively correlated to biomass in the PNA (though not significantly so, Fig. 37a). On average though, biomass declined

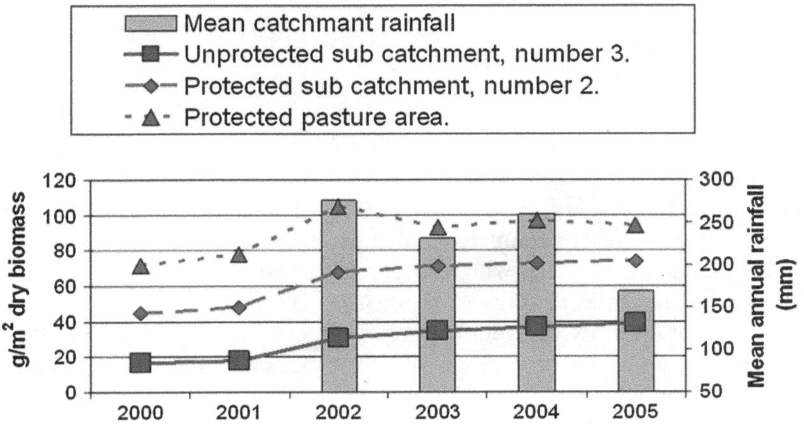

Fig. 36 Effect of rainfall and of protection from grazing on non-woody vegetation in different sub-watersheds, the Kiziloz microcatchment, Turkey

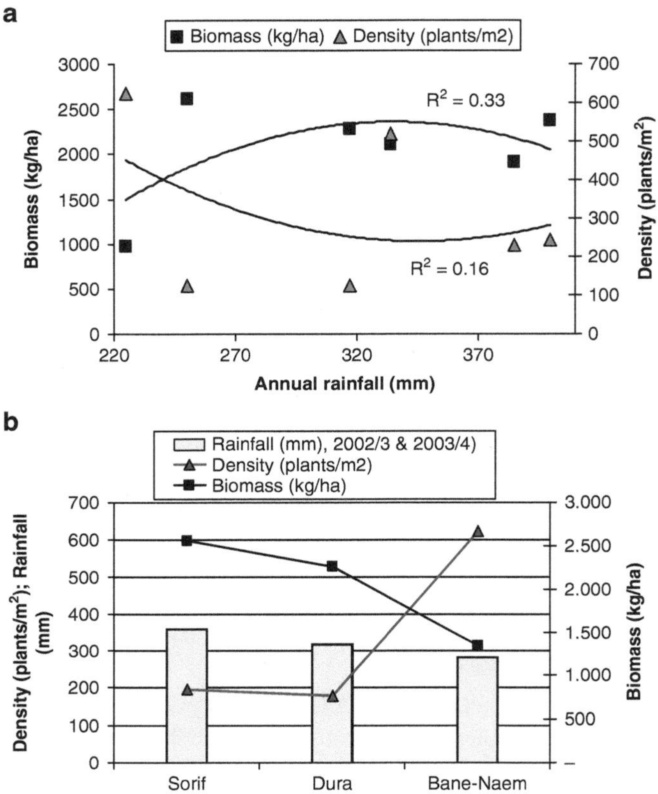

Fig. 37 Vegetation aggregates – plant biomass and density, South Hebron Mountains, PNA: (**a**) Regressions of biomass and density of each site in each of two rainy seasons 2003/2004 and 2004/2005 on rainfall; (**b**) Means of 2002/2003, 2003/2004 and 2004/2005, the differences in biomass and plant density between the driest site and the other two sites are significant (5% level, Fisher LSD test)

with the PNA sites' aridity (Fig. 37b), and in Turkey it was much reduced by grazing, being up to three times higher in protected than in non-protected plots (Fig. 36). Dry mass of the non-woody biomass within the Yatir forest was only 22% of that of the adjacent non-afforested watershed ($p < 0.001$; Ariza 2004). Biomass was significantly higher in mounds as compared to depressions and flat surfaces (Fig. 38, left; Wang 2005), but did not differ with respect to slope aspect in either afforested or in non-afforested sites in Yatir (Fig. 38, right; Ariza 2004).

The apparent disparity of overall biomass and overall density can be explained by the differences in the proportions of functional plant groups, like shrubs vs. grasses and forbes (or herbs). Indeed, the difference between the wettest site (Sorif) and the driest (Bane Noem) is that in the mesic one, Sorif, shrubs contribute to more than 70% of the biomass (Fig. 39, top) but their share in surface cover is the smallest compared to the other groups (Fig. 39, bottom); and in the driest site (Bane Noem)

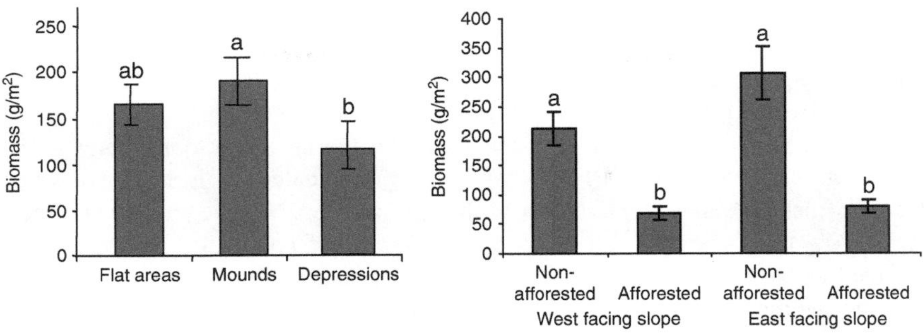

Fig. 38 Mean plant biomass, Yatir, Israel. Bars indicate 1+ standard error, the means of groups with the same letter are not significantly different at $\alpha = 0.05$

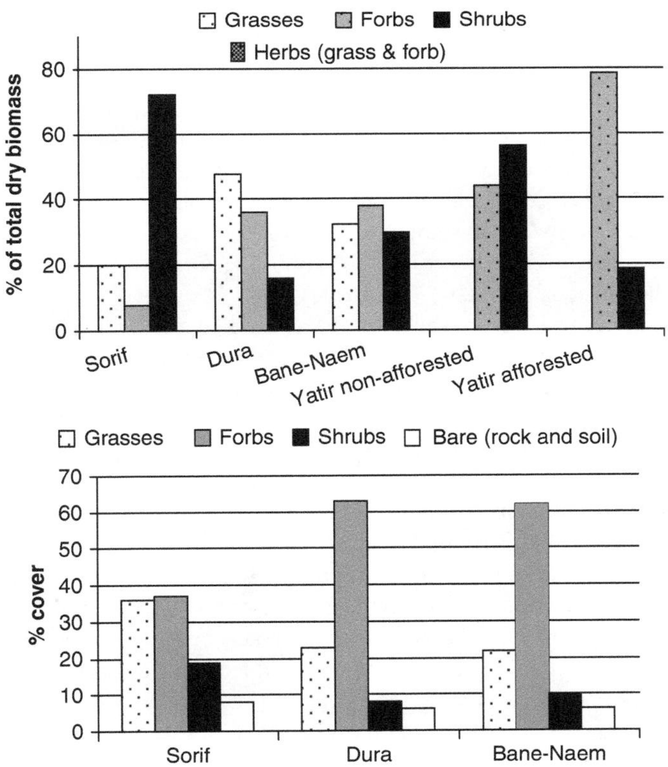

Fig. 39 Plant functional group biomass and cover. *Top*: Percent of total biomass (grasses, forbs and shrubs in the PNA sites in 2004/2005, and herbs and shrubs in Yatir, 2002/2002); *Bottom*: % of surface cover, in the PNA sites, 2004/2005

shrub biomass is only 29%. Altogether the share in biomass and in cover does not correlate, and forbs take the lead in cover of the more dry sites. In Yatir the forest non-woody plant cover is taken up mostly by shrubs, whereas outside the forest – by herbs (forbs and grasses combined, Fig. 39 top, Ariza 2004).

No correlation was found in Yatir between the average daily total Photosynthetically active Flux Density (PPFD) below canopy (calculated for the month of March from the fisheye photographs taken at the central site of the vegetation transects; Ariza 2004) and the herbs' harvested biomass at that point, thus suggesting that reduced herbaceous cover is not due to light-competition with the trees. Also, depth of pine needle litter layer on the Yatir forest soil was not correlated with the herbaceous biomass, suggesting that herbaceous growth in the forest was not constrained by allelopathy of pine needles.

Finally, employing an analysis of co-variance (involving microtopography and stone cover as covariates) and linear regression analyses (between shrubs and herbs), no effect of shrub presence and biomass on herb biomass was detected. Hence it was suggested that facilitation of herb growth by shrub presence can not be invoked as instrumental in the biomass differences between afforested and non-afforested sites (Ariza 2004).

As to plant density, however, no decrease in the density of herbs in the afforested area was detected, yet it was found that a herb individual in the non-afforested sites yields almost two times the biomass of an individual growing in the forest. This in-spite of the afforested sites having similar rainfall and relatively reduced evaporation and high nutrient, compared with the non-afforested areas. These observations suggest that competition for water and nutrients between the herbs and the trees (which have a shallow root system) is responsible for the decrease in herbs biomass and in the number of shrubs established within the forest, resulting in the overall low forest herbaceous biomass (Ariza 2004).

The effect of grazing on vegetation was studied in the Palestinian and the Turkish sites. Using grazing-protected plots within their watersheds the researchers regressed the ratio of the biomass in the grazed plots relative to that in the protected plots on the annual rainfall. Biomass ratio was positively correlated with rainfall, while density and cover were negatively correlated. Namely, with increased aridity grazing pressure declines (high ratio of the measure means a small biomass consumed). But plant density and the overall cover relative to the protected plots, were reduced as rainfall increased (Fig. 40).

Assuming that overall productivity increases with rainfall (both within site and between sites, which differed in their degree of aridity), the findings suggest that livestock preferred annuals, that are of low biomass hence do not contribute much toward overall biomass, but contribute more to the overall density and cover. Thus the lowest percentage of "remaining" density (i.e. highest grazing pressure that reduces density) was at low precipitation, and highest percentage, i.e., lowest grazing pressure occurred at low aridity sites. Also the lowest the rainfall, stronger was the reduction in plant cover due to grazing. Assuming also that plant cover rather than plant biomass is instrumental in soil conservation, it is suggested that with reduced rainfall (either due to droughts or due to the site's inherent aridity)

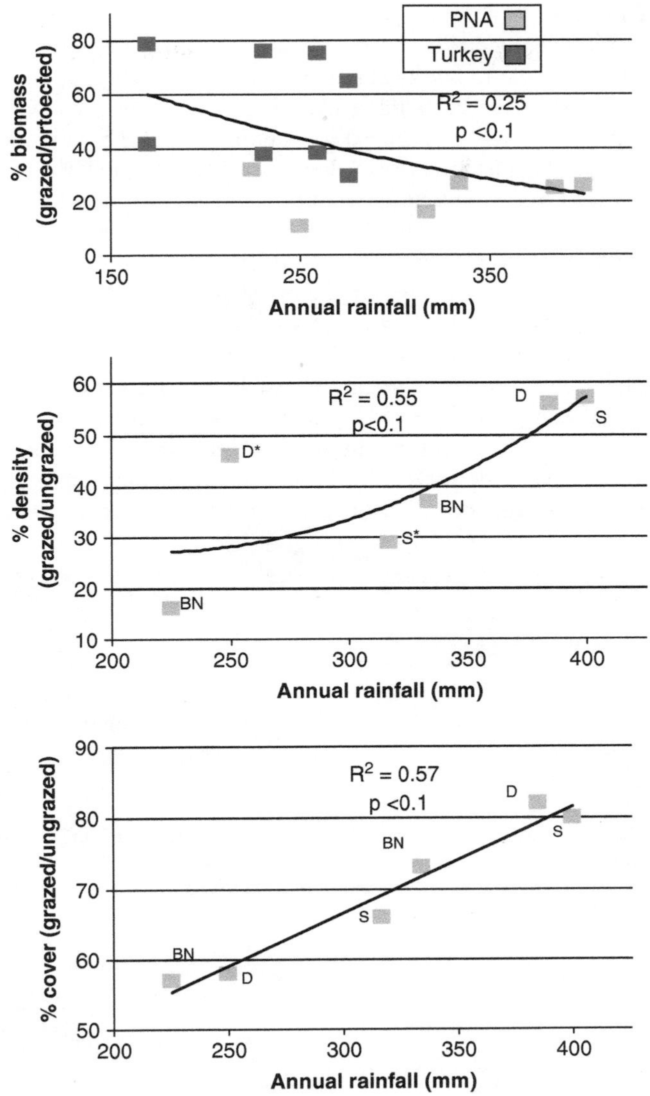

Fig. 40 The effect of grazing on aggregate vegetation measures regressed on annual rainfall. Each measure value (biomass, density and cover) is the percent of the measure in the grazed plots relative to that measure in protected plots, representing the biomass, density and cover relative to those existing prior to grazing. This value is treated as % forage left following grazing. Each data point represents a significant difference at the 5% level between the grazed and the non-grazed plot, except the asterisked value. *Red points* are for Turkey, in two replicate sites of Kiziloz for years 2002–2005, and *green point* are for the PNA sites, *BN* Bane Noem, *D* Dura and *S* Sorif, for rainy seasons 2003/2004 and 2004/2005. *Top* – biomass, *Center* – density, *Bottom* – cover

grazing may increase its detrimental effect on soil conservation. However, though values were significantly lower in the grazed than in the non-grazed plots, the regressions on rainfall were hardly significant statistically (Fig. 40).

It was also found in the Palestinian sites (Fig. 41) that in the low-rainy season (2003–2004) grass biomass was grazed upon more than shrub biomass, and in the least dry site the forb biomass increased due to grazing. This may indicate that some forbs were released from competition with grasses (due to grazing on grasses) and that some of the forbs are not preferred by grazers. In the high-rainy season (2004–2005) not only forb biomass increased in the least dry site due to grazing, but also the shrub biomass increased in the driest site, when and where grazing on forbs was more intense than grazing on grasses.

The effect of the runoff-harvesting/soil conservation structures installed in the Palestinian study sites on the aggregate measures of non-woody vegetation was explored too. The semi-circular bunds in Sorif increased plant biomass in 7% and 23% relative to control in the low-rain and high rain year, respectively (Fig. 42, left). Conversely, plant density, which was much higher relative to control, decreased when biomass increased (142% and 104% increase for the low and

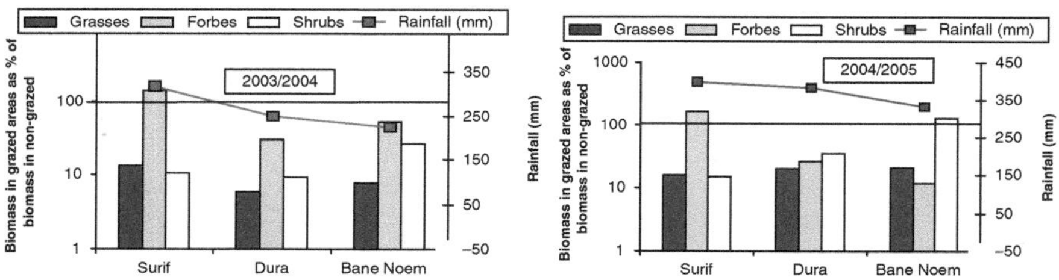

Fig. 41 The effect of grazing on biomass of grasses, forbs and shrubs, in 2 different years. Sites of South Hebron Mountains, PNA: biomass in grazed areas as percentage (log scale) of biomass in non-grazed sites. Horizontal line at 100% denotes that values above it are of functional groups in which grazing resulted in biomass increase

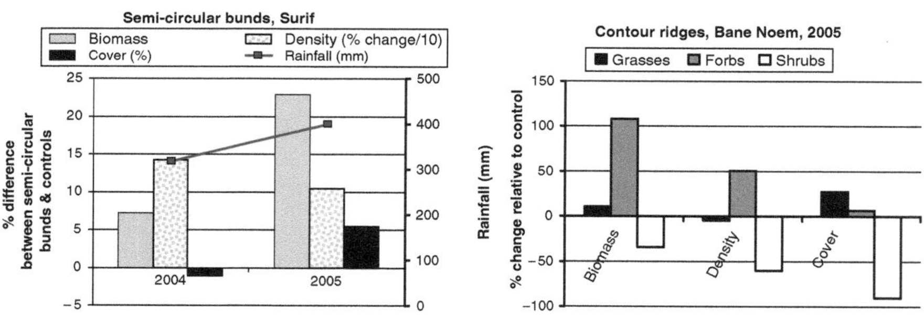

Fig. 42 Aggregate measures of non-woody vegetation in two runoff-harvesting structures, expressed as percent difference of the structures relative to controls, South Hebron Mountains, PNA. Note that density on the left-hand figure is expressed in units which are one order of magnitude lower than those of the other measures

high rainy year, respectively). The semi-circular bunds had a negligible effect on plant cover. Altogether the effect of this water harvesting practice was smaller in the low-rainfall year than in the high-rainfall one, but the difference between the 2 years might have been due to the disturbance caused when the bunds were dug, in 2004, the low-rain year. In the contour ridge practice, average biomass, density and cover in plots did not differ significantly (Fig. 42, right).

The increase in forbs was associated with decrease in shrubs, relative to controls. Furthermore, as the increase in forbs decreased (from biomass to density to cover), the decrease in shrubs increased. A small increase in biomass was associated with a small decrease in density and a somewhat higher increase in cover of grasses. To conclude, the semi-circular bunds of Surif, the least arid site, resulted in higher productivity than that of the contour bunds of Bane-Noem, the driest site. The difference in productivity between the two structures, practiced in sites differing in their aridity can be related to these difference in aridity, yet only small differences between them in the physical parameters could be detected (Table 10). Finally, it was found that the semi-circular bunds altered the relative density of functional groups – from highest density of grasses in controls, to highest density of forbs.

Attending just the contour trench practice of runoff harvesting in Yatir, the differences in biomass production in the different microhabitats along transect of such trenches was explored (Fig. 24; Wang 2005). In all its microhabitats as well as in the control, biomass was lowest in the least rainy year (2004, Fig. 43). On the other hand in the swales, acting as sink for the harvested runoff, biomass steadily

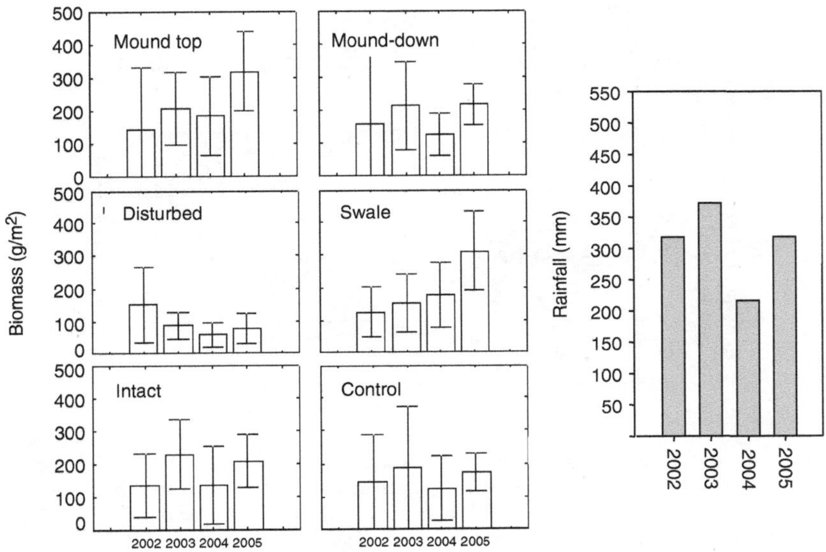

Fig. 43 Biomass in microhabitats of a contour trench's transect, Yatir, Israel. Note (framed figures in the center) the opposing trends in the disturbed microhabitat (disturbed during the creation of the swale) and the swale

increased during the 4 years of the study (the trenches were dug 2 years prior to initiating this study). However, the areas scraped and disturbed when the trench was created experienced a steady biomass decline during this period. Finally, the top of the mound created on the downslope side of the trench generated by the last year of the study the highest biomass, similar to that of the swale (Wang 2005).

3.18 Community Measures

Intensive vegetation surveys in which the floristic composition of the study sites resulted in identification of 206, 147 and 122 species in the Jordanian, Palestinian and Israeli sites, respectively. These were subjected to various analyses aimed at characterizing these plant assemblages and relating these characterizations to environmental variables. It was found (Fig. 44) that species richness (number of species in unit area) and species diversity (expressed by the Shannon-Wiener's information theory-based index and encompasses both species number and species' relative abundance) peaked in the mid section of the aridity gradient (Dura) along which their three sites were positioned. The highest value of the similarity index (38%) was also for Dura vegetation relative to the other two sites, with lowest similarity (32%) between these two. Finally, it was also found that the significance

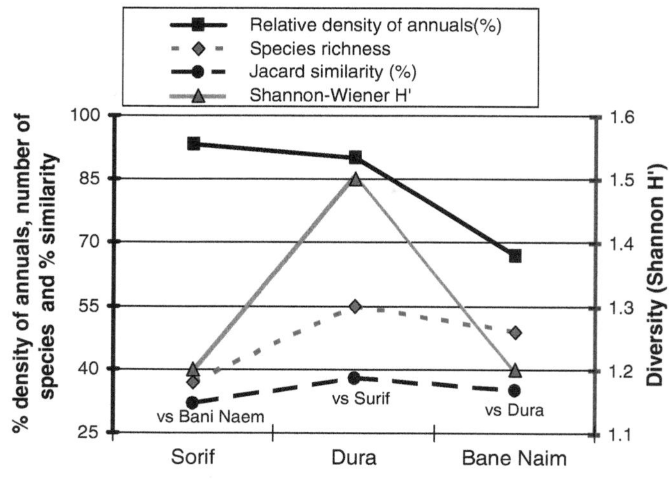

Fig. 44 Indices of plant community structure, South Hebron Mountains, PNA, 2005. *Red curves* – species richness (the cumulative number of species in 15 quadrates 0.24 m^2 each, i.e., number of species in 3.75 m^2), and diversity (H' = p*i* ln p*i* , where s is the number of species, P*i* is the proportion of individuals found in the *i*th species); Jacard index of similarity in species composition is *a*/*a* + *b* + *c*, where *a* is number of species present in both sites, *b* is number of species present in site A only and *c* is number of species present in site B only. The relative density of annuals is the number of individuals of all annuals combined per unit area divided by the number of individuals of all plants per unit area, times 100

of annuals as expressed in their relative density declines with aridity, probably since annual germination is opportunistic (Fig. 40).

A study of the vegetation community structure along an altitude/rainfall gradient of the Jordanian study watershed revealed a correlation between rainfall and number of species at each site, but species diversity, as measured by Shannon's H' did not change much and did not demonstrate relations to the rainfall gradient (Fig. 45). Also, % vegetation cover increased where number of species decreased (not considering the low cover of the grazed site), suggesting that high plant diversity does not necessarily provide the best soil protection.

Plant community features under different "treatments" were explored in the Palestinian and the Israeli sites. It was found that species composition similarity (expressed by the Jacard index), between water harvesting plots and control plots was 43% for the semi-circular bunds and 54% for the contour ridges thus demonstrating that these practices favor some species and disfavor others. The difference between the two figures, however, may be due not just to the difference in the water harvesting structures but also to the differences in aridity between the two sites. Thus, in the least arid site the effect of the water harvesting structure on the plant community was small (as evidenced by the high similarity in plant composition between treatment and control), whereas in the driest site the similarity was small, possibly indicating a stronger effect of the water harvesting intervention.

More species were found to inhabit the non-afforested than the afforested plots of Yatir, as expressed by total number of species in each of the two areas and in the index of species richness, namely the number of species per unit area (Table 11, top; Ariza 2004). Many species living off the forest were not found within it, meaning they were excluded by the afforestation. On the other hand, the forest had species that were not recorded out of the forest, meaning that the ecological conditions created by the afforestation enhanced the invasion of species alien to this ecosystem. Yet, the similarity between the two floras was relatively high (60%, using the

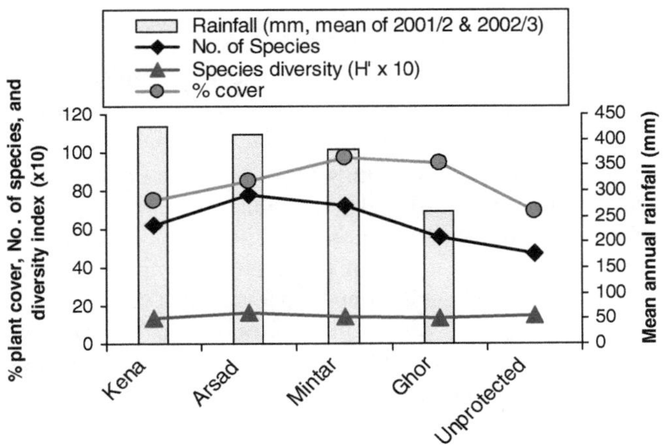

Fig. 45 Vegetation community indices in Humrat Es-Sahin, Jordan

Table 11 Species richness (number of species per unit of habitat) and number of species in Yatir, Israel, comparing afforested with non-afforested rangeland

	In transects				
	Richness	Total number of species			
	Identified and not-identified			Identified only[a]	
	(No./3 m^2)	In each area	Exclusive to area	In each area	Exclusive to area
Total (afforested and non-afforested)		122		88	
Afforested area	76.3	79	27	62	17
Non-afforested area	87.3	95	43	71	26
Shared by afforested and non-afforested			52		45

	In soil seed bank				
	Richness	Total number of species			
	Identified and not-identified				
	(No./8.4 × 2 cm soil sample)[b]	In each area	Exclusive to area	Only in seed bank	Also in transects
Total (afforested and non-afforested)		155		59	96
Afforested area	22.7	109	55		
Non-afforested area	38.3	100	46		
Shared by afforested and non-afforested			54		

Top: information of the plant survey, carried out by vegetation sampling transects; Bottom: soil seed bank data

[a]Identified to species level (6 species identified only to genus level, and are included in the "not-identified")

[b]Dimension of a soil seed bank sample is 8.4 cm diameter of the sampling cylinder, removing the top 2 cm of the soil

Sorensen similarity index). Whereas the differences in species richness between afforested and non-afforested plots were statistically significant, the differences between the diversity and evenness estimates were not (Fig. 46; Ariza 2004).

On top of aggregate vegetation measures in the microhabitats at transects across the contour trench water-harvesting structure used in the Yatir watershed for the newly afforested slopes (Figs. 24 and 43; Wang 2005), species-specific features there were also studied. It was found that the commonest functional group in these runoff-harvesting structures was that of the grasses, comprising 60% of the total biomass except for swales, where it dropped to less than 40% (Fig. 47; Wang 2005). The abundance of legumes was double in the Swale microhabitat, compared to the Intact and Control ones. As to shrubs, most of them were removed during the construction of the contour trenches except for the Intact and Control microhabitats. Their species slowly recovered during the 4 years of the study and though their biomass proportion was still low, the abundance of shrub individuals substantially increased over time (Wang 2005).

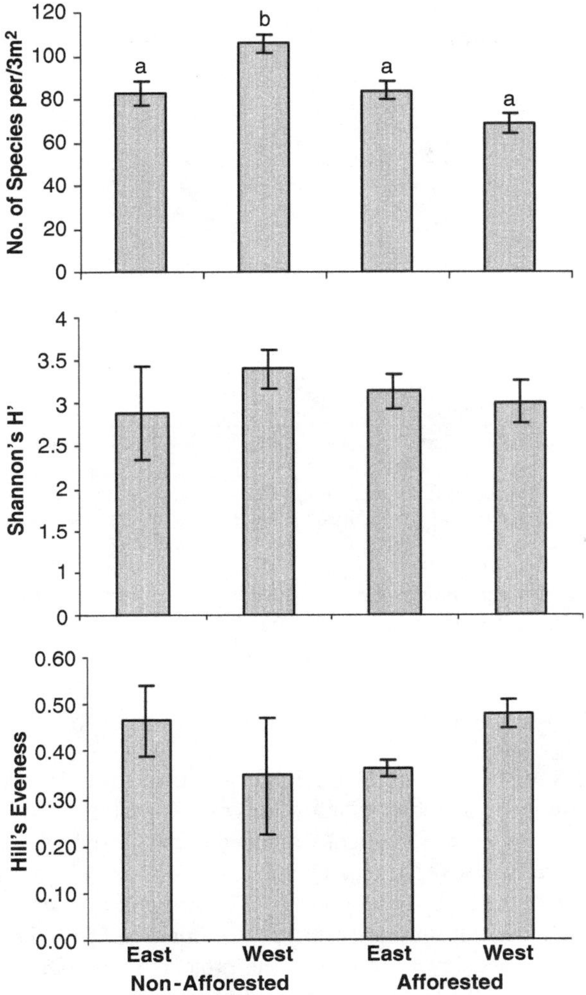

Fig. 46 Species richness, diversity and evenness in Yatir, Israel, 2002, comparing "treatments" (afforested and not-afforested) and slope aspects. Columns of different letters are significantly different from others at the 5% level

The impact of grazing on plant species composition in the PNA and the jordanian sites was explored. The researchers noticed that the five species (*Sarcopoterium spinosum*, *Asphodelus aestivus*, *Echinops polyceras*, *Eryngium creticum*, and *Thymelaea hirsute*) that dominated the grazed plots had lower dominance in the protected ones, whereas other six species (*Medicago* spp., *Bromus* spp., *Hordeum* spp., *Aegilops* spp., *Poa bulbosa* and *Avena sterilis*) showed the opposite trend, some

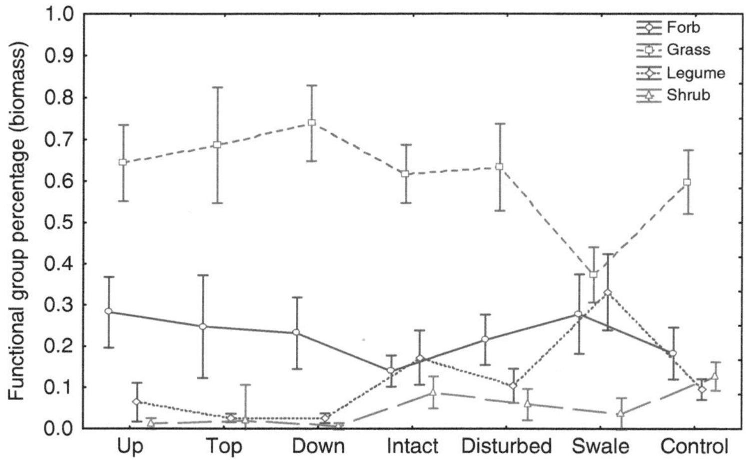

Fig. 47 Percentage biomass of functional groups in microhabitats across transects of the contour trench runoff-harvesting structure, Yatir, Israel (vertical lines – one standard error)

of them often thriving under *the Sarcopoterium spinosum* thorny shrubs. The researchers suggested that these differences were due to "non-palatability" and "palatability" of these two species groups, respectively. It was found that 17 of the most common species of the Yatir site are known by range managers as prime forage species, thus establishing the range quality of this site, in-spite of the apparently intensive grazing there.

A protected site within Humret Es-Sahin watershed was compared with four other sites within this watershed which were grazing-protected and found too that four palatable species were dramatically reduced in the grazed area. Furthermore, it was also found that whereas the mean height of plants was 81 cm in the protected sites, in the unprotected one it declined to 33 cm. Finally, it was found that the unprotected site's percent similarity in species composition using the Sorensen Index (68%), was most similar to that of the protected site, but of the inherently lowest productivity (Ghor), whereas the similarity to the other, less arid but protected sites was lower (65%).

The global geographic distribution of the species identified in the Palestinian and the Israeli sites was consulted, and the plant communities in each site were classified by chorotypes (the global phytogeoraphical region to which a species belongs). Since their territories are positioned in a global climatic transition zone, several global phytogeographical regions meet and overlap within Israel and the PNA. It is, therefore, not surprising that the teams found within their sites a high diversity of chorotypes: the geographical distribution of most of the species in each site was either Mediterranean or Irano-Turanian (of central and western Asian steppes). Furthermore, many species were of a wider distribution as they occur in more than one phytogeographic region (a combination of the two above, as well as including also the mesic Euro-Siberian region on the one hand, or the Sahara-Arabian deserts on the other hand). In each of the four sites (3 Palestinian and one Israeli) more than half of the species belong to only one chorotype, but in the driest site, Yatir, the proportion of this category was much higher (Fig. 48), suggesting that species in high aridity are specialized, what reduces their global geographical extent (Ariza 2004).

When the proportions of the different chorotypes in each site are plotted against the phytogeogrphical regions, which is arranged along an axis of increased aridity (Fig. 49) it is evident that only a very small proportion of the species are of a geographical distribution that includes also the Euro-Siberian region. The peak in representation is that of the Mediterranean, Mediterranean/Irano-Turanian, and Irano-Turanian chorotypes. The curve then rapidly descend towards the highest aridity, as expressed in low proportion of species in whose global distribution the Sahara-Arabian deserts are included. Moreover, as one moves from the mesic sites (Sorif and Dura) to the dry sites (Bane Noem and Yatir), fewer species are of the mesic chorotypes, and the representation of the Irano-Turanian and Sahara-Arabian and their combinations in the local flora increases (Fig. 49). This highlights the potential to detect and express sites' aridity by the relative distribution of their plant species chorotype (Ariza 2004).

Studying the relative representation of chorotypes in the afforested and the non-afforested sites of Yatir, depending on the aridity of the phytogeogrphical regions to which the species in both sites belong, it was found (Fig. 50) that though both afforested and non-afforested sites are exposed to identical macro-environmental conditions, there is a higher preponderance of more mesic chorotypes and lower incidence of arid chorotypes in the afforested compared the non-afforested sites. This suggests that afforestation of this rangeland is instrumental in modifying the micro-environmental conditions, thus making them favorable to species adapted to lower aridity, as suggested by the pattern of their global geographical distribution (Ariza 2004).

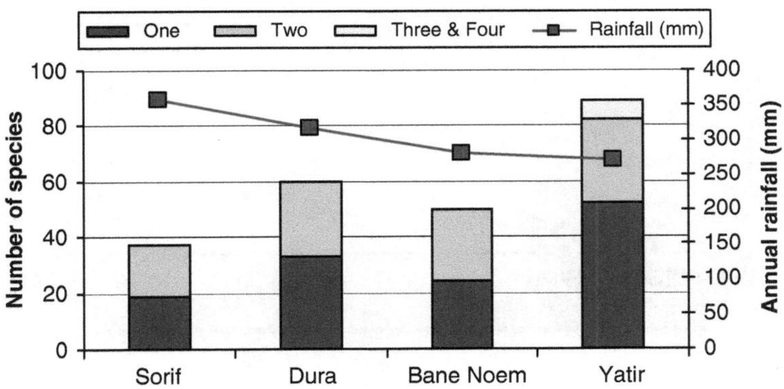

Fig. 48 Number of species of geographical distribution covering one, two and three or four global phytogeographical regions in the PNA sites (in 2004/2005 growing season) and in the Yatir watershed, Israel (in 2001/2002 growing season)

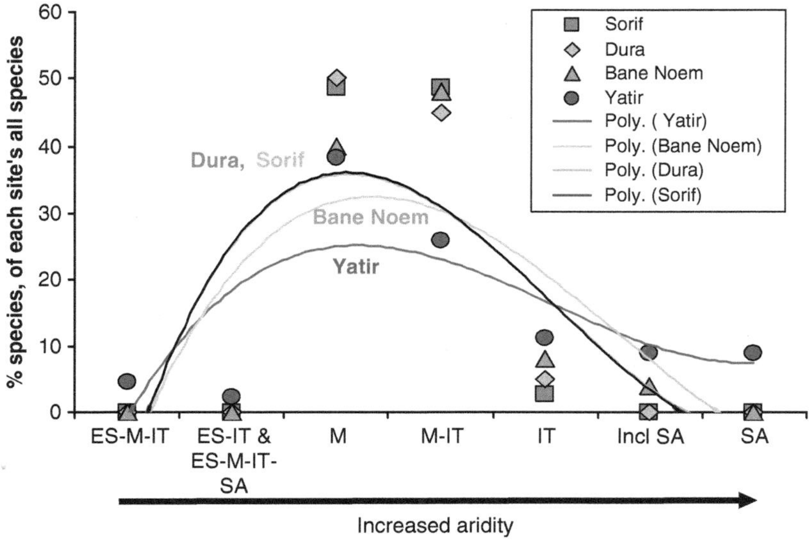

Fig. 49 The number of species in each chorotype (expressed as percentage of all species in each site, the ordinate), plotted against the phytogeographical regions arranged on an increased aridity gradient. The curves of the more arid sites (Yatir and Bane Noem) flatten towards the right more than those of the less arid sites (Dura and Sorif). *ES* Euro-Siberian, *IT* Irano-Turanian, *M* Mediterranean, *SA* Sahara-Arabian, "*Incl SA*" this category is of species for which the SA region is only a part of their global geographical distribution

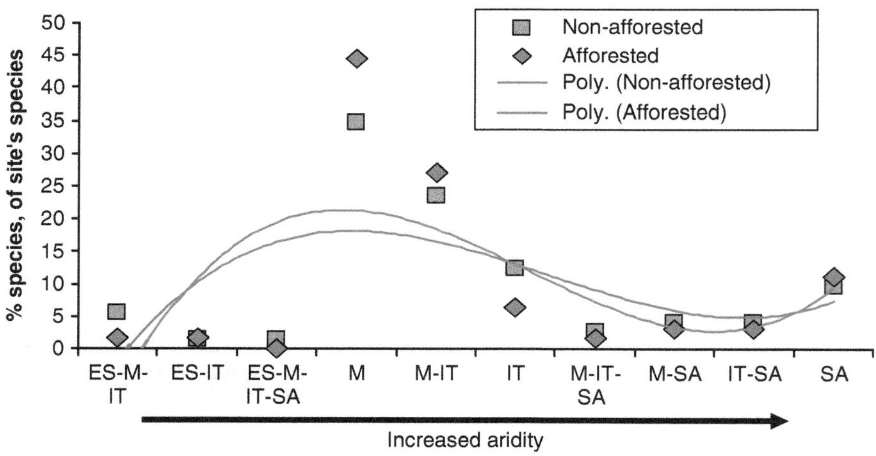

Fig. 50 The proportion of chorotypes along the aridity gradient (following Fig. 49), in the afforested and non-afforested sites of Yatir, Israel. For acronyms of the *X*-axis (see Fig. 49). In the afforested site the proportion of chorotypes of medium aridity is higher and of those of higher aridity are lower compared to the chorotypes of the non-afforested site

The distribution of chorotypes was also studied in the microhabitats within the runoff harvesting practice of the contour trenches in Yatir. In all habitats, the percentage of the Mediterranean biomass and species were highest compared to the Irano-Turanian and Saharo-Arabian species groups (Fig. 51). Furthermore, a trend of increased Mediterranean component with time was observed in both species

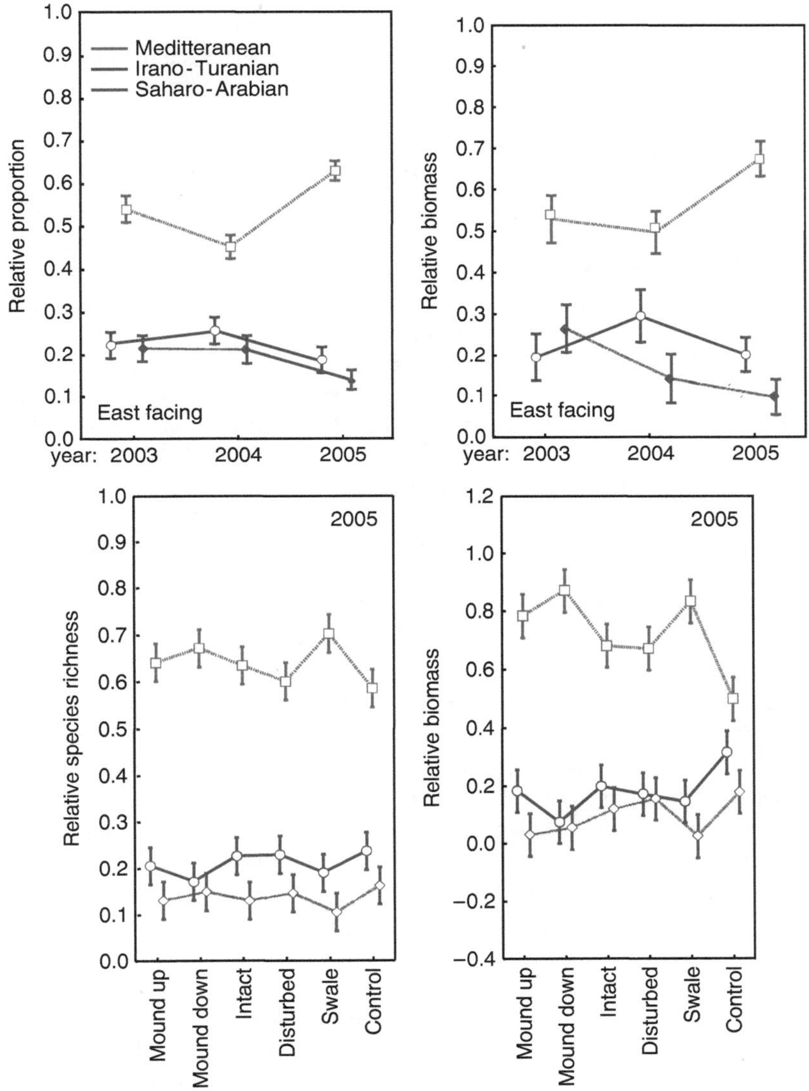

Fig. 51 Chorotypes in microhabitats, contour trench transect, Yatir, Israel. *Top*: There were no significant differences between east-facing and west-facing slopes hence only east-facing is presented; *Bottom*: 2005 was a year with average rainfall for Yatir, hence this is the one chosen for this presentation

richness and biomass during the studied years. In contrast, the Irano-Turanian and Saharo-Arabian components decreased with time (Fig. 51, top; Wang 2005).

The microtopography of the trench's transect significantly affected species chorortype identity; the Mediterranean chorotype had the highest species richness and biomass in swales and on the downslope of the mounds, whereas the other two more arid chorotypes were low in richness and biomass in these microhabitats. Also, the disparity in richness and biomass between the Mediterranean and the other two chorotypes in the controls, i.e., sites without water harvesting, was much smaller than within the runoff harvesting structure (Fig. 51, bottom). Thus, the added runoff due to the forester intervention aimed at collecting runoff water for the tree saplings planted in these contour trenches increased the preponderance of the Mediterranean flora in this micro-scale too (Wang 2005).

The discovery that species of the Mediterranean chorotype were doing better under watershed manipulations such as contour furrows and afforestation, and that species not found under the "natural" conditions in Yatir were detected in afforested sites brought up the question whether these species, apparently new to the indigenous Yatir plant community, are invaders or they are found in the seed bank but germinate only under favorable conditions, which did not occur during the studied years outside the afforested and newly-afforested areas, but were prevalent in the old afforestation site. Therefore a study of the Yatir soil seed bank was conducted, by collecting soil samples and let their seeds germinate both in marked plots in Yatir, and under promoted germination conditions (mainly irrigation), in the greenhouse (Ariza 2004).

A total of 155 species germinated from the entire soil seed bank samples combined, which is 27% more than the 122 species detected in all study transects in Yatir. Furthermore, 59 species that germinated in the experiments from the soil seed bank were not found in the afforested vegetation transects (Table 11). Thus, a sizeable number of species found within the potentially germinable soil seed bank of Yatir, was not detected in the vegetation surveys during the study years (Ariza 2004). Other comparisons relate to species richness which was higher in the non-afforested than in the afforested areas in both the "active" plant community surveyed by vegetation transects and in the germinable fraction of the dormant soil's seed bank (Table 11). However, whereas this was the case regarding the overall number of species in transects, the seed bank had slightly more species in afforested than in non-afforested areas. The number of species exclusive to each of the two habitats also behaved in a contrasting manner-more of these in the non-afforested transects but fewer of them there in the seed bank, as compared to the afforested habitat (Ariza 2004).

The differences found in the species richness of the soil seed bank's germinable fraction between afforested and non-afforested sites, agree with the results found for the present growing vegetation: the species richness was lower in the afforested sites, regarding both the active vegetation and the dormant seeds in the soil. However, in the transects no difference was found in diversity or dominance between the afforested and non-afforested plant communities, whereas the Shannon diversity was higher and the Simpson evenness was lower in the seed bank of

Table 12 Mean number of species germinated in the field and in the greenhouse, from seed bank samples in Yatir afforested and non-afforested sites, Israel

Habitat/Germinated	Afforested	Non-afforested
In the field	25.9	34.2
In the greenhouse	20	42.4

All differences are statistically significant, $P < 0.05$

non-afforested areas as compared to that of the afforested ones (Ariza 2004). More importantly, species exclusive to the afforested area were represented in the seed bank of the non-afforested area, and vice versa. All these indicate that the environmental conditions, more than the availability of seeds, determine which species will germinate where; species new to the afforestation area had seeds also in the non-afforested area, but they apparently did not germinate there (Ariza 2004).

Contrary to expectation that the enhanced greenhouse conditions would induce more species to germinate, it was found that this was true only for the non-afforested seeds (Table 12), though seedling density was similar in samples from both field and greenhouse replicates, thus suggesting that the afforestation did not negatively affect the germinability of the soil seed bank. Furthermore, experiments demonstrated that pine needle litter, expected to reduce germinabily due to its allelopathic effect, actually increased it (average of 22 species per sample germinated with pine needles' litter as compared to 18 species when litter was removed). This result can be attributed to the litter cover's positive effect on soil temperature (Ariza 2004). The forest's canopy cover too was not instrumental in changing overall germinability, though it increased community dominance measures, by considerably favoring a few species, probably due to improved topsoil moisture conditions. Altogether most findings pointed at appropriate germination conditions within the forest, in-spite of the forest having fewer species than the non-afforested areas (Ariza 2004).

4 Discussion and Conclusions

To address the risk of desertification, mainly expressed by soil erosion, management measures (afforestation and runoff harvesting) in selected Middle Eastern/ Mediterranean dryland watersheds (in Israel, the PNA, Jordan and Turkey) were explored.

Research in an arid Israeli watershed traditionally used for centuries as rangeland but currently partly covered by a 35-year-old Aleppo pine forest (the Yatir Forest area) yielded the following results: (a) the older afforested area had no runoff while the surrounding non-afforested, apparently overgrazed rangeland discharged much runoff and sediment; (b) an adjacent newly-afforested rangeland fitted with machinery-constructed contour trenches was very effective in harvesting runoff and reducing soil erosion (by more than a one order of magnitude) and the natural vegetation demonstrated rapid recovery following trench construction: only 2 years after the initial disturbance plant biomass and species richness regained their pre-disturbance values; (c) non-overgrazed and non-afforested areas (clearings within the afforested area) generated very little if any runoff; (d) the forest floor and the non-forest area hardly differed in the amount of rainfall they received and in the dimension of their soil water storage, which has been virtually depleted in both areas by the end of the dry season; (e) the older forest ameliorated soil conditions by doubling the soil organic matter content and together with the new contour-trenches intervention facilitated the growth of mesic flora (mostly of Mediterranean Basin origin) at the expense of the more xeric species (mainly of Asian origin).

Overall herbaceous plant biomass was contemporaneously one half of the herb biomass and one eleventh of the shrub biomass of non-afforested sites, compared with the non-afforested areas (likely resulting from competition for water and nutrients with the trees); (f) in-spite of the forest bringing "new" species to the Yatir area (27 of the 122 identified species found in the forest were new to the local natural flora) its overall species richness (number of species per unit area) declined relative to that of the non-afforested sites; (g) plant distribution across the contour trenches was rearranged following the local distribution of soil moisture caused by this intervention, but the contour trench not only maintained the native vegetation biomass but also increased its diversity.

Given these observations, it follows that the dryland can be used in one of three different ways: as rangeland, as (planted) woodland, or for a combination of both. In either uses loss of water by surface runoff and its associated soil loss is regarded as detrimental. However, the controlled grazing (the first option) and afforestation (the second option) may equally reduce runoff and erosion, while their effects on the soil water storage is similar.

Approximately 30% of the stored soil water is lost through direct evaporation in the afforested site, loss that can be reduced by increasing canopy density. Yet, increasing tree density may further decrease the herbaceous biomass and diversity within the forest. As the presence of the forest increased the species pool of the Yatir region, the combination (third option) through a spatial patchwork of forest stands in a matrix of rangeland may increase the overall regional biodiversity. Another option is to create within the forest tree-thinned patches, which may increase evaporation yet promote herbaceous species richness while not reducing its biomass. This notion can be expanded into forming landscape heterogeneity with added patches of stands in the surrounding non-afforested areas, which might increase overall species richness, due to the patches "edge effect". Regarding the practice of afforestation in contour ditches, increasing the distance between trenches, thus leaving more area of the runoff-harvesting structures intact would reduce soil erosion and increase plant regeneration since the intact area is the source of seeds for reconstructing the disturbed sections of the trench structure. In addition, less severe soil striping (leaving more than 10 cm thick soil layer) at the disturbed area next to the swale may allow swift recovery of the original vegetation instead of a dead-end phase-shift to sterile patches still common under the current construction practices. Somewhat reducing the amount of runoff collected in the trench and allowing more runoff to infiltrate prior to reaching the trench would also increase the diversity of the herbaceous plant community without compromising water supply for the trees.

Whereas in the study years both trees and herbaceous vegetation equally depleted the soil storage, since water did not infiltrate deeper than the herbaceous root zone, the option that soil moisture may extends below the rooting depth of the herbaceous vegetation during very high rainfall years should be investigated. This is because in this case trees might have a greater effect on water resources than of the herbaceous vegetation outside the forest. Also, the effects of increasing tree density on the fraction of the soil water that is evaporated in the forest need to be further explored.

Regarding the non-afforested rangeland, the optimal stocking rates for maximizing the amount of soil water by the end of the dry season and minimizing runoff generation is worth studying. The optimal spatial dimension of the created patchiness of forest and non-forest is unknown. Also to be considered are the maximization of overall plant biodiversity (due to edge effect), but to the point that animals species are not constrained due to their habitat fragmentation, and in a way that the plant community does not become dominated by edge species. With respect to the contour trench afforestation two issues require further attention: (a) the extent of competitive interactions between the native vegetation and the planted trees in the contour trenches, and (b) the relative importance of the contour trench "intact" habitats as a repository of sources for the regeneration of the trench plant communities.

Studying an arid watershed protected from grazing in Jordan revealed a very high species richness of plants (400 species) and demonstrated that a simple runoff harvesting practice, contour furrowing, can dramatically reduce surface runoff, soil erosion and sedimentation, increase soil moisture storage both in volume and in depth and encourage rangeland productivity. Furthermore, using three different furrow alignments differing in the runoff contributing area between furrows, the Jordanian study identified a space of 3 m between successive furrows as most productive. This simple water harvesting technique is unique in that it does not target agricultural productivity, but natural range productivity and its interference with the environment is relatively small.

Testing a number of runoff harvesting measures, including different types of terracing and bunds as well as "reforestation" with shrubs in arid watersheds of the PNA, revealed the efficacy of these practices in reducing runoff and sedimentation, increasing soil moisture and plant productivity, be it olive trees or indigenous, natural range vegetation. Furthermore, such water-harvesting practices were found to not only benefit agricultural and forage production, but also add a few new plant species to the local flora (as also found in Yatir).

The Palestinian studies re-asserted that the traditional practices of slope terracing not only creates more space and farmer's convenience for cultivation, but also dramatically improve soil properties, including its moisture profile, resulting in higher productivity. Furthermore, the more recent practices and the less labor-intensive bunds are also very efficient in conserving soil and promoting productivity. Interestingly, it was found that the effectiveness of all the tested measures, as expressed in vegetation response in biomass and density, declined with aridity, as evidenced by the strong response in the least arid site (Sorif) compared to the very low one in the most arid site (Bane Noem). The researchers suggested that the low response in the driest site is due to the thin and low-structured soil there, typical of arid dryland and expressed in the inherent low productivity of arid drylands. It is therefore likely that the low water holding capacity of drylands' soils does not enable them to respond to the improved water input provided by the water-harvesting measures.

Also notable is the use of shrub planting as a measure for rangeland rehabilitation, which unlike the tested "conventional" runoff-harvesting measures, it increased soil organic carbon storage. Thus, though arid afforestation as the one in Yatir also increased carbon soil, the Palestinian practice of shrub planting constitutes a milder ecosystem change. Finally, it was also asserted that the role of grazing in not only reducing overall plant density but also changing plant species composition, as well as the role of the Mediterranean shrub *Sarcopoterium spinosum* in providing grazing refuge for many species, and in improving soil moisture conditions at the microhabitat level.

The improved carbon storage and the protection from evaporation provided by the planted shrubs in Bane Noem were expressed in higher soil moisture in the driest site. These findings suggest that measures like dryland range rehabilitation by planting shrubs, given enough time, would prove more successful in increasing productivity in such arid areas than other runoff-harvesting structures and manipulations. Regarding orchards of fruit tees, especially olives, the comparative study of the several structures clearly recommend planting trees within the semi-circular bunds. Finally, given that some of the measures tested in the Palestinian sites can be instrumental in the rehabilitation of overgrazed arid and semiarid rangelands, more research is required for developing means to avoid further degradation under increasing population pressure and the demand for livestock products.

The Kiziloz watershed in Turkey is one of 17 watersheds contributing sediment to some of the largest dams on the Euphrates River, thus reducing their efficiency. The researchers found that soil erosion annually discharged from sub-watersheds of the Kiziloz watershed amounted to about 0.088 kg/m^2, regarded as a low erosion rate, though the annual discharge from the whole watershed was still above the average for Turkey. In addition, areas to which soil protection measures were applied (terraces, plantation, grazing exclusion) released less sediment than others did. On the other hand, the satellite image analysis, the trees, and vegetation ground sampling provided an interesting, rather unexpected insight: the natural vegetation areas such as natural forest and natural rangelands exhibited higher values in soil conservation and productivity measures than reforested and afforested areas. It is therefore concluded that the protection of natural areas is more effective and much cheaper than the reforestation and afforestation ventures. This distinction between conservation and rehabilitation is portrayed by the finding that plantations of the local oak species carried out by the national rehabilitation project, had a lower survival and growth rates as compared with those of orchard species (apparently due to insufficient attention given by the foresters to the soil type preferred by this oak species). Thus, protecting existing oak stands seems indeed to be preferred on oak plantations.

Acknowledgements The initiative for this joint project came from the Monitoring Unit of the US Forest Service in Boulder, CO, USA, with professional connections to researchers in the three countries. The project owes its existence and achievements to the vision and the inspiring leadership of Dr. Tom Hoekstra and to the continuing enthusiasm, dedication, logistic and processional support of Dr. Chuck Troendle, throughout the lifetime of the project. The authors thank the research students of the Israeli team, Clara L. Ariza, Ilana Wener-Frank, Matan Chocron, Michael Sprintsin, Yafei Wang and Yonit Schachnowitz, who carried out their studies under the umbrella of the project, and their theses constituted an invaluable source for the compilation of this chapter. Special thanks to Yulia Alexandrov for her invaluable work and consultation in the study of rain-runoff-sediment interrelationships. Thanks are also due to US-AID staff, especially Dr. Adam Reinhardt of the Washington DC office, who followed the project implementation and Mr. Boaz Ayalon of the Tel-Aviv office for his assistance and support. Much support for this project has been also provided by Dr. Chuck Lawson of the US State Department. Warm thanks are due to Jennifer Peterson and Christopher Soriano for their encouragement and continued support during the final stages of the project.

References

Adeel Z, Safriel U (2005) Ecosystems and Human Well-being: Desertification Synthesis. The Millennium Ecosystem Assessment, World Resources Institute, Washington, DC.

Alexandrov Y, Balaban N, Bergman N, Chocron M, Krispil S, Laronne JB, Powell DM, Reid I, Wener-Frank I (2008) Differentiated suspended sediment transport in headwater basins of the Besor catchment, northern Negev. Israel Journal of Earth Sciences. 57: 177–188. Doi:10.1560/IJES.57.3–4.177.

Ariza Clara L (2004) Vegetation Monitoring in a Semi-desert Afforestation Project. M.Sc. Thesis, Albert Katz International School for Desert Studies, Ben-Gurion University of the Negev.

Chocron M (2009) Runoff and Sediment Supply from Small Gullied and Unsullied Basins in a Semi-arid Grazed Environment. Unpublished M.A. Thesis, Ben Gurion University of the Negev, 81 pp.

Cohen S, Svoray T, Laronne JB, Alexandrov Y (2008) FuDEM: a temporally dynamic catchment scale soil erosion model based on fuzzy logic. Journal of Hydrology. 356(1–2): 185–198. DOI:10.1016/j.jhydrol.2008.04.010

Frank-Wener I (2006) The effect of Afforestation on Runoff and Sediment Yield in a Semiarid Area. Unpublished M.A. Thesis, Ben Gurion University of the Negev, Israel.

Powell MD, Reid I, Laronne JB, Frostick LE (1996) Bedload as a component of sediment yield from a semiarid watershed of the northern Negev. Int'l Assoc. Hydrological Sciences, Publ. 236: 389–397.

Safriel U, Abu-Sliman S (2007) Building Bridges of Confidence Through Technical Dialogue. Middle East and North Africa Regional Initiative for Dryland Management. ICARDA-FAO-World Bank.

Schachnowitz Y, PR Berliner, Bar P (2008) Rainfall interception and spatial distribution of throughfall in a pine forest planted in an arid zone. Journal of Hydrology. 349: 168–177

Schachnowitz Y (2006) The Effect of Arid Land Afforestation on Local Water Balances: The Yatir Case. Ph.D. Thesis, Ben-Gurion University of the Negev, Israel.

Schiller G, Cohen Y (1998) Water balance of *Pinus halepensis*, Mill.afforestation in an arid region. Forest Ecology and management 105:121–128.

Sprintsin M, Karnieli A, Berliner P, Rotenberg E, Yakir D, Cohen S (2007) The effect of spatial resolution on the accuracy of leaf area index estimation for a forest planted in the desert transition zone. Remote Sensing of Environment. 109: 419–428

Sprintsin M, Karnieli A, Sprintsin S, Cohen S, Berliner P (2009) Relationships between stand density and canopy structure in a dryland forest as estimated by ground-based measurements and multi-spectral spaceborne images. Journal of Arid Environments. 73: 955–962
Wang Y (2005) Vegetation Monitoring of Contour Trenches in Yatir. M.Sc. Thesis, Albert Katz International School for Desert Studies, Ben-Gurion University of the Negev.

Soils of the Mediterranean Region, Their Characteristics, Management and Sustainable Use

Pandi Zdruli, Selim Kapur, and Ismail Çelik

Abstract Twenty-two countries surround the Mediterranean Sea and a few more experience similar climate and soil conditions. The diversity of landscapes and natural conditions has created tremendous opportunities for economic development but has increased environmental concerns too. Human-induced pressures are on the rise throughout the region. They are particularly acute on the coastal areas. There are many competing interests especially for soil and water resources and the long term sustainable development of the region is far from being achieved. This paper offers a general overview and description of major soils of the Mediterranean region, explain their extent and characteristics and offer suggestions for their use and management. We argue that due to increased anthropic pressures throughout the region, but most importantly in North Africa and the Middle East, the endorsement of long-term policies for soil protection are not a choice but a prerogative to enhance sustainable development, reach food security objectives and meet the goals of poverty alleviation. It is concluded that unless there is political will to confront these realities chances for success may be limited and the region may continue to accelerate his path towards instability with consequences for all the countries in the region.

Keywords Mediterranean · Soil descriptions · Classification · Use and management · Human pressures

P. Zdruli (✉)
International Centre for Advanced Mediterranean Agronomic Studies, Mediterranean Agronomic Institute of Bari, Italy Via Ceglie 9, 70010 Valenzano (BA), Italy
e-mail: pandi@iamb.it

S. Kapur and I. Çelik
Department of Soil Science, Faculty of Agriculture, University of Çukurova, 01330 Adana, Turkey

S. Kapur et al. (eds.), *Sustainable Land Management*,
DOI 10.1007/978-3-642-14782-1_4,

1 Introduction

The Mediterranean is placed at the crossroads of three continents and yet continues to have a strategic importance at global scale. In itself, is a unique fragile ecosystem, which includes many natural features and a great variety of landscapes, soils, vegetation, geology, climate, water and biodiversity. Most importantly, at the centre are its people, a family of 428 million (Plan Blue 2005) members that call the region home. Predictions are for 517 million in 2020.

Twenty-two bordering countries surround the Mediterranean Sea (Fig. 1), even though there are other ones like Portugal and Jordan for instance that rightly consider themselves as Mediterranean too. The Mediterranean type of climate domains on the other side, have a much larger geographical distribution including areas as far as the southern parts of California, and limited regions of Australia, New Zeeland or Argentina (Eswaran et al. 1995).

One of the most common used criteria to establish whether a certain area "qualify" for being classified as Mediterranean climate is the "xeric" (Gr. *xeros*, dry) Soil Moisture Regime (SMR) of the USDA Soil Taxonomy (2006) that identifies as: "Soils of temperate areas that experience moist winters and dry summers, in areas of xeric SMR the soil moisture control section, in normal years, is dry in all parts for more than 45 or more consecutive days in the 4 months following the summer solstice and moist in all parts for 45 or more consecutive days in the 4 months following the winter solstice,...the mean annual soil temperature is

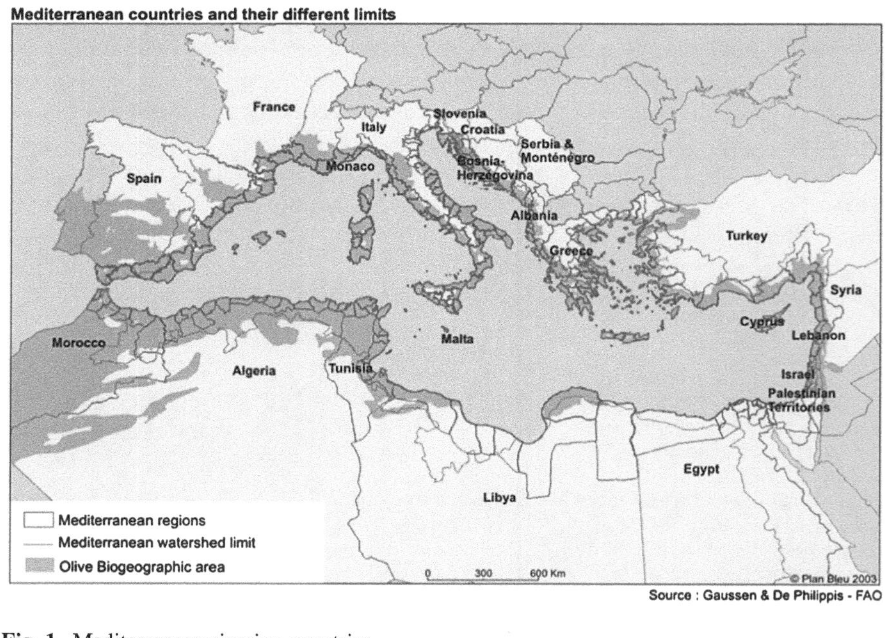

Fig. 1 Mediterranean riparian countries

lower than 22°C, and the mean summer and mean winter soil temperature differ by 6°C or more at a depth of 50 cm or wherever shallower... ." This paper deals with soils surrounding the Mediterranean Sea.

The stakes for the very existence of the Mediterranean people have been continuously the same: scarcity of water and land resources. Confronting these realities required for instance the development of some of the most antique and efficient irrigation systems of the world (Kapur and Akca 2003). Only 13% of the Mediterranean land is considered fit for agricultural use (FAO 2000) and around 5% of the land resources included in the North African and the Middle Eastern countries are arable lands; the rest is made of pastures, wetlands, forests, shrubs, urban zones, badlands, rocky areas, and deserts. Due to increased population pressure and continued pace of land degradation it is expected that the available arable land per capita in the whole Mediterranean would decrease from 0.48 ha per person as it was in 1961 to 0.22 ha per capita in the year 2020 (Zdruli et al. 2007). However, that ratio is around 0.20 already in North Africa and the Middle East. Still today's agriculture relies on these precious natural resources to provide food and fibber for the rapid growing population and the tourism industry that provide both important revenues for the local economies, but put as well a strong pressure especially on natural resources. Sustainable land management and land use planning has yet to become a common practice in the region and endorsing international agreements such as the United Nations Convention to Combat Desertification (UNCCD) or the Biodiversity Convention are important prerequisites for the future.

Pressures are particularly high on the coastal zones and various estimates predict that 50% of the Mediterranean coastline could be cemented by 2025. Recent studies have begun to quantify also the cost of degradation of the environment in several countries. Environmental degradation in Egypt was estimated to cost between €2.7 and €5.1 billion per year (or 3.2–6.4% of GDP), €1.5 billion per year (or 3.6% of GDP) in Algeria and €1.2 billion per year (or 3.7% of GDP) in Morocco (Montanarella 2007).

In particular the UNCCD recognises the intensity and the risk of desertification not only in the southern Mediterranean but also in the northern countries. The Annex IV of UNCCD includes a number of European Mediterranean countries affected by desertification and require the establishment and implementation of National Action Plans (NAPs) to fight against this major threat.

2 Description of Major Soils

Mediterranean soils are characterized by a wide diversity reflecting differences in climate, landscape, vegetation, time and especially, the long-term influence of the human activities. Erosion has been a dominant factor in carving landscapes and influencing soil distribution. Some of the best-known examples of Mediterranean soils are the famous "*terra rossa*" or the Rhodic and Chromic Luvisols (WRB 2006) or the Rhodoxeralfs of Soil Taxonomy (Soil Survey Staff 2006). However, there can be great confusion when one tries to classify these particular soils, as they possess variable soil depths, evidences of clay translocation within the soil profile, and age of soil development. Thus, their classification may vary from Regosols or Leptosols to Cambisols and Luvisols (Fig. 2).

Understanding the formation and behaviour of Mediterranean soils is a challenge that yet requires the inputs of many pedologists. The dominant parent materials of southern Europe and the Middle East are the limestones and dolomites that gave rise to the development of the typical *karstic* ecosystem of the Mediterranean. Furthermore, special attention is needed to understand the past and present role of wind-borne materials from the Sahara desert in the Region's soil formation. Following are brief descriptions of major soils using the World Reference Base for Soil Resources (WRB 2006) terminology and classification system. Short descriptions of soil characteristics and issues related to soil use and management are also provided (Fig. 3).

2.1 Histosols

Histosols cover relatively small areas throughout the region. They are formed from organic soil material and are continuously influenced by the presence of shallow groundwater. The parent material is made of partially decomposed plant remains with or without the addition of sand, silt or clay. Histosols cover the lower parts of fluvial, lacustrine, and marine landscapes and are often associated with Fluvisols, Gleysols, Solonchaks and Vertisols. In a natural way, Histosols act as the highly valuable wetlands that maintain natural biodiversity. The depth of peat layers is usually shallow but when a combination of tectonic lowering and accumulation of

Terra rossa formation in limestones: Leptic Cambisols and Luvisols (Skeletic Rhodic)
Photo: P. Zdruli, Eastern Turkey

Fig. 2 Terra rossa

Fig. 3 Typical Mediterranean landscape, South-eastern Albania (Photo: P. Zdruli)

peat occurs, the depth may reach much as 300 m as reported in the Drama Plain, Greece (FAO 2001). Pockets of Histosols are found in Albania, Spain and, very occasionally, in the Middle East where they follow the Antakya Fault (the northern extension of the Rift Valley) in Southern Turkey.

Histosols represent very fragile soil ecosystems. Studies have shown that reclamation, land management and crop production can drastically affect them. Naturally, they are suitable for grazing and forestry. However, under good drainage systems, they are very productive soils for horticulture. If mismanaged, their cultivation could accelerate mineralization of organic matter, enhance soil subsistence, and waterlogging. Histosols act as organic carbon pools and play an important role in carbon sequestration. Zdruli et al. (1995) report on study conducted in a Histosol in Albania where the average soil organic carbon (SOC) loss due to land reclamation, drainage and cultivation in only 38 years was as much as 80.6% for the top 0–30 cm depth or a SOC rate decrease of 26 g m^{-2} yr^{-1}. Therefore Histosols should be properly managed; preferably left in their natural conditions. If cultivated, well-managed and functional drainage systems have to be fully operational to keep the soil profile at optimal wetness.

2.2 *Anthrosols*

These are the so-called man-made soils that have been profoundly transformed by human interventions either by long-term cultivation or manuring or by drastic and immediate actions. Their name derives from Greek *Anthros* (human being) and is very connotative to their formation and genesis. Despite the long history of cultivation, overall Anthrosols in the Mediterranean are rather limited and are found mainly within the *Irragric*, *Hortic*, and *Escalic* qualifiers. These last are

Fig. 4 Anthrosols used for grape cultivation under drip irrigation in Apulia, Italy (Photo: P. Zdruli)

present throughout the region. One of the most famous places is in the Ligurian coast in north-western part of Italy in the area known as Cinque Terre where still millenary terraces are well-functioning for grape production in slopes as steep as 70%.

Very unique man-made soils exist also in Apulia Region in the south of Italy in about 20,000 ha. Otherwise the practice is locally known as *frantumazione* and could be described as the "grinding" of the rocky topsoil to increase by mechanical power the effective soil depth. The process occurs on shallow soils overlying limestone bedrock mixed with rock outcrops (Zdruli 2008), even though is common for farmers to add additional calcareous rocks just before the process starts. The benefits of these expensive interventions are twofold: increased rooting zone and release of available Ca, a much-preferred nutritional element especially for grapes' cultivation (Fig. 4).

Such process has expanded also in the slopping areas for the cultivation of other crops like cereals for instance, however, numerous erosion problems were reported afterwards (Canora et al. 2003; Trisorio-Liuzzi et al. 2005).

2.3 Leptosols

There are rarely other soils that could characterize better the Mediterranean environments than the Leptosols. These widespread shallow soils of the undulating lands and steep slopes located mainly on the northern Mediterranean Basin occur

Fig. 5 Ancient Roman terraces on Leptosols in Turkey used for the cultivation of olives, grapes and cereals (Photo: M. Serdem, University of Çukurova, Turkey)

predominantly as *Calcaric*, *Mollic*, *Rendzic*, *Eutric* or *Haplic* soil units. The *Mollic* and the *Rendzic* Leptosols are dominant in Turkey and southern France on partly degraded forested lands. The *Calcaric* Leptosols occupy large areas of the Dalmatian Coast (Croatia), the Spanish and Greek islands and much of the nearby coastal zone. Extensive areas of such soils occur on eastern and southern Turkish coastlines, the highlands of Cyprus and large areas of Syria, Lebanon, and Israel. They are shallow soils of the karstic landscape, highly susceptible to erosion due to the conversion of widespread forests to Mediterranean shrub-lands and degraded *maquies*. This process, initiated thousands of years ago, is still continuing.

Although Leptosols are known to be a potential resource for grazing and forestry, their terracing has created the traditional Mediterranean landscape, which is a unique and productive agro-ecosystem suitable for water harvesting with long adapted indigenous crops, dating back to the Roman period, providing sustainability to the system itself (Fig. 5).

2.4 Vertisols

Vertisols cover large parts of the deltas of the Mediterranean region as dominant and/or associated soils with Fluvisols, Cambisols, Calcisols, Luvisols and occasionally Leptosols. Extensive areas are found in Turkey, the western Balkans, Spain and Italy. Some particular types of Vertisols in Morocco are locally called *Tirs* while the magnesium-rich Vertisols in the Balkans, are known as *smonitsa* soils (Stebut 1923; Bogatyrev 1958; Zdruli 1997).

Their swelling and shrinking properties lead to the development of an undulated surface micro topography that requires a careful land management approach. In the relatively wetter parts of the Mediterranean, Vertisols can be maintained as areas of traditional use, such as improved rangelands. The use of green manure (legumes) as part of soil improvement technology has also been found to be highly beneficial in improving their physical properties. Some of the Vertisols from the sub-basins of eastern Turkey have been used as indigenous rangelands for centuries, whereas the Vertisols in the Tavoliere delle Puglie in the Apulia Region in Italy are used for the cultivation of irrigated tomatoes, sugar beet, asparagus and non-irrigated durum wheat.

2.5 Fluvisols

Fluvisols are frequently located along past or present riverbeds, as in the alluvial basins in Spain, southern France, Greece, Italy, Turkey, Albania, Middle East and especially in Egypt. An old Egyptian proverb depicts this, as "Egypt is the gift of the Nile." Occasionally, they are present in the deltas of coalescing alluvial fans, which extend from mountains towards the Mediterranean Sea.

Due to their setting in the landscape, Fluvisols are susceptible to the build up of salinity especially in Spain, Turkey, Egypt, Morocco and Tunisia and extra care is needed in irrigated areas. They are highly productive soils under irrigation and extensive fertilisation and are allocated mainly to cash crop production, such as vegetables, cotton, maize, sunflower and for rein fed cereal cultivation.

2.6 Gleysols, Solonchaks and Solonetz

Gleysols are found only in small areas throughout the region often in association with Solonchaks and Solonetz soils. Their typical characteristic is the presence of *gleyic properties* within the upper 50 cm topsoil layer of the soil profile. Drainage is a land management prerequisite for their use and cultivation.

Solonchak soils occur in the semi-arid parts of southern Europe. In Turkey and Spain, they are mostly associated to dominant Fluvisols overlying marls of lacustrine origin. As salinity builds up, the rising water table follows deep seepage from lower elevations as well as spring waters reaching the alluvial surfaces, and is ultimately enhanced by high evaporation of the semi arid environment. *Gleyic* Solonchaks are widespread within the context of waterlogged soils as back swamps. They are estimated to cover some 50,000 million hectares throughout the Mediterranean region. In areas with very high salinity content or more than 15 dS m^{-1} ECe only a few salt tolerant crops (halophytes) could grow. The presence of a *Salic* horizon is essential to differentiate Solonchaks from other soils such as Gleysols. *Mollic*, *Calcic* and *Gypsic* Solonchaks may be associated with Kastanozems, Phaeozems, Calcisols and Gypsisols. *Saline* Vertisols occur very occasionally in the semi arid parts of southern Europe (Fig. 6).

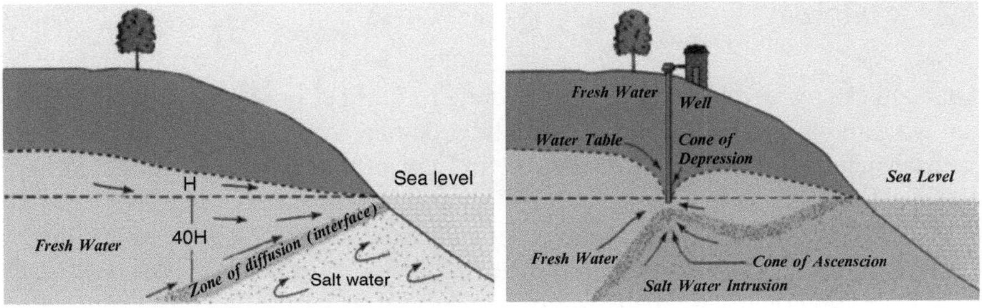

Fig. 6 Seawater intrusion and salinity built up through ground water pumping

Fig. 7 Artiplex cultivation in Marrakech, Morocco (Photo: P. Zdruli)

Solonetz soils are rich in exchangeable sodium and magnesium ions and are confined to flat lands found in association with Gleysols and Solonchak. In coastal areas they could be associated with Fluvisols. The transition of Solonetz soils to Gypsisols and Calcisols occurs due to the accumulation of gypsum and calcite below the diagnostic *Natric* horizon, as found in semi arid conditions of Spain and Turkey.

Extensive areas of saline soils occur in Morocco, Tunisia, Algeria, Egypt and throughout the Middle East. One dominant factor for the expansion of saline soils in the Mediterranean is the introduction of irrigation and the intensification of crop production, especially along the coasts as shown in Fig 7.

Solonchaks and Solonetz soils are very difficult to manage and require extensive effort to be productive. However, they are widely used for arable farming and extensive grazing. The production of salt tolerant crops (i.e. Atriplex) on unused grasslands seems to be a highly profitable and environmentally friendly practise used in several countries of the Mediterranean region, especially in Morocco.

2.7 Andosols

Andosols occur in volcanic regions of southern Europe, primarily in Italy, France and with minor distributions in the undulating mountainous areas of Turkey. They are associated with Cambisols, Podzols and Luvisols in France, Cambisols in Italy whereas in Turkey with Calcisols, Leptosols and adjacent Fluvisols. Determination of Andosols has been highly controversial within the Mediterranean context, due to the lack of the bulk density requirement (0.9 g cm^{-3}), organic matter content as well as morphological-field characteristics, such as *thixotropy* (the smeary feeling that develops due to the sol-gel transformations under pressure).

Andosols are cultivated for a variety of crops. They represent soil ecosystems that are vulnerable to disturbances, especially to soil sealing, which may lead to the development of landslides (as it has happened in southern Italy and eastern Turkey) and erosion. The unavailability of phosphorous is their main chemical constraint of soil fertility.

2.8 Kastanozems and Phaeozems

The distribution of Kastanozems in southern Mediterranean Europe is very limited and mainly associated with Leptosols, Cambisol, Calcisols and Fluvisols. They occasionally associate with Gypsisols, Solonchaks and Solonetz in depressions of warmer and drier areas. Some Kastanozems are covered by forests or are used for grazing. Other areas are cultivated for cereals and occasionally for irrigated crops. Salinity build up may be a problem for these fertile and productive soils if water is not well managed during irrigation.

Phaeozems are soils with distinct dark surface colour and rich in organic matter content. In the Mediterranean region *Leptic*, *Vertic*, *Gleyic*, *Luvic*, *Pachic*, *Calcic*, *Skeletic*, *Siltic*, *Chromic*, and *Haplic* represent the most common soil units. They are found in Spain, France, Italy, the Western Balkans, Greece, Turkey, Israel, Lebanon, Syria, and Morocco. Phaeozems in rainfed flat to undulating environments become very productive soils especially for cereals. Under irrigation, a great variety of crops, including sugar beet, vegetables or fodder, may be cultivated. In sloping mountain areas, they are suitable for forestry or managed grazing. If mismanaged, Phaeozems of the mountainous regions could be seriously threatened by erosion. The lack or loss of the *mollic* horizon in many parts of the Mediterranean mountains is the best evidence of historic soil erosion.

2.9 Umbrisols

Umbrisols occur mainly in the northwest of Portugal and Spain with smaller patches in the Balkans. They are associated with soils of moist but free-draining conditions, namely Phaeozems that ultimately may integrate by cultivation to Anthrosols. Cambisols and Leptosols may also be associated with Umbrisols as well as being adjacent in the landscape.

Regions of cleared forests, converted to short grasses, widely used as grazing lands, dominate Umbrisols in the north-western part of the Mediterranean. Thus, the management, in this context, has aimed the introduction of improved pasture and correction of soil acidity by liming. Cambisols, Luvisols and Leptosols at higher elevations face similar problems as Umbrisols.

2.10 Gypsisols, Durisols and Calcisols

These are typical soils that form under arid or semi-arid conditions. With the exception of Calcisols, their extension is limited in the northern Mediterranean areas but highly visible in the southern countries. In the arid regions of North Africa and the Middle East, mud and debris floods formed the so-called "*wadis*" located in the areas of low relief. Many *wadis* were formed during the Late Pleistocene and Early Holocene period between 13,000 and 8,000 years ago (FAO 2001). Consecutive mudflow occurrences have been responsible for the development of Calcisols with variable contents of secondary carbonate (lime) accumulation, leading to the development of *calcic* and *petrocalcic* horizons within a depth of 100 cm or deeper and often grading to "*calcrete*." *Petric* Calcisols or Calcretes are widely distributed in the Mediterranean and are used primarily for grazing and agroforestry.

Calcisols occupy large areas in Spain while in Turkey and Cyprus they are ranked the second most dominant soil after Leptosols. Minor distributions are also found in Italy, France, Greece, and Albania. The major soil units include the *Vertic*, *Luvic*, *Petric*, and *Haplic* categories. *Petric* and *Luvic* Calcisols represent the dominant calcretes of Turkey, Cyprus, Spain as well as North Africa and the Middle East. Calcisols are associated with Leptosols at higher elevation and with Gleysols, Vertisols, and Solonchaks at lower depressions.

Gypsisols are characterised by the substantial accumulation of secondary gypsum ($CaSO_4$). The natural vegetation is made of xerophytic shrubs and ephemeral grasses. The largest concentration of Gypsisols in the Mediterranean occurs in the Libyan Desert, Jordan, Syria, areas in Central Anatolia and Cyprus. The agricultural use of Gypsisols depends on the gypsum content in the topsoil. If the gypsum content is low, they could be used for the production of small grains, cotton, and forage crops. In the Gypsisols of the alluvial plains, irrigation and drainage are necessary for the cultivation of crops and fruit trees, including grapes. However, large areas of these soils are used only for extensive grazing.

Durisols contain cemented secondary silica (SiO_2) in the upper one meter of the soil. Their typical feature is the presence of a hard-cemented layer identified as the "*duripan*" phase. In the Mediterranean, they occur on level and slightly sloping alluvial plains, terraces and gently sloping piedmont plains mainly in Jordan, Syria, Morocco, Tunisia, and Algeria. They are often found in association with Gypsisols, Calcisols,

Solonchaks, Solonetz, Vertisols, Arenosols, and Cambisols. Pedogenetically, they are characterised by the presence of *duric* and *petroduric* horizons that practically limit the effective soil depth available to plant roots.

In fact, the *petroduric* horizon can vary between 10 cm and 4 m in depth. In addition, Durisols could have an *argic*, *cambic* or *calcic* horizon above the *petroduric* horizon. The agricultural use of Durisols is limited to extensive grazing (Fig. 8).

Calcisols in the Mediterranean basin are productive soils with moderately deep profiles. However, effective soil depth could be reduced by the presence of a *petrocalcic* horizon. Occasionally, shallow *Petric* soil units (with relatively high water holding capacity) are allocated for the traditional production of tree crops like olives, almonds, vineyards, apricots and figs. In eastern and central Spain and Turkey, these soils have been used extensively for grazing creating erosion prone areas especially at the vicinities of active fault lines.

Extensive areas of Calcisols in the Mediterranean are also used for production of irrigated winter wheat, melons, and cotton. Calcium tolerant crops such as sorghum, rhodes grass and alfalfa, are best suited for Calcisols. A number of vegetable crops have successfully been grown on irrigated Calcisols, if fertilization of nitrogen, phosphorus and trace elements (Fe, Zn) are provided.

Fig. 8 A striking example of a calcrete. Massive secondary accumulation of soft lime forming a calcic horizon in the hilly areas of Central Albania (Photo: P. Zdruli)

2.11 Luvisols

Luvisols are well-developed soils whose main pedological characteristic is the formation of the *argic* sub surface horizon through the illuviation and translocation of silicate clay from the surface layer.

The Luvisols of the Mediterranean basin are widely distributed throughout Portugal, Spain, Italy, Greece, Albania, Croatia, Turkey and Cyprus and are often associated with Cambisols, Leptosols and upland Regosols under forest cover. The *Gleyic* and *Haplic* units dominate the Luvisols of Italy and France. *Chromic* and *Rhodic* units (terra rossa soils) are widespread in Greece, the southern part of Portugal, central Spain, western, southern and south-eastern Turkey and throughout Italy. Luvisols are observed to be associated with Alisols and Acrisols in the humid parts of the Mediterranean requiring further research to understand the relations between the soil typologies and the fluctuating past and present climates of the region. Luvisols are also present in Morocco, Algeria, and the Middle East (Jordan, Syria, Israel, Lebanon).

They are fertile soils suitable for a wide range of agricultural uses (except for the shallow *Leptic* units). On sloping land they require measures, such as man-made terraces, and careful control of erosion and deforestation. The *Chromic*, *Calcic* and *Vertic* Luvisols of the Mediterranean are commonly used for cereals and sugar beet cultivation while the upper slopes are best fitted for fruit trees, vineyards, olives, and grazing. The *Haplic* Luvisols of central north Italy are used for high quality wine production.

2.12 Arenosols

Arenosols are soils having sandy or coarser texture in the upper 50 cm soil layer. In southern Mediterranean European countries, they occur mainly in the lacustrine, fluvial and marine environments of the contemporary and ancient coastal areas including sand dunes along the coast. In more humid environments Arenosols, are suitable soils for horticulture provided efficient systems of irrigation and drainage are put in place, complemented by fertigation (the application of nutrients through irrigation systems). In arid areas where rainfall is less than 300 mm per year, Arenosols are predominantly used for extensive and nomadic grazing. Uncontrolled grazing and clearing for cultivation without appropriate soil conservation measures can easily make these soils unstable and revert the land to shifting dune systems. Dry farming (the profitable production of crops without irrigation) is possible where the annual rainfall exceeds 300 mm. Some Arenosols in the coastal areas of southern Turkey have been anchored by agroforestry through planting of stone pine (*Pinus pinea*) (Yaktı et al. 2005), eucalyptus (*Eucalptus camaldunensis*) and acacia (*Acacia cynophylla*) (Fig. 9).

Fig. 9 Pine tree reforestation programmes in coastal sand dunes of Adana, south-eastern Turkey. (Photo: P. Zdruli)

2.13 Cambisols

Cambisols represent soils of young age in a continuous process of pedological maturation, as revealed by the presence of the *cambic* horizon. In pedogenic terms they stand between Fluvisols and Luvisols. Cambisols are the most widely distributed soils of the Mediterranean delimiting diverse climatic areas. The major Cambisol units are *Eutric* and *Dystric* which occur in France, Italy, Spain, Portugal, Croatia, Bosnia-Herzegovina, Montenegro, Albania and Greece, whereas the *Vertic* ones are mainly found in Italy, Greece, Turkey and Cyprus. Other units such as *Endogleyic*, *Fluvic* and *Andic* only cover minor parts of the Mediterranean. Cambisols are widespread in North Africa and the Middle East as well.

Cambisols with moderate to deep soil profiles are among the most productive soils of southern Europe. They associate with Luvisols and Vertisols and with Fluvisols and Gleysols in wet depressions or fluvial deltas. The *Eutric* Cambisol is an excellent soil used for all types of crops while *Dystric* Cambisols are used for mixed arable farming, forestry and grazing. The *Calcaric* and *Chromic* Cambisols on steep slopes are often covered with forests. The *Vertic*, as well as the *Eutric*, *Calcaric* and *Chromic* Cambisols in irrigated or rainfed alluvial plains are dominantly used for cereals and seed oil crop production throughout the region. Where the climate is wetter, many Cambisols are allocated for Mediterranean indigenous agro-ecosystems comprising carobs, figs, olives, and grapes.

2.14 Regosols

Regosols are weakly developed soils found on unconsolidated materials. They are typical of the mountainous regions of Spain, Italy, Albania, Croatia, Montenegro,

Greece, Turkey, Syria, Jordan, Lebanon and Israel. Regosols occur primarily on eroded land associated with Cambisols, Leptosols, Luvisols, Phaeozems, Umbrisols, and Calcisols. The *Dystric* suffix qualifier is the main characteristics of the Regosols in the Mediterranean region followed by the *Calcaric* and *Eutric* soil units.

Regosols have little value for agriculture. In Spain and Turkey some of them are used for capital-intensive irrigated farming systems. However, they are better suited for forestry, managed grazing and recreation. The continuous threat of erosion as well as forest clearing badly damages the fragile natural ecosystem of these soils.

3 Soil Degradation and Soil Protection in the Mediterranean

Despite continued efforts from both local, regional and international institutions operating in the region, soil degradation continues to be a very threatening problem. Zdruli et al. (2007) reports water and wind erosion, soil sealing and urbanisation, loss of organic matter and biodiversity decline, nutrient mining, chemical pollution and contamination, floods and landslides, salinisation, overgrazing and degradation of vegetation cover as well as unsustainable irrigation practices as the main land degradation factors in the Mediterranean. Another threatening problem is the "littoralisation" or the cementation of the coastal areas with grave repercussions on the sustainability of fragile coastal ecosystems as well as inland areas.

Figure 10 intends to establish relationships between the soil type and the major soil degradation problems occurring at any particular part of the landscape. These types of assessments are necessary for defining soil protection strategies and sustainable land use and management plans.

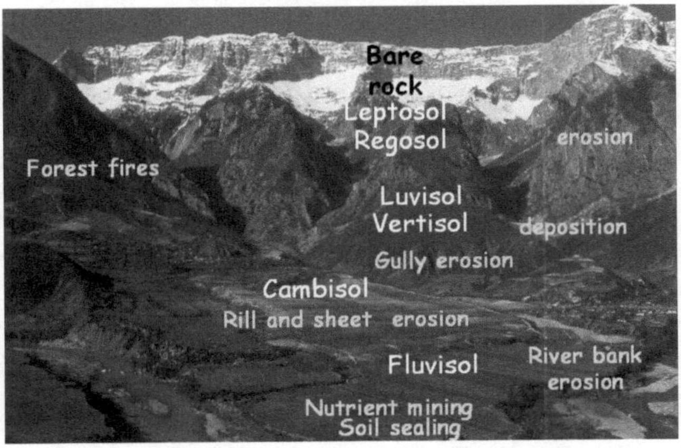

Fig. 10 Schematic correlation between the soil name and soil degradation problems

The main shortcoming s that region-wide studies are rare or missing and due to lack of harmonisation of methodologies for estimating the extent of the problem comparisons between countries are difficult, even within short distances. Such weaknesses raise again the need for establishing Mediterranean reliable and up-dated soil databases. The MEDCOASTLAND project outcomes (Zdruli et al. 2007) could be used as a platform for further enhancing collaboration in this field. It is hoped also that the newly established Union for the Mediterranean could support activities in these domains as well.

Most analysts argue that widespread land degradation and desertification, inefficient and inequitable use of water lie at the root of the problem, for which effective solutions are still pending. Almost one million hectares of the irrigated area are salt-affected in Egypt alone (Gomma 2002) due to poor quality water used for irrigation. Throughout the Mediterranean saline soils are estimated at some 50,000 million hectares. It is equally clear that increased food and water security through equitable, productive and sustainable utilization of land and water resources would provide a measure of relief, and increase especially the well being of the most vulnerable people in the region.

The situation may get worst due to projected climate change impacts. Almost all climate models predict that the Mediterranean may get hotter and drier and the intensity of extreme rainfall events may increase (Giannakopoulos et al. 2005). Such scenarios require preparation and adaptation, as the consequences may be devastating. For instance, due to sea level rise countries like Egypt may experience devastating consequences (Plan Blue 2008).

4 Conclusions

Soil's multifunction needs to be better understood and studied. The diversity of Mediterranean soils requires intense soil surveys at larger scales. While appreciating the development of remote sensing technologies in support of digital soil mapping, it is strongly suggested for the need to continue field soil surveys and that the biodiversity component should be an essential part of all the studies dealing with soil degradation as an important indicator of the soil quality and its trends.

Endorsing soil protection policies and sustainable soil management is not a choice but a paramount prerogative for the Mediterranean region. Soil resources are limited, water is lacking more than ever and anthropic pressures both by population growth and tourism are increasing. Chances for expanding arable land area are almost nil, unless alternative water resources (like groundwater in the case of Egypt and Libya) become available, so that cultivation may be expanded in desert like environments. Even, if this becomes reality, soil protection should be considered as the only long lasting measure to secure sustainable development and meet the goals of food security in the Southern Mediterranean and food safety in the North. The eco-system approach is the best method to reach these goals.

Land degradation and desertification have no political borders, thus transboundary assessments are needed to estimate their extent region wide and consequently similar remediation practices should be applied across and beyond national borders. Collaboration among all the Mediterranean countries should be strengthened.

The Mediterranean is the birthplace of European civilisation and the place where the skills of growing crops started. The *Mediterraneans* were ingenious to construct terraces and adapt to climatic hardships and overall survived even at Sahara oasis receiving only 50 mm or less rain per year. Present day conditions require additional societal efforts to succeed. However, success will ultimately depend on widespread political support translated into a clear determination to deploy all the necessary resources for soil protection.

References

Bogatyrev KP (1958) "Smolnitsy" (Smonitsa) of Albania (Cinnamon-brown meadow and meadow cinnamon-brown dark-coloured magnesium-like soils). Soviet Soil Science. 14, No. 4, pp. 358–364

Canora F, Ferrigno L, Fidelibus MD, Spilotro G, Straziuso K (2003) Modificazinoni antropiche delle tessiture dei suoli carsici del'Alta Murgea e finalizzazione agricola: implicazioni idrogeologiche. Riassunti FIST, Bellari, pp. 476–78 (In Italian)

Eswaran H, Van den Berg E, Reich P, Almaraz R, Smallwood B, Zdruli P (1995) Global Soil Moisture and Temperature Regimes. World Soil Resources Report. Natural Resources Conservation Service, United States Department of Agriculture, Washington

FAO (2000) Land Resource Potential and Constraints at Regional and Country Levels (World Soil Resources Report 90). FAO, Rome

FAO (2001) Lecture Notes on the Major Soils of the World. In: (Eds. Driessen PM, Deckers JA, Spaargaren OC, and Nachtergaele FO) World Soil Resources Report 94. FAO, Rome

Giannakopoulos C, Bindi M, Moriondo M, Le Sager P, Tin T (2005) Climate change impacts in the Mediterranean resulting from a 2°C global temperature rise. Report for WWF, 1 July 2005. http://assets.panda.org/downloads/medreportfinal8july05.pdf [last accessed 13 March 2009]

Gomma M (2002) Land and water resources information system in Egypt. In: Land Resources Information Systems in the Near East, Cairo, Egypt, 3–7 September 2001. Workshop proceedings. FAO, Rome

IUSS Working Group WRB (2006) World Reference Base for Soil Resources 2006. World Soil Resources Reports No. 103. FAO, Rome

Kapur S, Akca E (2003) Technologies, Environmentally Friendly Indigenous. Encyclopaedia of Soil Science. DOI: 10.1081/E-ESS120006648. Marcel Dekker, New York

Montanarella L (2007) The EU Thematic Strategy for Soil Protection and its Implications in the Mediterranean. In: "Status of Mediterranean soil resources: Actions needed to promote their use"(Eds. Zdruli P and Trisorio Liuzzi). Conference Proceedings, Tunis, Tunisia 26–31 May 2007. MEDCOASTLAND series of publications 6. CIHEAM-IAM Bari, Italy, 368pp. ISBN 2-85352-371-3 pp15–30

Plan Blue (2005) A Sustainable Future for the Mediterranean, The Blue Plan's Environment & Development Outlook (Eds. Benoit G and Comeau A). Earthscan, London

Plan Blue (2008) Climate Change and Energy in the Mediterranean. Regional Activity Centre. Sophia Antipolis, France, July 2008. European Investment Bank and EUROMED

Soil Survey Staff (2006) Keys to Soil Taxonomy, 10th ed. USDA-Natural Resources Conservation Service, Washington

Stebut AI (1923) Tipovi zemalja Shumadije (Types of land in the Shumadije area). Glasnik Srpske Kralyevske Adamedye, SU Belgrade. (In Serbian)

Trisorio-Liuzzi G, Ladisa G, Grittani R (2005) The effects of two human activities on land degradation in a typical area of Southern Italy (the Lama Balice-Tiflis watershed in the Apulia region). In: Determining an Income-Product Generating Approach for Soil Conservation Management (Eds. Zdruli P, Trisorio-Liuzzi G). Workshop Proceedings, Marrakech, Morocco 12–16 February 2004. MEDCOASTLAND Publication 2. IAM Bari, Italy, pp. 297–315

Yaktı S, P Zdruli, E Akça, M Pekel, S Yılmaz, A Düzenli, M Serdem, B Kapur, S Kapur (2005) Management of coastal sand dunes by agroecosystem approach. In: "Determining an income-product generating approach for soil conservation management". (Eds. Zdruli P, Trisorio-Liuzzi G). Workshop proceedings, Marrakech, Morocco 12-16 February 2004. MEDCOASTLAND series of publications 2. CIHEAM-IAM Bari, Italy. ISBN 2-85352-311-X.

Zdruli P (1997) Benchmark Soils of Albania: Resource Assessment for Sustainable Land Use. International Conservation Division, USDA Natural Resources Conservation Service, Washington; 2 Volumes, 293pp

Zdruli P (2008) Anthropic interventions for increasing orchard cultivation areas in rocky calcareous shallow soils of Apulia Region in South-eastern Italy. Geophysical Research Abstracts. Vol. 10. EGU2008-A-01696, 2008. Sref-ID: 1607-7962/gra/EGU2008-A-01696, EGU General Assembly 2008, Vienna, Austria

Zdruli P, Eswaran H, Kimble J (1995) Organic carbon content and rates of sequestration in soils of Albania. Soil Science Society of America Journal. 59, pp. 1684–1687

Zdruli P, Lacirignola C, Lamaddalena N, Trisorio Liuzzi G (2007) The EU-funded MEDCOAST-LAND thematic network and its findings in combating land degradation in the Mediterranean region. In: Climate and Land Degradation. Environmental Science and Engineering. (Eds. Sivakumar MVK and Ndiang'ui N). International Workshop on Climate and Land Degradation, Arusha, Tanzania 11–15 December 2006. World Meteorological Organisation and United Nations Convention to Combat Desertification. Springer-Verlag, Heidelberg, pp. 422–434

Mountain Anthroscapes, the Case of the Italian Alps

Franco Previtali

Abstract After an attempt of a definition proposal, some typical and frequently occurring anthroscapes in the Italian Alps are here examined. By referring to the original landscapes, such man-made landscapes are divided into positive/ameliorative and sustainable, on one hand, and worsening/negative and unsustainable on the other hand, depending on the type of impact that should have generated them.

Keywords Anthroscapes · Italian Alps · Soil · Landscape

1 Introduction

In the broadest sense of the term, an Anthroscape can be defined as a landscape clearly and prevalently bearing visible features imprinted by human activities. According to Kapur et al. (1999, 2004) this concept must be reserved to conditions where marked differences or deviations from the normal and natural landscapes are observed. The anthroscape model can allow to understand the underlying processes which created these landscapes.

A frequent question arises, when resorting to the concept of anthroscape, concerning its age of formation. Some have proposed to link up the anthroscapes with a specific historical or geological period. So, for instance, Crutzen and Stoermer (2000), have stated that, basically, the Anthropocene is the epoch when "impact of human activities on ecosystems has exceeded that resulting from natural forces", and that should have occurred, in their opinion, since the invention of the steam engine in 1784. This is the period when data retrieved from glacial ice cores show the

F. Previtali (✉)
Department of Environmental and Territorial Sciences, University of Milano-Bicocca, Piazza della Scienza, 1-20126 Milano, Italy
e-mail: franco.previtali@unimib.it

S. Kapur et al. (eds.), *Sustainable Land Management*,
DOI 10.1007/978-3-642-14782-1_5, © Springer-Verlag Berlin Heidelberg 2011

beginning of a growth in the atmospheric concentrations of several "greenhouse gases", in particular CO_2 and CH_4. Furthermore, according to these Authors, the Anthropocene becomes the temporal context in which anthroscapes develop.

However, as argued by Eswaran et al. (2005), it seems that it may not be necessary to assign a precise date to the onset of the Anthropocene. For the purposes of the discussion on the impacts on land, the beginning of the eighteenth century is adequate as a milestone, because the recent centuries have documented information on changes in the ecology and on the land. But we think that the creation and appearance on the Earth of the anthroscapes and the beginning of the so-called Anthropocene epoch can be not necessarily coincident, also because many different definitions of this epoch have been proposed. In any case, it must be noticed that many anthroscapes developed much before of the latter part of the eighteenth century.

It is the case of many archaeological anthroscapes dating to the late Pleistocene-early Holocene, aged from 15,000 to 10,000 years BP. Almost continuous human presence, under changing climatic, cultural, economic conditions, different dominations, etc. deeply reshaped the landscape several times, in particular after the Roman period, moving the equilibrium lines between men and the natural environment up and down the valleys and the mountains. These processes have produced in many sites real layered anthroscapes. Such "stratifications" can be decoded by means of archaeology and historiography, but they become more difficult to solve and interpreted when occurring as juxtapositions and interpenetrations.

Hence, it should be remembered that the reference of Archeological Anthroposols, recently introduced in the Référentiel Pédologique (Baize and Girard 2009), implicitly recalls the concept of anthroscape. However, many remarks carried out on the different anthroscape types refer to concepts partially existing in the different soil taxonomies, even if the last ones aim prevalently at classifying soils and not landscapes.

Furthermore, it must be taken into consideration that the time-wise developments of the anthroscapes are very different from each other. Some of them quickly materialize and develop: within few minutes, in case of accidents or man-induced catastrophes; years, in case of building, mining, industrial, and so forth activities; ages and millennia, in case of agriculture, progressive deforestation, and so on. In addition to the ages, a further aspect to be considered is the size of the various anthroscapes, which also vary from one to the other: from region to district, municipality, catchment area, and farm. The visible effects of the works and human actions sometimes do not correspond to the entity of their environmental impact on terrestrial and aquatic ecosystems. For example, small-size waste incinerators or waste landfills have strong chemical, biological, atmospheric, and landscape impacts on the land. Moreover, on the long run, they cause by intensity, dimension, seriousness of their frequently indirect effects, abrupt drops in crop quality and quantity, depletion in fish resources, worsening of human and animal health, etc.

A particular aspect of the anthroscapes is represented by their possible invisibility, because not always the impacts of the human interventions are visible and corresponding to the true and visible anthroscape. It concerns immaterial phenomena, which men can not perceive unless by the use of surveying tools or by measuring their biological impacts on the living organisms. As an example, the outcomes of the slow chemical and radioactive contaminations of water, soil and biota and the varied types of air pollution (by CFC, CO_2, ozone, SOx, NOx, etc.) may be possible to determine at long periods via their intensity, dimensions and seriousness of their frequently indirect effects (e.g., drop in quality and quantity of crops, depletion of fish resources, worsening of human and animal health, etc.).

Finally, some anthroscape degradations could be defined potential, because of their dramatic visibility only in case of huge technical accidents. This would be the case, for instance, of the nuclear power stations and the big arch dams with high upright development. Such plants occupy relatively small portions of the territory, but in case of serious accidents can deeply modify the environmental characters of the entire regions. Some of the most marked examples of the anthroscapes of the Italian Alps facing the problems mentioned above will be examined in the following sections.

2 Negative Human Impacts Creating/Reshaping Anthroscapes

Some characteristic landscapes have been clearly created in the mountainous areas by recent human activities. These landscapes show the evident marks of the many negative impacts, which are valid for the anthroscapes originating from:

- The winter resorts, with their civil settlements, the ski-lift facilities and the ski slides
- The mountain off-roading
- The mining dumps
- The industrial and civil settlements along rivers and valley floors
- The stone quarry districts
- The waste disposals.

2.1 The Impacts of Winter Resorts: the Case of Ski Slides and Ski-Lift Facilities

At a first and superficial observation, the landscape of many winter resorts appears composed by green meadows and grazing land, with scattered shepherd huts (or so presumed) and few innocuous ski-lifts (Fig. 1).

But a closer examination, mainly in the more popular and better equipped localities of winter sports, would reveal the serious negative modifications of the local ecosystems and social structure.

Climate change as inducing a frequent decrease in snowfall is a primary threat to the Alpine region (Watson 1985; Ritter 2004; De Jong 2007; Edwards et al. 2007). In addition to the direct and consequent decrease of snowfall, ice and water supply,

Fig. 1 Ski-lift facilities, shepherd huts, meadows of a typical winter resort in the Italian Alps (Photo by F. Previtali)

an indirect effect is encountered by the shift in the altitude of ski-lift facilities, ski slides and related plants, by progressively being pushed to higher locations by the anthropic impact, with markedly amplified effects on the environment (air pollution, water capitation, noise and wastes).

On the other hand the recourse to artificial snow-making is growing as a technique in need of a large supply of water (about 300,000–400,000 L to cover one hectare with 30 cm of snow). But the outlet water returned to the drainage is not purified and contains all the primary chemical additives (silver iodide, kaolin, detergents and soaps) and further the frequently added organic compounds as well as the ones that have developed in the meantime (fungi and lichens and the organic compounds) (Ritter 2004), where all, ultimately, lead to the creation of an invisible and highly degraded anthroscape. The other important environmental modifications on the Alpine anthroscapes are:

- The relief modifications of the streams, water tables and spring captures and diversions, that are enhanced on karstic morphologies
- The degradation of the forests and pastureland, inducing decrease in rainfall and rise of temperature triggering erosion (Fig. 2)
- Slope reshaping
- Fauna disturbance and expulsion
- Soil consumption, sealing (soil compaction reducing its porosity, permeability and water holding capacity), and contamination due to the effects of artificial snowing
- Increasing air pollution, waste and noise
- Biodiversity loss or degradation
- Massive building of holiday homes.

Fig. 2 A low altitude ski resort in the Pre-Alps. Deforestation, accelerated water erosion, artificial snow-making and uncontrolled buildings involve heavy environmental consequences (Photo by courtesy of C. Zucca)

Incidentally, it must be stated that the growing number of high altitude resorts suffer from water shortage and pollution, despite the large local water sources. This is a clear indicator of an excessive water uptake and the exceeding the limit of sustainable tourism. In addition, such resorts, commonly cause the disruption of the architectural and cultural heritage, with non-aesthetic and gigantic buildings, lacking the local and traditional style (Fig. 3).

By the socio-economic and cultural point of view, another aspect to be noted is the abandonment of the traditional activities (grassland, pastoral, and dairy farming, silviculture, craftwork, etc.), induced by the transformations in tourist resorts, and the replacement by the unskilled activities (waiters, doorkeepers, shopkeepers, etc.) or seasonal jobs (ski instructors, ski-lift keepers, etc.). In spite of the income increase, this has made the local economies weaker and more vulnerable, exposed to the consequences of climate change and the fickleness of passing tourist fads.

2.2 The Mountain Off-Roading: A Growing Environmental Problem

Some individual vogues and habits having a considerable impact on the environment are becoming widespread in the Alps. This concerns the motorized sleighs, the helicopter ski, and the motorcycle and car off-roading (Fig. 4).

Fig. 3 A typical locality of the central Italian Alps, located at about 1,900 m a.s.l., mainly exploited as a winter ski resort. The primitive needle leaved woodland has been removed and replaced with artificial meadows for ski slides. Peat bog, pastureland and woodland have been almost totally destroyed and replaced by buildings of questionable style, not integrated to the environment

Fig. 4 The overlapping of motorcycling (see tyre prints) off-roading and overgrazing trigger off intensive erosion processes (Central Italian Alps, 1,900 m a.s.l.). (Photo by F. Previtali)

Pollution of snow, ice, water, and air, together with waste and noise are transferred up to 3,000–4,000 m a.s.l. Moreover, the risks of accident occurrences are always impending. Motorcycles and cars tear the soil and their print tracks become weakness paths along which the runoff progressively digs rills and gullies. At some locations, the impacts of overgrazing and off-roading overlap, causing accelerated erosion. These sites do not normally cover large areas, and thus could not be considered as typical anthroscapes in the Alps. Whereas, such phenomena, if uncontrolled, could generate large eroded areas extending to the different parts of the anthroscapes.

2.3 *Mining Dumps*

Nearly 1,000 mining sites have been recorded in the six northern regions of Italy, with part of their territory within the Alpine mountain range (A.P.A.T. 2006). The majority are in Piedmont (375), 37 in the Aosta Valley, 294 in Lombardy, 80 in Trentino-Alto Adige, 114 in Veneto and 32 in Friuli-Venezia Giulia. These numbers indicate not only the past economic relevance of this activity but also its environmental impact. The list of minerals quarried from the Alpine ore bodies is very large. Some of the major ones are: blend, galena, barite, kaolin, feldspars, pyrite, fluorite, siderite, talc, calamine, cinnabar, gold, hematite, magnetite, asbestos, cobaltine, nickel, uraninite and others.

The extractive industry nowadays has practically ceased, mostly for metalliferous ores, owing to their non-remunerativeness as regards the world market. However, serious problems concerning these industries are still valid even today despite the abandonment of these sites.

Overburden and tailing of abandoned mineral dumps, although having low contents of extractable ore and trace elements, slowly and continuously release them to surface and ground-waters through run-off and leaching (Fig. 5). Despite of the somewhat small size of such bodies, the pool of trace elements threaten the human health via the food chain: from waters to vegetation, animals, and man. In addition to spoiling the landscapes, such deposits, in the absence of drastic reclamation, constitute a serious and long-term environmental risks.

2.4 *Industrial and Civil Settlements Along Riversides and Valley Floors*

The widespread and increasing occupancy of the Alpine valley floors with buildings and roads are responsible for the enhanced hydrogeological, urbanistic, and ecological problems. Such settlements undergo high destruction risks from rivers and streams, and constitute on their side a physical obstacle to their regular flow (Fig. 6). The rivers, being thronged and lacking natural "expansion boxes", in flood events, overflow and submerge lands causing loss of lives. Furthermore, such land-use removes the fertile alluvial soils, lying in the confined space of narrow valley floors and between very steep slopes.

The natural landscapes thus assume the usual features of the urbanized sites, and become a kind of ectopic urban anthroscapes.

Fig. 5 Mining dump of magnetite overburden, periodically undergoing water erosion processes under intense autumn rain. Chromium oxides, sulphides and sulphates, heavy metals and trace elements slowly flow into the river below. (1,500 m a.s.l., Western Italian Alps) (Photo by F. Previtali)

Fig. 6 Industrial and commercial settlements of an Alpine valley in the Central Alps. The buildings throng the alluvial riverbed and obstruct the discharge of the almost ten-yearly floods. *Arrows* indicate the direction of the river flow (Photo by courtesy of IRPI-CNR)

2.5 Stone Quarrying

The stone quarry exploitation stands for a relevant economic resource along all the Alpine and pre-Alpine ranges. The extractive industry constitutes a centuries-old tradition and significantly contributes to local income. Such activity has been strictly connected with mining, which is nowadays progressively disappearing. The quarries of these ranges provide building and raw materials as well as ornamental stones.

Even though, as prescribed by law, each quarry must be submitted to preliminary evaluation for approval, the environmental impacts of these activities are very intense and evident. The alterations to the hydrogeological systems and landscapes are strong and only in part mitigable (Fig. 7). Furthermore, the damages and inconveniences to the human and animal health and wellbeing are in some localities very heavy (Fig. 8). The landscape is strongly and irreversibly modified and assumes, mainly in the wider districts, the typical features of a degraded and derelict anthroscape, devoid of vegetation, soil, and water. Naturally, cautious planning and management of these sites and accurate restoration can mitigate the negative impacts, but in any case the reshaping, the re-vegetation, and/or the natural resilience will give back the landscape its original aspect only in decades or centuries.

The annual production of 46,621,000 metric tons of lithoid materials exploited just in the region of Lombardy (ISTAT 1990), points effectively out the extent of this activity in Italy. These materials include granite, gneiss, marble, slate, sandstone,

Fig. 7 Aerial view of a quarrying district of limestone. The visual and functional impacts on soils and environment are highly evident (Photo by http://www.miramarmi.it/it/cave.html).

Fig. 8 Awkward and unsustainable contiguity and interpenetration between limestone quarry, cement factory and built-up area are evident. The quarry front is undergoing reshaping and re-vegetation works (Photo by F. Previtali)

limestone, dolomite, calc-schist, serpentinite, gypsum, quartz and quartzite, porphyry and siliceous rocks. The loose materials are the clays, sands, and gravel, all quarried in lowland or low hill and constituting about 77% of the exploited materials. This could suggest that their impact on environment and landscape should concern just marginally the Alpine lands. But the remaining 23% on the Alps and Pre-Alps, represented by concrete, foundry, covering, and paving stone quarries, exploited in opencast or tunnels, has a heavier impact both visual and functional on the ecosystems. As a matter of fact, the strong upright growth of the quarries in the mountains and hilly regions increases strongly their visibility and the magnitude of their damages to the hydrogeological systems.

2.6 Waste Disposal in Mountain Areas

If in the lowlands the quest for suitable sites to disposal and inactivation of the urban solid wastes is a problematic question, the problem becomes even more complicated in the mountain areas. Actually the weakness' of the ecosystems, i.e., the scarcity of the flat land, the high risk of contamination of surface and ground water, the unstableness of the slopes, the hazard of landslides and avalanches, even the beauty and the singularity of the landscapes, are the drawbacks in the selection of the suitable sites, turning the problem into a real emergency.

The presence of a dump lends a different look to the concerned land, which even though within a relatively limited radius assumes an unmistakable derelict

Fig. 9 Aerial view of a municipal garbage dump in a valley of the western Alps. The complexity of the interrelationships of this body with the environment is evident. Roads, dwellings, rivers and streams, valuable vegetation cover, and the uneven relief make awkward the coexistence with the slump (Photo by Google Earth)

appearance and maintains it for a long time, before a complete (even though partial) re-naturalization of the site. Notwithstanding the accuracy of the technical restoration actions, the landscape inevitably assumes the features of a characteristic and degraded anthroscape (Fig. 9).

3 Positive Human Impacts and Inheritances Creating Anthroscapes

Some human activities and settlements create anthroscapes which have positive impacts on the territory and somewhere constitute relevant cultural heritages. It is the case, for example, of:

- Mountain terracing
- Archaeological and historical settlements
- Land cover pattern typical of traditional and sustainable pastures and rangelands.

3.1 Mountain Terracing

The stone terraces built by the Alpine peoples over the centuries, still nowadays, carry out multiple functions. Primarily they consolidate and bear the slopes against landslides, mudflows, erosion and excessive runoff. They have been constructed on alluvial fans, at footslopes, and on midslopes. The deterioration of the traditional terracing systems, which constitute structural elements widespread in the Italian rural landscape, is one of the most evident consequences of the abandonment of farming in mountain lands (Piussi and Pettenella 2000). Actually this process is followed by land degradation problems, because the collapse of these hydrogeological artificial infrastructures is accompanied by the cessation of their protection function towards soil erosion and overland flow (Dunjo et al. 2003). The collapse of terraces triggers a dangerous process of soil degradation and the break up of the buttress walls can favour the formation of landslides (Baldock et al. 1996).

Furthermore, an important and historic role played by terraces has been to enable soil-less cultivations, of mainly vineyards along with orchards, chestnut trees, potatoes, officinal plants, soft fruits, and many other types of vegetables. The soil in the course of time, and even nowadays, has been collected from the valley floor, transported, suitably manured, and then unloaded on the tread-boards. These soils may be considered as typical man-made soils and named as Anthrosols in the World Reference Base for Soil Resources (IUSS Working Group WRB 2006) and Anthroposols in the Référentiel Pédologique (Baize and Girard 2009).

The cultivated terraces (Fig. 10) contribute to create typical and frequently scenic landscapes in many Alpine valleys (Fontanari and Patassini 2008; Scaramellini and Varotto 2008). Over the centuries, construction, preservation, and maintenance of terracing have been closely connected with viticulture. The mountain vineyards,

Fig. 10 The typical features of the vine-clad terraces which frequently adorn the toe slope of the south-facing sides of the Alpine valleys. The three functions carried out by the terraces are evident: agricultural production, slope consolidation and runoff control along with the addition of the unintentional effect of landscape embellishment (Photo by courtesy of G. De Giorgi)

in the majority of cases just lying on terraces, have recently been estimated to cover a total surface area of 93,000 ha in Italy (Stanchi et al. 2008).

This particular slope arrangement in the Alps dates back to many centuries ago. For the vineyards on terraces of Valtellina (Central Italian Alps) a Roman or at least Longobardic origin has been assumed. The total vineyard surfaces were covering 900 ha in this valley around 1642 A.D, whereas in 1968 this area was risen to about 3,800 ha, dropping to 854 ha in 2001 (Zoia 2004).

In the Aosta valley (Western Italian Alps) the viticultural traditions date back from at least 2,000 years and certainly both the Romans and the Franks were vine growers. At the end of the nineteenth-century the vineyards were covering 3,000–4,000 ha of this valley, but today this area amounts to 500 ha only (Moriondo 1999).

The scaling down of the vine surfaces, mainly due to screening and improvement of the best species and cultivars, is a process that has occurred all over Italy, where between 1996 and 2005 more than 50,000 ha planted with vines, with an 8% of the total being lost (Vania 2009).

While the landscape remains irreversibly marked by the terracing, the decline of their productive uses for agriculture, accompanied by depopulation of the mountains, could in future increase the risk of weakening of the involved stability of the steepest slopes.

3.2 *Archaeological Imprints: From Palaeolithic to the Roman Ages and More*

From the Late Pleistocene already, throughout the Holocene the Alps were inhabited by different peoples leaving there innumerable and peculiar imprints of their presence. The Alps are dotted with countless remains of dwellings, rock shelters, and features of various human activities, such as hunting, breeding and farming, burial and sacred ceremonies, lithic artefacts, graffiti, etc. Figure 11 illustrates one of these sites, in which lake dwellings were built around 2100 B.C. (Early Bronze Age) on one of the several ponds of the moraine amphitheatre of Garda lake. The settlement caused the transformation of the small lake into a marsh and induced deep changes of the surrounding natural environment, through deforestation and creation of pastures and crops (De Marinis et al. 2005).

The vestiges of historic and pre-historic human settling and attendance occur somewhere at altitudes even greater than 2,000 m a.s.l. Figure 12 shows a case of morphology and partial modification induced by the habitual visiting of the Palaeo-Mesolithic Man, which hunted, celebrated rites, and likely spent there some seasonal stays. Eventually, the Palaeo-Mesolithic man partially modified the morphology of the area, there leaving marks and artefacts (Fedele 1981, 1999).

The mountain landscape and environment have progressively acquired over the centuries quite different shapes up until the present conditions. Of some transient

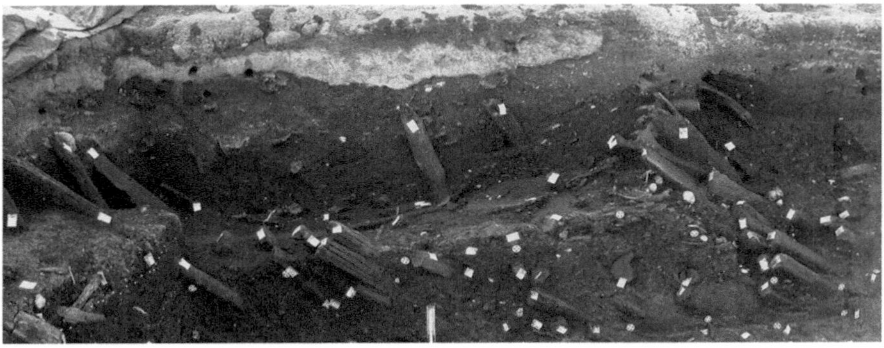

Fig. 11 Lake dwellings of the Early Bronze Age (2100 B.C.) at Lavagnone (Desenzano del Garda, Brescia) (Photo by courtesy of R. De Marinis)

Fig. 12 Pian dei Cavalli (Sondrio). The presence and the activities of the Palaeo-Mesolithic Man have partially moulded the morphology of the area, leaving several marks and artefacts (Photo by courtesy of E. Minotti)

and ephemeral Alpine anthroscapes, created in the last centuries, only iconographic marks remain or else the less old of them survive in mind of the oldest inhabitants. They are frequently incorporated in the present anthroscapes and so they are at this point invisible but represent a historic and cultural heritage of great significance.

3.3 Pastureland, Overgrazing, Forestry

Only a few of the current highland pastures and meadows of the Alpine land could be considered as "natural", since most of them were degraded by cutting and thinning out the native forests and woodlands. Nevertheless, pastures and meadows can be considered at present a semi-natural component of the Alpine environment. Over the centuries the grazing and fodder requirements have driven people to clear the wooded areas located at different altitudes, with low up to steep sloping topography (Fig. 13). Frequently, meadows and pastures are located on weakly sloping and loamy textured Wűrmian moraine deposits.

Between 1979 and 1997 the total meadow and pastureland surfaces have suffered in the Alpine Italian provinces a drastic decrease (−11.4%) (Conti and Fagarazzi 2004).

According to some estimates about 800,000 ha of meadows, grazed meadows and pastures have been abandoned since 1960, which means that 45% of the former area is today essentially gone (Chemini and Gianelle 1999). Such trends are also proceeding in parallel with the decrease in the utilized agricultural area (UAA), which dropped from 17.8% of the total area of the Alpine range in 1990 to 13.0% in 1980. This has favoured the constant increase of forests, reaching now the 22.8% of the land surface of Italy (ISPRA 2009). The progress of woodland presents

Fig. 13 Grazed meadow in a karstic zone of the Julian Alps at an altitude of about 700 m a.s.l. In the middle of the picture some dolines are visible (Photo by courtesy of E. Casati)

“lights and shades”, being linked not only with the activities of afforestation and reforestation, but also, in the last years, mainly with its invasion of marginal agricultural areas, both hilly and the mountaunous, abandoned by man. Depopulation causes mostly the cessation of farming, wood cutting and gathering, and grazing. Weeds and bushes, poor in species and highly inflammable replace, for at least some decades, the woodlands and pastures. When abandoned these lands are further under the risk of fires. Vegetal and animal biodiversity are in turn endangered, while erosion and runoff are intensified (Garcia-Ruiz et al. 1995).

Mowing and grazing are the key factors of preventing the natural sequences and the re-colonization processes, which actually start again only after the cessation or reduction of such perturbing factors. The abandonment of semi-natural meadows trigger off the prevailing expansion of arboreal and shrubby formations in secondary pastures, but also the development of dwarf arboreal formations in the primary meadows above the tree-line (Chemini and Rizzoli 2003; Ziliotto et al. 2004). Even if, the end of the mowing and grazing periods in the alpine meadows cause a risk for increasing the natural disasters in short terms, in the long period the development of an arboreal cover often gives the slopes a higher stability and a consequent lower risk (McDonald et al. 2002).

On the other hand, the expansion of woodlands in the Mediterranean regions is hindered by two serious and increasing events: the fires (mainly arsons) and overgrazing (ISPRA 2009).

With more specific regard to the deliberate anthropic actions on the development and conservation of woodland, it must be noted that their partial destruction has practically corresponded with the prehistoric development of agriculture and farming. The natural arrangement of woods and forests has been over the centuries deeply modified because of the use of wood as timber and fuel, together with fires, more and more turned from natural into arsons.

Frequently it is at the lower and the upper limits of the conifers belt that the anthroscape features become more evident: the abrupt boundary between woodland and meadow; spread of larch even at low altitudes, below the spruce belt; replacement of the upper natural belt of larch by meadow. At middle-low altitudes, further evidences of the anthropic influence are represented by the chestnut wood introduced instead of the broad-leaved wood. Furthermore, the marks carved by the First World War (bomb craters, grenade fragments in the trunks, trenches and so on) on the woodlands and meadows, should not be disregarded on the matter of nature degradation.

Among the anthroscapes generated by agricultural activities, overgrazing (Fig. 14) should not be overlooked, which can be charged with accelerated erosion, soil fertility loss, soil sealing, and ultimately land degradation.

4 Conclusions

The term “Anthroscape” has very recently been introduced into the scientific literature (Eswaran et al. 2005) and a legitimate and an internationally accepted definition of this entity probably does not exist yet. The concept of anthroscape

Fig. 14 Mountain pastures on steep slopes. The generalized overgrazing, combined with snow melting and water erosion, has produced different types of land degradation. On the right side, a landslide marked by tree tilting and falling along with shear lines; in the middle, eroded pathways by cattle trampling and a watering pond at the top of the hill are visible (Photo by courtesy of C. Zucca)

even though is a not univocal context, leads back to the geo-chronological unit of “Anthropocene”, which came into use not long ago (Crutzen and Stoermer 2000).

In mountain environments the anthroscapes assume absolutely peculiar features, since most of the anthropic impacts have different origins and dynamics in comparison with the hilly and flat regions.

Furthermore, in this paper we briefly mentioned the existence of invisible anthroscapes, only detectable by analytical means, which can occur jointly or separately in regard to the material and visible anthroscapes.

Some Anthroscapes can be considered as created by anthropic activities, having positive impacts on the land, e.g. the mountain terracing, the archaeological settlements, and the pastures and rangelands. These all are consistent with the related natural environments and anthroscapes, which were aesthetically and functionally improved and where some constitute relevant cultural heritages.

In other cases some characteristic landscapes show the evident marks of the negative impacts deriving from human settlements and activities. It is the case of many winter resorts, of the mountain off-roading, of the largest dumps, of the industrial and civil settlements along valley floors, of the widest stone quarries, and of the waste disposals.

Many of the above-mentioned anthroscapes strictly require firm precautions against the contemporary land-use trends, and not only in the Italian Alps, but also in many mountain ranges of the world. Actually, the alpine ecosystems must be preserved not only because they are fragile as valuable and unique by themselves, but also because the quality of life in the vast plains (i.e., water resources, rivers, air, climate, vegetation, environmental hazards, etc.) strongly depends on the state of conservation and protection of the above stretching mountain lands.

Acknowledgements The author would sincerely like to thank: Prof. S. Kapur and Dr. C. Zucca for additional reviews and feedback of this paper; Prof. F. Rodeghiero for the information on mining sites; Drs E. Sturani, A. Cavallo, C. Zucca, E. Casati, G. De Giorgi, R. De Marinis, E. Minotti for kindly providing several photos; IRPI-CNR, Google Earth, and Miramarmi S.r.l. for authorizing the use of some images.

References

A.P.A.T. Agenzia per la Protezione dell'Ambiente e per i Servizi Tecnici – Min. Ambiente Tutela Territ (2006) Censimento dei Siti Minerari Abbandonati. I siti minerari italiani (1870–2006).

Baize D, Girard MC (Eds) (2009) Référentiel Pédologique 2008. Quae Éditions, Paris.

Baldock D, Beaufoy G, Brouwer F, Godeschalk F (1996) Farming at the Margins: Abandonment or Redeployment of Agricultural Land in Europe. Institute for European Environmental Policy, London/The Hague.

Chemini C, Gianelle D (1999) Pascolo e conservazione della biodiversità. In: Presente e futuro dei pascoli alpini in Europa. Franco Angeli, Milano. pp. 36–48.

Chemini C, Rizzoli A (2003) Land use change and biodiversity conservation in the Alps. Journal of Mountain Ecology, 7:1–7.

Conti G, Fagarazzi L (2004) Sustainable Mountain Development and the key-issue of Abandonment of Marginal Rural Areas. Planum, XI:1–20.

Crutzen PJ, Stoermer EF (2000) Anthropocene, IGBP Newsletter, No 41.

De Jong C (2007) Melting mountains: focus on the European Alps and beyond. Mountain Forum Bulletin, 2:8–9.

De Marinis RC, Ravazzi C, Arpenti E, Deaddis M, Perego R (2005) Lavagnone (Desenzano del Garda): new excavations and palaeoecology of a Bronze Age pile dwelling in northern Italy. In: Della Casa P (Ed) Proc. Int. Conf. on Wetland Economies and Societies. Zurich, 2004.

Dunjo G, Pardini G, Gispert M (2003) Land use change effects on abandoned terraced soils in a Mediterranean catchment, NE Spain. Catena, 52: 23–37.

Edwards AC, Scalenghe R, Freppaz M (2007) Changes in the seasonal snow cover of alpine regions and its effect on soil processes: A review. Quaternary International, 162–163:172–181.

Eswaran H, Kapur S, Akça E, Reich P, Mahmoodi S, Vearasilp T (2005) Anthroscapes: a landscape unit for assessment of human impact on land systems. In: Yang JE, Sa TM, Kim JJ (Eds) Application of the Emerging Soil Research to the Conservation of Agricultural Ecosystems. The Korean Society of Soil Science and Fertilizers, Seoul, Korea, pp. 175–192

Fedele F (1981) Il popolamento delle Alpi nel Paleolitico. Le Scienze, 160:22–39

Fedele F (1999) Le ricerche del Pian dei Cavalli nel contesto del popolamento preistorico della Valchiavenna. Proc. II Conf. Archeol. Prov., 3:17–34, Grosio.

Fontanari E, Patassini D (Eds) (2008) Paesaggi terrazzati dell'arco alpino. Marsili, Venezia

Garcia-Ruiz JM, Lasanta T, Martí C, Gonzales C, White S, Ortigosa L, Ruiz Flaño P (1995) Changes in runoff and erosion as a consequence of land-use changes in the central Spanish Pyrenees. Physics and Chemistry of the Earth, 20:301–307.

ISPRA (2009) Annuario dei dati ambientali 2008. Agenzie regionali e delle Province Autonome per la Protezione dell'Ambiente-Sistema Statistico Nazionale, Roma.

ISTAT (1990) Annuario statistico italiano. Istituto Nazionale di Statistica, Roma.

IUSS Working Group WRB (2006) World Reference Base for Soil Resources 2006. World Soil Resources Reports No. 103. FAO, Rome.

Kapur S, Eswaran H, Akça E, Dinç O, Kaya Z, Ulusoy R, Ba Y, Yılmaz T, Çelik İ, Özcan H (1999). Agroecological management of degrading coastal dunes in southern Anatolia. In: Özhan E (Ed) The Fourth International Conference on the Mediterranean Coastal Environment, Vol. 1. Medcoast 99, Antalya, Turkey, pp. 347–360.

Kapur S, Zdruli P, Akça E, Arnoldussen AH, Kapur B (2004) Anthroscapes of Turkey: sites of historic sustainable land management (SLM). In: van Asselen S, Boix Fayos C, Imeson A (Eds) 'Briefing Papers' of the Second SCAPE Workshop in Cinque Terre (Italy), 13–15 April 2004.

McDonald MA, Healey JR, Stevens PA (2002) The effects of secondary forest clearance and subsequent land use on erosion losses and soil properties in the Blue Mountains of Jamaica. Agriculture, Ecosystems and Environment, 92:1–19.

Moriondo G (1999) Vini e Vitigni Autoctoni della Valle d'Aosta. Imprimerie E. Duc, Ed.IAR, Aosta.

Piussi P, Pettenella D (2000) Spontaneous afforestation of fallows in Italy. In: Weber N (Ed). NEWFOR – New Forests for Europe: Afforestation at the Turn of the Century. Proc. Sci. Symp., Freiburg, 16–17 February 2000, pp. 151–163

Ritter SK (2004) Artificial snow. Science and Technology, 82:3.

Scaramellini G, Varotto M (Eds) (2008) Paesaggi terrazzati dell'arco alpino. Atlante. Marsilio, Venezia. Sondrio, Sondrio.

Stanchi S, Freppaz M, Galliani C, Revelchion V, Zanini E (2008) Problematiche produttive ed ambientali delle aree terrazzate. L'Informatore Agricolo, Anno 24–n.1.

Vania E (2009) Diminuiscono superficie vitata e numero di aziende vitivinicole, crescono invece le denominazioni d'origine. Teatro Naturale, Anno 7, n. 33.

Watson A (1985) Soil erosion and vegetation damage near ski lifts in Cairngorm, Scotland. Biological Conservation, 33:363–381.

Ziliotto U, Andrich O, Lasen C, Ramanzin M (2004) Tratti essenziali della tipologia veneta dei pascoli di monte e dintorni. Regione Veneto, Accademia Italiana di Scienze Forestali, Venezia.

Zoia D (2004) Vite e vino in Valtellina e Valchiavenna. L'officina del libro, Sondrio.

Development and Challenges of the Anthroscapes in the Clay Soil District of Eastern Norway

Arnold Arnoldussen

Abstract Anthroscapes are dealing with the adaption of land management to soil- and terrain conditions. Under human influence specific soil and landscape conditions can develop over long periods of time. Due to the low population pressure and historical developments other systems of Anthroscapes could develop in Norway.

In this chapter the landscape, soil and land-use situation in the Landscape Region "The Clay Soil District" in SE Norway is described. Attention is paid to the original situation, on the negative impact of land use and land management change and which measures were taken to redirect the negative developments into a positive direction. Understanding the relation between ecosystem functioning, landscape development, land use and soil development is important to develop a sustainable land management. Due to human influence a modern Anthroscape is under development.

Keywords Agro-environmental schemes · Anthroscapes · Land levelling · Land use change · Soil resources · Sustainable land management

1 Introduction

Prevention of land- and soil degradation is in focus. Several developments in society influence the development of degradation. Land use and land management should be in balance with soil, terrain and climate factors. If this balance is lacking and certain thresholds are passed, over longer or shorter time, processes of degradation start. The consequences are often irreversible.

Soil development is influenced by many factors as geology, climate, vegetation, animals and human influence. An important decisive factor is time.

A. Arnoldussen (✉)
Norwegian Forest and Landscape Institute, PB 115, 1431 Ås, Norway
e-mail: Arnold.Arnoldussen@skogoglandskap.no

S. Kapur et al. (eds.), *Sustainable Land Management*,
DOI 10.1007/978-3-642-14782-1_6,

Human impact on soil development can both be negative and positive. In the case of a negative impact soil degradation will develop and soil quality decreases. The contrary may happen in cases where human impact has a positive influence on soil quality. The development of Anthrosols over many centuries time is an example. These soils developed under influence of agriculture over centuries of time. By using manure and plaggen soil quality improved and resilience against disturbing factors increased. This is the situation sustainable land management is aiming at. Anthrosols are may be an extreme example, but we see at many places that soil quality increases under influence of human impact over long time.

In Norway Antrosols cover 60 km^2 (1%) of the soil mapped area and they are confined to the areas which are cultivated over centuries time. Often these areas are on a south slope and close to the place where the original farm buildings were established. In Norway these places are often quite spread in the landscape.

However due to the introduction of unsustainable land management practices situations of imbalance may develop, resulting in land degradation. Examples are land levelling operations, the introduction of heavy machinery to get a more productive and effective agriculture or a change to a land use or cropping system which is not longer adapted to the local soil terrain and climate conditions. For reasons of economical survival and due to globalisation processes farmers are often obliged to introduce less sustainable land management methods.

Land management has not only impact on soil development, but influences also the development of landscape and biodiversity. Both aspects are now also in focus. Modern land management techniques can damage the historical cultural landscape and over time these landscapes can disappear. Old historical landscapes represent ethical, aesthetical and economical values. The disappearance of the cultural landscape can damage for example the possibilities for tourism as an economical activity.

Many species, varying from plants, insects, mammals and microbes, are bound to specific living conditions in the landscape. Disappearance of these conditions or the fact that they get isolated cause the disappearance of populations or the degeneration of their genetic quality. As many species do have direct influence on ecosystem development and biological processes, negative impact on ecosystem quality and functioning may be the result.

Development and functioning of Anthroscapes should be related to landscape and biodiversity development. A holistic view is required when discussing Anthroscapes.

Degradation of the land resources cause that the soil is not longer able to fulfil one or more of its functions. Soil functions consider the eco-services that the soil is providing. When looking at soils you can ask the question: "What does the soil for you?" The most important soil functions are according the European Commission (2002), Dorren et al. (2004) and Imeson et al. (2006):

- The production function (food and biomass).
- The regulation function of the soil: the cycling of water and nutrients. The soil functions as a chemical factory by storing and transforming minerals, organic matter, other chemical substances and energy.

- Habitat and gene pool: soil performs essential ecological functions and harbours an important gene pool.
- Physical and cultural environment for mankind: soil is a platform for human activity and is an element of landscape and cultural heritage.

In addition the soil has a communication and an aesthetic, scientific and carrier function. It contains geological and archaeological treasures, enabling mankind to understand better its history and development.

Many of the functions are related and soils are often acting multifunctional. Overexploiting some functions may damage automatically other functions.

Soil functions are threatened by the following threats (European Commission 2002): soil erosion, soil contamination, decline in organic matter, soil sealing, soil compaction, decline of soil biodiversity, salinisation, and floods and landsides.

The different soil threats are often related. Soil compaction can increase erosion and results in a decrease of organic matter content. Soil threats are influenced by many different developments in society, varying from a local to a national to a global scale. In the globalisation process farmers can be forced to take measures which are not sustainable in the long run.

Knowledge about the development and functioning of Anthroscapes may help to understand the ongoing processes and support the development of a more sustainable land management. Often Anthroscapes represent old cultural values.

In this chapter the case of Anthroscape development in the Clay Soil District in Eastern Norway is taken as an example. In a historical perspective and due to low population density developments of Anthroscapes should be seen in another light.

The development and characteristics of the landscape, soils, land-use and land-management are described for the Landscape Region "Clay Soil District." Focus will be on the developments and consequences of the land use change and the measures taken to establish a more sustainable situation. How can knowledge about the characteristics and processes of Anthroscapes help with the implementation of a more sustainable land management?

1.1 Development and Characteristics of the Landscape Region Clay Soil District in Eastern Norway

During history Norway was several times covered with thick layers of ice. The heavy weight of the ice caused that the earth crust was pressed down. During the Young Dryas (12,800–11,500 years ago) the ice from the last ice age started to melt quickly down. At that time the sea level in Southern Norway was around 200 m above present sea-level. The melting of the glaciers caused both a raise of the sea- and land level. The redraw of the glaciers was not a constant process. During periods the glaciers could advance and redraw again. This process of a changing advance and redraw is still visible in the landscape. End moraines pushed up during the advance periods form clear ridges in the landscape. After the last ice age large parts of SE Norway were covered by sea. Here thick packets of clay were deposited; in some parts of Eastern Norway 150 m of clay was deposited (Ramberg et al. 2006). As the land level rise was bigger than the sea level rise large parts of the area became dry. As clay is an erosive material a landscape with steep and deep ravines could develop during the last 10,000 years.

The clay was deposited in a salt environment. After the land rise the salt was washed out and replaced by fresh water. This change caused a destabilising of the clay structure. Reason why there have been many small and bigger land slides in the area. These land slides are still visible in the landscape. Also today land slides occur now and than.

The melting of the glaciers went relatively very fast and enormous amounts of sediments were produced. In the northern part of the district a large packet of glacier deposits were deposited. These deposits contain today one of the biggest ground water reservoirs (Ramberg et al. 2006); the national airport of Oslo is build on these deposits.

The former Norwegian Institute for Land Inventory, now Norwegian Forest and Landscape Institute, developed a national landscape reference system. This system is hierarchical, dividing the country into 45 landscape regions and 444 sub-regions. The landscape regions are recognised based on major landform, minor terrain form, water and water courses, vegetation, agricultural land and built-up areas/technical installations (Puschmann 2005).

The Landscape Region "Clay Soil District" (Fig. 1) is characterised by the clay deposits which were sedimentated directly during and after the melting of the glaciers. The landscape is characterised by a mosaic of plateaus, mainly build up by marine sediments and cut by small and bigger ravines. The ravines may be steep and border sharp to the surrounding plateaus (Fig. 2).

The big rivers in the region (Glomma and Vorma) are following cracks in the earth crust and are not able to meander. The other rivers in the area are generally streaming slowly and could develop meanders.

The forests in the region are characterised by coniferous forest. Along the Oslo fjord more requiring deciduous trees like oak and beech can be present. These trees can form stands at more protected places; often on the steep slopes of the ravines with a southern exposition. During historical times most land suitable for agriculture was brought under agriculture. The forests are generally present on steep slopes and on soil types not or less suitable for agriculture (very stony moraine deposits). Sandy parts of the big moraine ridges are now used for vegetable production.

The Clay Soil District is in Norway the most cultivated area. In total 30% of the area is used for agriculture and covers nearly 1,800 km^2. At national scale only 3% of the area is under agriculture. In the district 70% of the agricultural area is owned, 25% is hired out and 5% is abandoned (Puschmann 2005). Abandoned land is often marginal for agriculture. Ca. 80% of the agricultural area is today used for cereal production. Dairy farming has been strongly reduced after the Second World War and is now limited to the areas with remainders of the old ravine systems.

The clay soil district is close to the most urban areas in Norway. This explains why prevention of soil sealing is here high on the political agenda. Focus is on the

Fig. 1 Landscape region 3: the Clay Soil District [from Puschmann (2005)]

Fig. 2 A ravine landscape. Photo: Norwegian Forest and Landscape Institute

protection of the most productive agricultural areas for food production. During recent times the protection of the cultural landscape became also an argument to protect areas for soil sealing. Municipalities are obliged to reduce the yearly sealing of their good agricultural areas by 50% within 2010.

In prehistoric times the Clay Soil District was already inhabited by man (Elgersma and Asheim 1998). As the transport went most by water many stone-age settlements were found in the surroundings of the old coast. Stone inscriptions were found in the southern part of the district.

Important graves were found from the Viking time. An important settlement from the Viking time (Kaupang) is found in the western part of the district in the county Vestfold.

In the clay deposits Stagnosols, Luvisols and Albeluvisols were formed. On flat areas with a bad drainage Histosols could develop. These areas are situated both on the top of the plateaus and in the river valleys. On the sandy moraine deposits Arenosols, Podzols and Umbrisols were developed. As clay deposits are dominant in the District the Stagnosols, Albeluvisols and Luvisols characterize the original soils.

In this naturally high erosive landscape land use became during the centuries more or less in balance with the soil and terrain situation. Arable farming was developing on the flat plateaus on the better drained soil types. Badly drained soil types were used for hay production. The steep slopes of the ravines were used for grazing purposes. Forests were growing on the sites unsuitable for agriculture; either because of the limited soil quality or a too steep slope. Clear evidence is that over many thousands of years the Luvisols developed into Albeluvisols. By the transport of the clay minerals downwards Albeluvisols gets slowly more badly drained and went over into Stagnosols. Agriculture can delay this process, because measures to improve drainage were taken. Application of manure on sandy soils has a positive influence on the development of Umbrisols in an original Arenosol. In the Clay Soil District Anthrosols are scarcely present and cover only 0.1% of the agricultural area.

In the area land slides occurred. Due to the geological development of the area land slides can be considered as a natural phenomenon. However agriculture and other human activities can trigger land slides if certain thresholds are passed. They can be catastrophic and in both past and present human lives were lost in these events.

Due to the natural conditions in the area Anthroscapes could develop over time characterized by Luvisols, Albeluvisols and Stagnosols.

1.2 Land Use and Land Management

As already mentioned the Clay Soil District was already inhabited during prehistoric times. With the increasing population in historical times more and more land became cultivated. In the district plateaus were used for arable farming, the slopes of the ravines were used for dairy farm production. All the farms were mixed. The forests were in the end only growing on the sites less or unsuitable for agriculture.

After the Second World War the land use situation changed drastically. In the first place mechanisation in agriculture started; resulting in larger fields and levelling of the land. The introduction of machinery, fertilizers and pesticides made it possible to increase agricultural production.

However political changes had an even bigger impact on the land use structure in the district. After the Second World War in many European countries the focus came on increasing food production, in order to reach a certain level of food self-sufficiency (Bullock 1999). In Norway it was decided to stimulate cereal production in areas where this climatologically was possible. Dairy farming was prioritised in the western and northern parts of Norway. Big land levelling operations were needed to make the areas with ravines suitable for arable farming. During this canalisation policy the biggest part of the old ravine systems were levelled; a process going on to the 1980s. The farmers got a subsidy to fulfil the levelling process. During the land levelling ravines were filled with soil; small streams were laid in pipes and the forests and shrubs on the slopes were transformed into agricultural fields. The total landscape structure changed. Arable land increased, pasture land decreased and in many cases forest land decreased. The landscape became more open with less heterogeneity. With the disappearance of the ravines the river system changed totally. Figure 3 shows as an example the change in stream pattern in the surroundings of Spydeberg from 1964 to 1992.

A fraction of the original old river system still exists. A consequence of laying a part of the river system in pipes is that in case of extreme precipitation the earlier storing capacity of water has been reduced strongly. A surplus of water reaches very quickly the main river system with as a consequence the possibility to develop

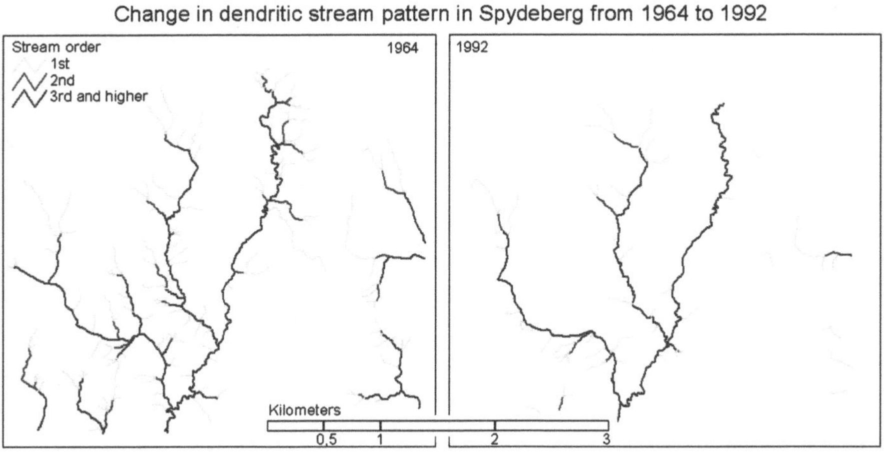

Fig. 3 Change in dendritic pattern in Spydeberg (1964–1992)

a flooding. In exposed areas the land management should be organized in such a way that the water stays as long as possible in the higher parts of the watershed. Considering the consequences of land levelling for the hydrological situation one can conclude that flexibility has decreased strongly.

In the beginning of the era of land levelling no attention was paid to the topsoil. The original underground could become topsoil after levelling. Due to bad soil structure and lack of organic matter this had big disadvantages for the establishment of new crops and the functioning of the soil ecosystem. To avoid this, the topsoil was later treated separately and placed back after finishing the land levelling operation. In some municipalities 40% of the total agricultural area is levelled.

Estimated is that the erosion increased with a 3–13 fold after a land levelling (Øygarden et al. 2006). In addition the slope length increased, resulting in a stronger down slope water flow.

Looking at agricultural production the process of mechanisation and canalisation was very successfully. Agricultural production increased and agriculture became more efficient.

However the other side of the medal became evident during the 1980s. During the land levelling process land originally marginal for arable farming became cereal land. These areas were generally much more erosive and erosion became a problem.

Compared to erosion in Southern Europe erosion in Norway is not that much a problem of transport of soil material, but more a problem of transport of nutrients. Fresh and salt water ecosystems were polluted with nitrogen and phosphorus; at the end resulting in the development of poisonous blue algae populations. The extent of the problems became clear when at the end of the 1980s sea life perished in Skagerrak and North Sea. Countries bordering the Botnic Gulf, East Sea, Skagerrak and North Sea agreed in the North Sea Declaration upon a substantial reduction of the pollution with nutrients. Besides the dead of water organisms blue algae made it impossible to use the water from severe polluted fresh water lakes for drinking water or use it for recreational purposes. Norway decided amongst others to prioritise the erosion reduction.

Another consequence of the land use changes in the Clay Soil District was the disappearance of the historical cultural landscape and the demising of biodiversity. Both elements are related with each other. During the land use change small landscape elements and border zones disappeared. These elements represent both an important living area for plants, insects and animals and are important for the aesthetic value of landscapes. Establishment of so called protected landscapes is on this moment on the agenda.

Several Agro-Environmental Schemes were at the end of the 1980s established to prevent and reduce erosion and to stimulate the development of a more sustainable land management.

In exposed watersheds the soil of all agricultural areas were mapped and the erosion risk maps were produced based on an erosion risk model (USLE) calibrated for Norwegian conditions. In the Agro-Environmental Schemes the erosion risk maps are used (Arnoldussen 2005a).

At a national scale on this moment nearly 50% of the Norwegian agricultural soils are mapped.

Today the same mistakes of destroying the existing Anthroscapes are still made in several parts of the world; examples are known from Spain and Portugal (Imeson et al. 2006). How may the lessons learned on one place contribute to prevent damage by the same mistakes in another place? Knowledge about Anthroscapes and their development may give a contribution.

The soil mapping was realized at the moment that nearly all the land levelling operations were finished. So no exact information is present which soil types disappeared with the land levelling. The most common soil types on agricultural land in the Clay Soil District today are given in Table 1.

Agricultural areas in the Clay Soil District cover nearly 1,800 km^2 and are today characterized by Stagnosols, Luvisols/Albeluvisols and Anthropic Regosol. Anthropic Regosols are due to the land levelling. Remaining soil types consist out Anthrosol, Fluvisol, Histosol, Leptosol, Phaeozem, Regosol and Umbrisol. Each of them is only covering a small area.

Table 1 Soil types (in % of the agricultural area) in the Clay Soil District

Soil type (WRB)	Percentage of agricultural area (%)
Stagnosol	32
Luvisol/Albeluvisol	30
Anthropic Regosol	16
Cambisol	10
Gleysol	4
Arenosol	3
Podzol	2
Rest	3

Fig. 4 Haplic Stagnosol (Ruptic). Photo: Norwegian Forestand Landscape Institute

All Stagnosols (Fig. 4) are badly drained and need artificial drainage. Stagnosols are often situated in flat or slow sloping areas, resulting in a relative low risk on erosion. In the Clay Soil District Stagnosols are mostly developed in sea deposits, but may also be formed in river and beach deposits. The clay content increases often with depth. The organic matter content in the A horizon varies generally from 3 to 6% (Solbakken et al. 2006). Stagnosols are generally suitable for growing cereals and potato, if a good drainage system is in place. In the Clay Soil District this soil type covers 32% of the agricultural area. For the total mapped agricultural area in Norway this figure is 22% (Solbakken et al. 2006)

The Luvisols and Albeluvisols are placed together in one group, because they are closely related to each other. Albeluvisols differs from Luvisols because they, due to leaching, developed albic tongues under the ploughing horizon. In the Clay Soil District 30% of the agricultural area has this soil type; at national scale this figure is 22% (Solbakken et al. 2006). Both Luvisols and Albeluvisols are characterized by downward transport of clay particles. These clay particles are sedimentated along pores longer down in the profile. Luvisols and Albeluvisols are generally formed in sea deposits; Albeluvisols are developed in the eldest depositions. Research showed that it took 4,500–5,000 years to develop Albeluvisols (Solbakken et al. 2006). In the top of the profile the texture is varying from a loam to a silty clay loam. In the underground texture is varying from a silt loam to clay.

The A horizon contains ca. 4–6% organic matter. These soils need artificial drainage when under agriculture. With a good drainage system they are suitable for cereal and potato production.

They are for 60% situated on slopes less than 6%. In a case of a steeper slope the risk on soil erosion will be a challenge and measures to reduce erosion will be necessary. Figure 5 shows a Stagnic Albeluvisol.

Anthropic Regosols cover 16% of the Clay Soil District. At national scale this is 9%. Anthropic Regosols are developed in the land levelled areas and in 90% of the cases slopes are steeper than 6% (Solbakken et al. 2006). Under the land levelling the topsoil was in most cases removed and the former underground is forming now in many cases the topsoil. Organic matter content in the topsoil is generally less than 3%, the soil has a bad structure and rooting possibilities for plants are limited. Seen the bad soil structure, the low amount of organic matter content and the steeper slopes the erosion risk is often a big problem and more or less extensive measures are needed. This soil type is not suited for potatoes and cereals. A cover with grass and the application of manure may help to improve the situation over time. Anthropic Regosols are an example where human impact had a strong negative impact on the ecological functioning of the soil and the potentials for agriculture. Figure 6 shows an Anthropic Regosol.

Cambisols cover only 10% of the agricultural areas in the Clay Soil District. At national scale this figure is 22% (Solbakken et al. 2006).

Cambisols are characterized by a relatively little developed B-horizon. Under Norwegian conditions Cambisols are developed in generally nutrient rich deposits.

Fig. 5 Stagnic Albeluvisol. Photo: Norwegian Forestand Landscape Institute

Fig. 6 Anthropic Regosol. Photo: Norwegian Forest and Landscape Institute

In the case of a nutrient poor substrate the soil develops in the direction of a podzol. In the Clay Soil District Cambisols are formed in beach, river and sea deposits. In many cases the natural drainage is bad and measures are necessary. Soil texture is generally varying from silt to fine sand. Organic material in the A horizon is around 5%. As most Cambisols are situated on flat or less sloping terrains erosion is not a problem. However on steep slopes measures are necessary. Generally Cambisols are suited for growing cereals and potatoes. Figure 7 shows a picture of a landscape with wheat growing on an Endostagnic Cambisol.

From 1991 to 2004 a national Agro-Environmental Scheme was regulating the reduction of erosion. Main focus was on decreasing autumn ploughing of agricultural fields, especially on sites vulnerable for erosion. The subsidy rates differed according the erosion risk of the field (Kollerud 2005; Arnoldussen 2005b) and were continuous evaluated and adjusted to be as effective as possible regarding the environmental benefit and cost-effectiveness. Besides "not ploughing in autumn" measures like the establishment of grassy zones in drainage ways, using catch crops and stopping arable farming on marginal agricultural land with a very high erosion risk were included in the Scheme. In later years the use of reduced tillage systems was included. In addition farmers could get a support to construct sedimentation ponds and to take hydro technical measures.

The scheme gave good results and in a few years the total autumn ploughed land increased to about 50% of the total agricultural area. Already in 2000 it was stated that Norway nearly had reached the tasks formulated in the North Sea Declaration (Snellingen Bye et al. 2000; Arnoldussen 2005a).

However it was felt that more could be reached by adapting the rules more to the regional and local situation. Especially in exposed watersheds more needed to be done. In the period 2003–2005 the whole system of the Agro-Environmental Scheme was reorganized and Regional Environmental Schemes were developed.

Fig. 7 Wheat growing on an Endostagnic Cambisol. Photo: Norwegian Forest and Landscape Institute

In the first place the farmers were in 2003 obliged to have an Environmental Plan for their farm to be eligible for full governmental support. This plan should cover aspects like erosion, nutrient runoff, pollution, cultural landscape and biodiversity. Needed actions should be planned and achieved goals should be evaluated. The Environmental Plan focuses the farmer on his natural resources and how he can influence the impact of his land management by taking the right measures.

The second step was to abandon the existing national Agro-Environmental Scheme and to establish Regional Environmental Plans. The counties were responsible for making these plans. In co-operation with farmer unions the regional environmental problems were analyzed and per county a priority plan with measures was formulated. Purpose is to direct the efforts to the places where the problems are most pressing. In 2004 the Regional Environmental Plans were implemented.

The implementation of the Water Frame Work directive from the European Union makes the approach of the Regional Environmental Plan even more important. In a few years surface and ground water quality should reach certain standards. In areas with erosion water quality is threatened. Measures are needed to improve the situation. Based on the soil and erosion risk data soil statistics are produced for the most exposed watersheds. The statistics accompanied by maps show the part of the watershed which is most vulnerable for erosion. Figure 8 shows an example from the river Rakkestadelva in the county Østfold. It is clearly to be seen which areas close to the river system has a high erosion risk. Here specific measures are on its place. In 2008 the Regional Environmental Plan is evaluated on its effects and efficiency.

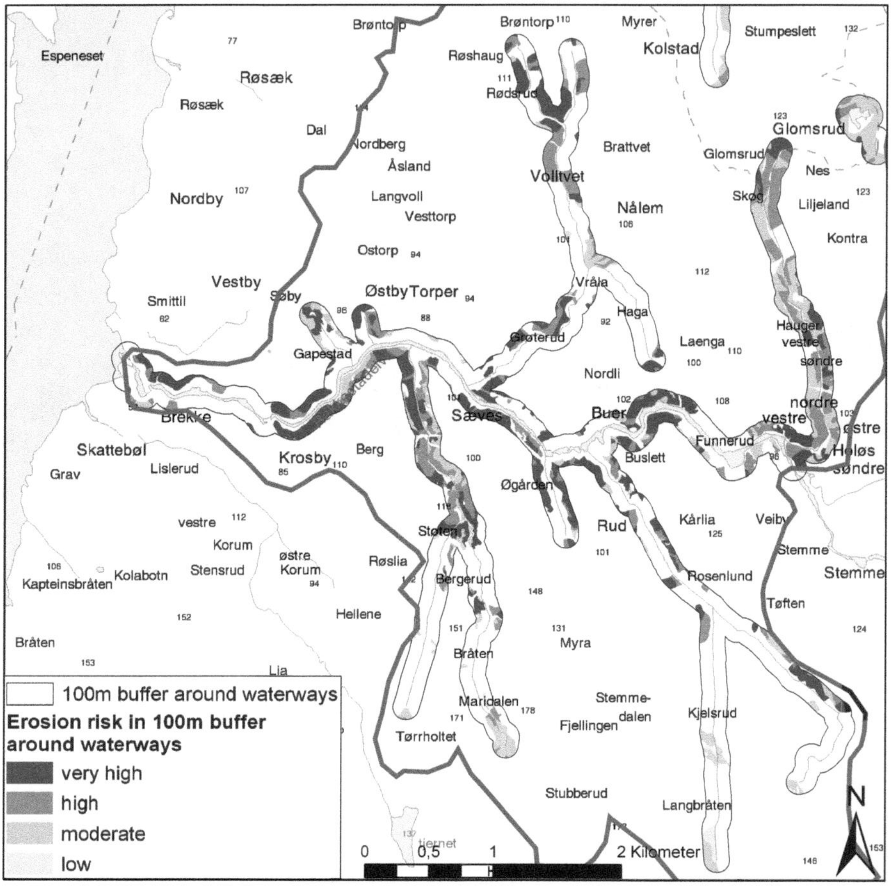

Fig. 8 Rakkestad river: erosion risk in a 100 m buffer around waterways

During the last few years the implementation of no-tillage and reduced tillage systems increased. Long term research learned that the time of ploughing (spring vs. autumn) had little effect on the yield levels. With spring ploughing the sowing period can be delayed. This result in a more varied annual yield compared with autumn ploughing (Øygarden et al. 2006).

No-till systems are generally successful on well-drained loam and clay soils under the relatively dry conditions in SE Norway. However these systems showed to be more problematic under wetter conditions and than especially on silty and sandy soils. Reduced tillage and direct drilling show yields of an equal level compared with autumn ploughing. In years with very dry weather in the beginning of the growing season these yields are even higher than with autumn ploughing. No-tillage and direct drilling shows a strong reduction in erosion risk compared with autumn ploughing (Øygarden et al. 2006).

1.3 Knowledge About Anthroscapes and the Implementation of a More Sustainable Land Management

In the Clay Soil District land has naturally a relatively high risk for soil erosion. Clayey and silty soils are dominating and steep slopes are present. In the old land use situation arable farming was limited to the flat well drained areas on the plateaus. Grazing and pasture land was present on the more badly drained soils on the plateaus and on the steep slopes in the ravines. The situation was much more balanced as it is today. Improving the situation means that knowledge of the impact of land use and land management types on the environmental conditions should be known under the given soil, terrain and climate conditions. Also knowledge should be available which cropping systems are economical feasible or could be made feasible.

For the implementation of a sustainable land use in the area this means that information is required about:

- Which areas are very susceptible for erosion and in particular which areas should not be used as arable land.
- Which land use types and crops will, under the given erosion risk class, not give any problems.
- Which land use types and crops will, under the given erosion risk class and under more extreme weather conditions, give problems. Which measures could be taken to reduce this risk down to an acceptable level?

In the original land use situation the most erosive areas were used as grazing and pasture land. In many parts of the levelled areas the soil structure is still very bad, resulting in low productivity, bad hydrological conditions and reduced rooting possibilities. Here the land use and land management operations should be focussed to increase organic matter content, to improve soil structure and to increase rooting depth. Establishment of a deep rooting grass vegetation and the application of manure will over a period of 10–20 years improve the situation a lot.

Anthroscapes could develop over time in balance with the terrain, soil and climate conditions. Knowledge about their development and functioning may help to develop modern land management systems adapted to modern economical requirements. As practice shows development of land management systems with a one-sided focus on economy will in many cases result in a degradation of soil and water resources. The costs in repairing the damage will in most cases be high and some damage will be irreversible.

This chapter shows how a negative development could be reversed. The practice from Norway learned that including farmers and land owners in the measures to be taken is very important. Knowledge and motivation are the start of a process in the right direction.

For the areas where the risk on erosion under more extreme weather conditions is too high specific measures should be taken to reduce the risk level down to an acceptable level. Measures may vary from changing land use type, to introduce specific soil elaboration methods or to take hydro technical measures.

The land use type and land management method should not decrease the soil quality, but should preferably increase it. The Soil Bank is the most important bank for a farmer!

References

Arnoldussen AH (2005a) The effectiveness of Agro-Environmental Schemes in reducing erosion. Land Degradation and Development, 16: 541–546

Arnoldussen AH (2005b) Soil survey in Norway. In: Jones RJA, Houšková B, Bullock P, Montanarella L (eds), 2005: Soil resources of Europe. Second Edition. European Soil Bureau Research Report no 9: pp. 257–262

Bullock P (1999) Soil resources of Europe: an overview. In: Bullock P, Jones RJA, Montanarella L (eds), Soil resources of Europe. European Soil Bureau Research Report no 6: pp. 15–25

Dorren L, Bazzoffi P, Sánchez Díaz J, Arnoldussen A, Barberis R, Berényi Üveges J, Böken H, Castillo Sánchez V, Düwel O, Imeson A, Mollenhauer K, De la Rosa D, Prasuhn V, Theocharopoulos SP (2004) Soil erosion. Task group 3 on impacts of soil erosion. In: Van Camp L, Bujarral B, Gentile A, Jones RJA, Montanarella L, Olazabal C, Selvaradjou SK (eds), Reports of the Technical Working Groups established under the Thematic Strategy for Soil protection. Volume II Erosion. European Commission Joint research Centre and European Environmental Agency. EUR 21319 EN/2: pp. 187–198

Elgersma A, Asheim V (1998) Landskapsregioner I Norge – Landskapsbeskrivelser. NIJOS Report 2/98

European Commission (2002) Communication from the Commission to the Council, the European Parliament, the Economic and Social Committee and the Committee of the Regions. Towards a Thematic Strategy for Soil protection. COM:179

Imeson A, Arnoldussen A, De la Rosa D, Montanarella L, Dorren L, Curfs M, Arnalds O, Van Asselen S (2006) Soil Conservation and Protection in Europe. The way ahead.

Kollerud J (2005) The reduction of soil erosion in Norway. Erosion in Norway – problem and goals, strategies, instruments, results. In: Van Asselen S (ed), 2005: SCAPE Soil Conservation and Protection for Europe. Briefing Papers of the 4th SCAPE Workshop in Ås (No), 9–11 May 2005:pp. 49–55

Øygarden L, Lundekvam H, Arnoldussen AH, Børresen T (2006) Norway. In: Boardman J, Poesen J (eds), 2006: Soil erosion in Europe: pp. 3–17

Puschmann O (2005) Nasjonal referansesystem for landskap. NIJOS Report 10/2005

Ramberg IB, Bryhni I, Nøttvedt A (2006) Landet blir til. Norges Geologi. Norges Geologisk Forening, Trondheim

Snellingen Bye A, Mork K, Sandmo T, Tornsjø B (2000) Result control agriculture 2000. Statistics Norway Report 2000/20. Kongsvinger, Oslo:82 pp (in Norwegian)

Solbakken E, Nyborg Å, Sperstad R, Fadnes F, Klakegg O (2006) Jordsmonnatlas for Norge. Beskrivelse av jordsmonn på dyrka mark i Vestfold. Viten fra Skog og Landskap 01/2006: pp. 168

Anthroscapes and Anthropogenic Soils in North-Western Sardinia: The Soils with Calcrete Horizon in the Alghero Area (Italy)

S. Madrau and C. Zucca

Abstract We describe the soils and landscapes studied in the area of the *Sella e Mosca* vineyard farm (Alghero, north-western Sardinia). In the study area the original soils and land use have been deeply modified, creating a new landscape and anthropogenic soils.

In recent years a wide area of the farms has been subjected to heavy reclamation works, especially through the removal of a shallow and thick petrocalcic horizon. The purpose of the intervention was the reduction of the negative effects of calcium carbonate on vine physiology and the increase of rooting depth and available soil moisture. The traditional rainfed olive and cereals crops were extensively substituted by vineyards.

Keywords Anthropogenic soils · Vineyard · Petrocalcic horizon · Calcrete, Sardinia, Italy

1 Introduction

1.1 Anthrosols and Anthroscapes

Soils whose principal characteristics are derived by human activities are defined as "anthropogenic". An Anthroscape can be defined as a landscape deeply modified by human activities and in particular showing marked differences from the natural landscapes (Kapur et al. 2004). The WRB Anthrosols Reference Soil Group

S. Madrau (✉) and C. Zucca
DIT (Department of Territorial Engineering), University of Sassari, V.le Italia 39, 07100 Sassari, Italy
and
Nucleo Ricerca Desertificazione, University of Sassari, V.le Italia 57, 07100 Sassari, Italy
e-mail: madrau@uniss.it

S. Kapur et al. (eds.), *Sustainable Land Management*,
DOI 10.1007/978-3-642-14782-1_7,

comprises "soils that have been modified profoundly through human activities, such as addition of organic materials, house hold wastes, irrigation and cultivation" (IUSS 2007). According to the WRB, the human influence is normally restricted to the surface horizons, and the horizons differentiation of a buried soil may still be intact at some depth.

Although the changes occurred in Anthroscapes (e.g., vegetation cover, land use, soil erosion rates, local morphology and topography, etc.) may have strongly affected soils. An anthroscape does not imply the presence of anthropogenic soils. However, in some cases human activities actively and purposely disturb or completely delete the original arrangement of soil horizons and features. In these situations, not only anthropogenic soils and anthroscapes coexist, but the properties of the rearranged soils may become the characterizing aspect of the reshaped landscape. This is the case, of some land reclamation sites.

Here we describe the soils and landscapes observed in the area of *Sella e Mosca* vineyard farm, Alghero's municipal area, north-western Sardinia, Italy. The study area is an interesting case in which the original soils and land use have been deeply modified; creating a new, man made landscape characterized by fully anthropogenic soils. In the last years a wide area of this vineyard farm has been actually subjected to heavy reclamation works, especially based on the removal of a shallow and sometimes very thick petrocalcic horizon, as explained below. The soils of the area had developed on old alluvial and aeolian deposits formations. Their most characterizing pedogenetic feature is the presence of strongly cemented calcic horizons, result of ancient, long lasting processes as briefly discussed in the following paragraph.

The aim of this study was to describe the changes made to the soils and landscapes by the reclamation interventions carried out in the study area, by comparing them with surrounding areas, still not subjected to such kind of transformation.

1.2 Calcrete Pedogenesis and Calcium Carbonate Accumulation in Soils

The working definition of calcrete is "a pedological altered limy material that contains more than 40% calcium carbonate equivalent and has sufficient accumulation of pedogenic carbonates to meet the minimum requirement of a calcic or petrocalcic horizon" (Wilking et al. 2001).

This material may be either cemented (Bkm horizon) or not (Bk horizon). It can have variable texture, thickness and origin, for example from aeolian sands or alluvial deposits characterized by a significant accumulation of pedogenic carbonate (more than 5% in volume). This definition excludes all non weathered materials, i.e. marl, chalk and hard or soft limestone.

The pedogenesis of calcrete requires prolonged subaerial exposition and the interaction of geomorphic, biogenetic and pedogenetic processes (Stokes et al. 2006).

Yaalon (1971) defines the calcic horizon as *a relatively persistent, slowly adjusting soil feature* and the petrocalcic horizon as *a feature produced by essentially irreversible self-terminating processes.*

The carbonate supplies in soils are strongly related to the internal water movements, either vertical (in relationship with groundwater) or lateral. So, the pedogenesis of Bk and Bkm horizons is related to the soil water flow and to evapotranspiration processes. The amount of calcium carbonate accumulation in soils is also a function of the bicarbonate concentration in the soil water. The origin of this may be:

- *From karst processes*; the bicarbonate is moving in soil through intrapedon or extrapedon illuvial transport (Sommer and Schlichting 1997).
- *From the hydrolysis of calcareous parent sediments and fine to very fine calcareous aeolian dusts*; the movement is in this case normally due to illuvial intrapedon transport.

The general ubiquity of calcareous Saharan dust depositions in Mediterranean areas (Kubilay et al. 1997) suggests that in these areas calcium carbonate accumulated in the Bk and Bkm horizons mainly derives from dust falls. Actually, it seems that the hydrolysis and illuvial intrapedon transport of aeolian fine dust material coming from North Africa had a particular influence on the genesis and evolution of calcic and petrocalcic horizons in Pleistocene aeolian sand deposits widespread in North-western Sardinia.

The accumulation of calcium carbonate in soils is related to: (1) the bicarbonate concentrations in soil water; (2) the duration of the process, and (3) the granulometry of parent materials. As a function of these three factors, different evolution phases can be identified.

In non gravelly deposits, commonly calcareous throughout but containing little or macroscopic carbonate, Gile et al. (1966) propose four progressive development stages.

- Stage I: *few filaments or faint coatings,* on sand grains on in the pores.
- Stage II: *few to common nodules*, commonly oriented in a diagonal or vertical mode. Their consistence ranges from slightly to extremely hard.
- Stage III: *many nodules and internodular fillings*. At this stage the entire horizon is impregnated (Bk horizon). In the last phase of this stage gravel or coarse sand grains are separated by carbonate and most pores are plugged. The horizon may be slightly to strongly cemented and can be indurated (Bkm horizon).
- Stage IV: *laminar horizon overlying plugged horizon.* An indurated laminar horizon of nearly pure carbonate or an indurated carbonate gravelly material are found at the top of calcic, petrocalcic or C horizons.

Machette (1985) proposed a classification of calcium carbonate accumulation processes according to six progressive stages based on carbonate content in the fine earth fraction, on nodules dimension and degree of coalescens and on cementation and induration stages.

2 The Study Area and its Land Use

The study area is located in Alghero's municipal land (north-western Sardinia), in the "Nurra" area (Fig. 1). The climate, strongly controlled by the closness to the sea and is typically Mediterranean. Hot temperatures in summer, with a maximum of 23.9°C in August,[1] associated with a minimum of mean monthly rainfall, 3 mm in July (Madrau et al. 2007).

The maximum of rainfall is in winter, during November and December and this corresponds to 33% of annual precipitations, associated with the minimum of mean monthly temperature of 9.6°C in January. The mean annual precipitation is 647 mm

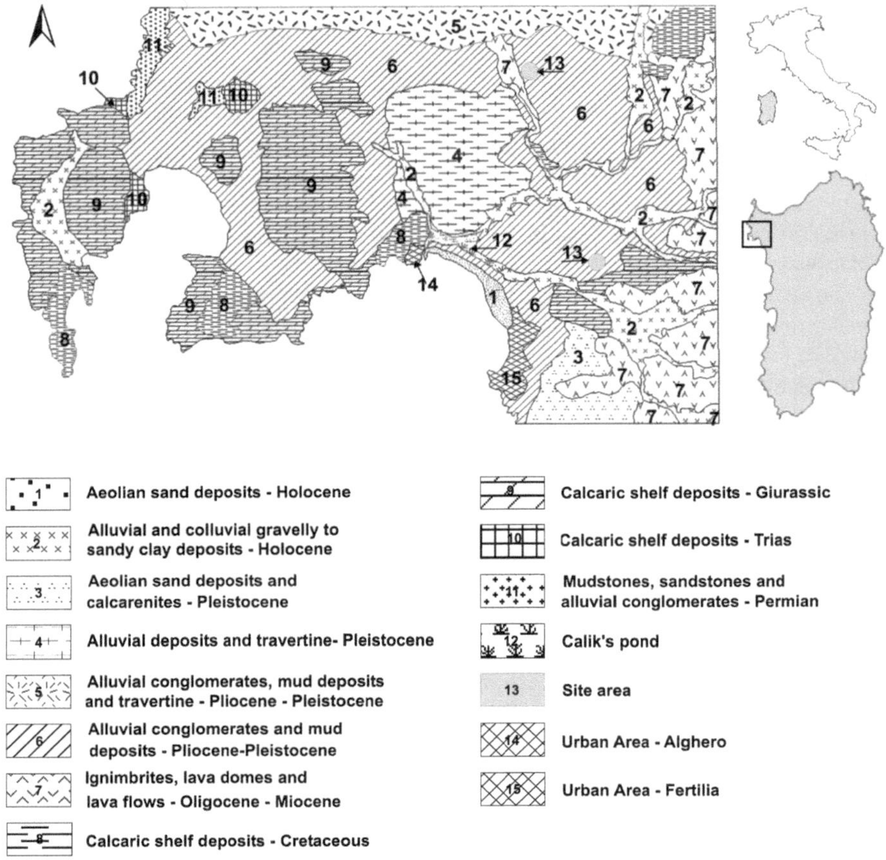

Fig. 1 The simplified geology of Nurra region [from Servizio Geologico d'Italia (1961); modified]

[1]Alghero' Airport weather data, 1951–1985 years.

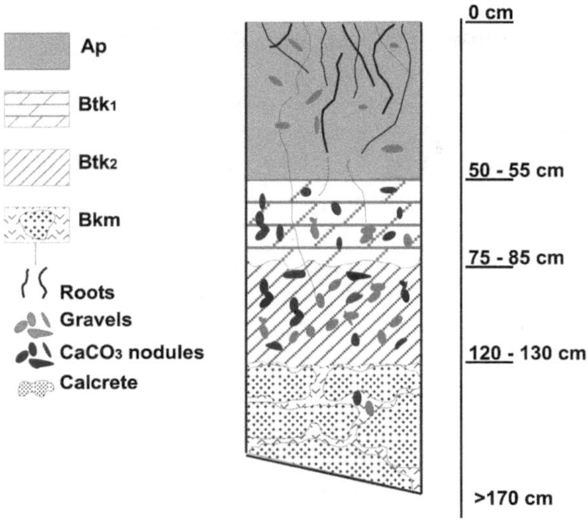

Fig. 2 Soil profile 1, on Pleistocene aeolian sand deposits. Mamuntanas Farm, Alghero (Italy)

and 16.4°C is the mean temperature. The soil moisture and temperature regimes are xeric and thermic.

The geological features of the area are described by the Geological Map of Italy at the 1:100,000 scale (S.G.I., 1961), *Alghero* and *Porto Torres* sheets (Fig. 2). The Nurra is a wide plain, mainly characterized by a sequence of Plio-Pleistocene fine to coarse gravelly alluvial deposits (Unit 6), interrupted by low Mesozoic limestone hills (Unit 9) (Fig. 2).

Plio-Pleistocene alluvial deposits are the dominant geological formation. They are made up of very thick sequences of buried soil horizons and layers constituted by quartz, metamorphic and sandstone gravels and boulders. These coarse layers are slightly to very cemented by silica, calcium carbonate or both. Miocene to Pleistocene soils are part of the sequence, characterized by calcium carbonate, silica or iron accumulation horizons (Baldaccini et al. 2002; Madrau 2004).

The south-eastern part of these alluvial deposits, between the sea, the Jurassic limestone hills and the Cenozoic effusive formations (Unit 7), is characterized by wide and thick Pleistocene aeolian deposits, cemented by clays and carbonates (Unit 4) (Fig. 2). Locally these sands cover Permian sandstone, Mesozoic limestone and Cenozoic lava flows.

Two wide areas with calcrete horizons can be observed in Alghero's municipality land: the Pleistocene aeolian sand deposits and the Plio-Pleistocene coarse to fine alluvial deposits. In the former land unit the depth of calcrete horizons is related to the palaeo-morphology, which strongly influenced deposition processes as well as present and past erosion processes.

In the study area, in addition to the processes described above and to the contribution due to Saharan dust depositions, decarbonation processes and subsequent lateral illuviation have also played a major role on the genesis of the Bk and Bkm horizons. They acted especially on the colluvial deposits at the feet of Mesozoic limestone hills, and on the flat to gently undulated Plio-Pleistocene alluvial deposits.

The Btk horizons observed in the study area, on aeolian sands and non gravelly Pleistocene alluvial deposits, may be classified as II or III stages as defined by Gile in relation to nodules dimension, while the Bkm horizons can be referred to the IV stage. According to Machette, if the calcic horizons have a carbonate content of above 4% and soft nodules within a range of 5–40 mm in diameter, they correspond to the second stage. The Bkm horizons, strongly cemented, indurated, thick and with no gravel contents belong to stage 6. The presence of deep or shallow Bkm horizons strongly affects the internal drainage, slowing down the deep infiltration of rain and irrigation water.

The prolonged period of soil moisture conditions during spring has favoured the development of traditional olive groves in these areas, now extended over about 2,300 ha. Depending on the local morphology, in the presence of shallow Btk and Bkm horizons, brief ponding periods are possible in winter.

The olive plants were established by means of a regular square breaking up, a traditional scheme allowing for a significant reduction in labour time. Even in the presence of shallow soils, this created favourable conditions for root growth and for increasing water reserves. This traditional approach also avoided the removal of the shallow calcrete horizons and soils conserved a high degree of natural condition.

The dug out calcrete material was often used as building material. After all, Alghero's historical town centre and his defensive walls were almost completely built up by means of aeolian sandstones bricks obtained from cemented layers of Pleistocene dunes.

2.1 *The Reclamation Interventions*

Founded in 1899, the *Sella e Mosca* company has 650 ha of total estate, 550 of which are under vine. Its wines are exclusively produced from the farm grapes. The canopy management, based on Tellis system and Guyot training, allows for high quality grapes and for a wide mechanization of cultural practices and harvesting.

For both the newly established vineyards, and for the old ones that reached the end of their productive cycle, the company, mostly starting from the last decade, has executed heavy works with the aim of creating a 100–120 cm deep continuous layer with more favorable edaphic conditions for the new productive cycle. In some years, all the farm area is expected to be subjected to such kind of intervention. As mentioned above, the intervention implies, if necessary, the removal of shallow calcrete horizons or ignimbrite layers. Most of the big blocks extracted so far are now piled up in the form of a huge wall, hundreds of meters in length (Fig. 3). The farm area can thus be considered as an important example of Anthrosols development process under irrigated tree crops in Mediterranean climate conditions.

Fig. 3 Blocks of calcrete horizons removed in Sella e Mosca's vineyard farm and piled up in the form of a "cyclopic" wall

Data allowing for a comparison of the productive levels before and after the intervention are not available so far. Actually the company appears to be not really interested to the quantitative aspect of production, trying to preserve a high quality level independently from grapes production.

3 The Original Soils and the Anthropogenic Soils

The soil profile 1 described below is representative for the soils developed on Pleistocene sand deposits. The profile is located in the Mamuntanas farm, property of the Regional Government of Sardinia. In the farm land, approximately 100 ha wide, the depth of calcrete horizon's top ranges from 20 to 60 cm or more. This has strongly conditioned the agricultural land use: typically, olive groves were grown in the shallow soils, cereals and vineyard in the deepest ones.

Profile 1	
Location	Mamuntanas Farm, Alghero (Italy)
Site coordinates	UTM32T 445043E–4495233N (ED50)
Map Sheet	459 IV Olmedo (IGMI 1:25,000)
Elevation	13 m a.s.l.

(*continued*)

Profile 1	
Morphology	Flat to gently undulating alluvial terraces. Slope: <2%
Parent material	Lower Pleistocene, fine to medium alluvial deposits weakly cemented by calcium carbonate
Rock outcrops	Absent
Surface stones	Very few boulders
External drainage	Good
Erosion	Active, very slight sheet water erosion
Land uses	Olive crops and sheep grazing

Ap	0 to 50–55 cm. Color (moist) 7.5YR 3/2 (dark brown). Common, very fine, rounded, weathered coarse fragments. Fine to medium, firm, not sticky and not plastic, strong subangular blocky. Common, fine pores. Non calcareous. Well drained. Few, fine roots. Common biological activity. Gradual and linear boundary.
Btk_1	50–55 to 75–85 cm. Color (moist) 7.5YR 4/2 (dark brown to brown). Common, fine and very few medium, rounded, weathered coarse fragments. Medium to coarse, firm, sticky and slightly plastic and strong subangular to angular blocky. Few clay cutans. Common, soft to hard, irregular to rounded, white, fine and medium carbonate nodules. Few fine pores. Calcareous. Slow drainage. Very few, fine roots. Common to few biological activity. Gradual and linear boundary.
Btk_2	75–85 to 120–130 cm. Color (moist) 10YR 4/3 (dark brown to brown). Common, fine to coarse, rounded, weathered coarse fragments. Medium to coarse, firm, sticky and slightly plastic, strong angular blocky. Few clay cutans. Common, soft to hard, irregular to rounded, white, fine and medium carbonate nodules. Few fine pores. Strongly calcareous. Slow drainage. Very few, fine roots. Few biological activity. Gradual and linear boundary.
Bkm	120–130 to 170 cm and more. Color (moist) 5YR 4/2 (dark reddish grey) Dominant, coarse, strongly weathered, rounded coarse fragments. Coarse to very coarse, massive blocky. Many, soft to hard, irregular to rounded, medium to coarse, white, carbonate nodules. Discontinuous, massive, strongly carbonate cemented. Calcareous. Roots and biological activity absent.
Soil Classification (Soil Taxonomy USDA, 2006): Petrocalcic Palexeralfs	

The pedological features of the Plio-Pleistocene alluvial deposits are very complex. This soil complexity is related to:

- Palaeo morphology and palaeo micro-topography; these have influenced, on one hand, the flooding frequency and the granulometry of depositions and, on the other hand, the ground water flow and the intrapedon and extrapedon transport processes of fine materials in solution and suspension.
- The presence of shallow layers of Mesozoic limestone or Oligocene ignimbrite which reduce, for example, the soil depth and generally increases its coarse gravel content.
- Human activities; during the 1940s and the 1950s of the twentieth century, relevant public investment was done in the Nurra area with the double purpose of land reclamation and reorganization of land tenure.

The impact of human activities on soil was determinant. The main aim of land reclamation works was to create favorable conditions for the introduction of a range of irrigated crops. So, in many sites the natural Mediterranean maquis cover was removed, as well as rock outcrops, large surface boulders and shallow calcrete horizons.

The subsequent deep tillage practice has reworked the soils disturbing the diagnostic horizons, whose fragments are now often mixed and arranged in a not discernible sequence. These soils may be classified as Alfic Xerarenths according to Soil Taxonomy and as Terric or Tecnic Anthrosols with reference to WRB Reference Soil Groups. In some of the Nurra irrigation district the complete removal of the calcrete horizon is still a practice carried out.

The profile 2 described below was opened outside the vineyard farm, in order to describe an almost undisturbed sequence of horizons developed on the Plio-Pleistocene aeolian deposits. The profile is located by the vineyard, at the boundary between a cereal field and the Anghelu Ruju archeological site.

The presence of three Bkm horizons in this soil profile suggests that the secondary carbonates accumulation processes are correlated to the change from aridic to humid climate conditions during the Pleistocene. These climatic changes have strongly influenced the micro-morphological formation processes,[2] and consequently, the intensity of intrapedon and extrapedon illuvial processes (Figs. 4 and 5).

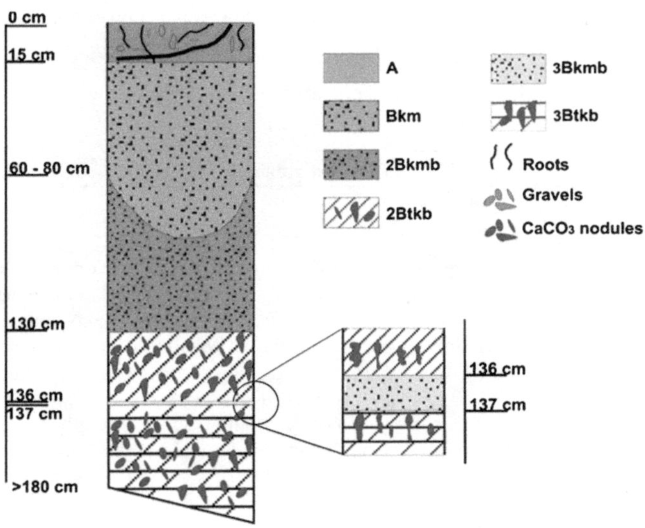

Fig. 4 Soil profile 2 on Plio-Pleistocene alluvial deposits. Anghelu Ruju archeological site, Alghero (Italy)

---

[2]Examples of buried palaeo fluvial beds can be observed in several sections of Plio-Pleistocenic aeolian or alluvial deposits in the Nurra region.

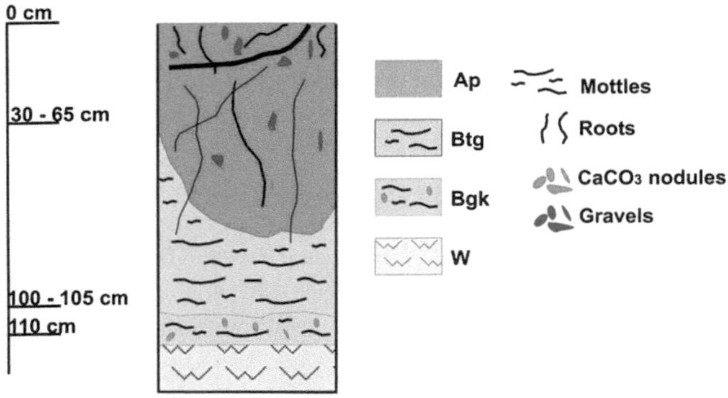

Fig. 5 Soil profile 3 on Plio-Pleistocene alluvial deposits. Sella and Mosca farm, Alghero (Italy)

Profile 2	
Location	Anghelu Ruju archeological site, Alghero (Italy)
Site coordinates	UTM32T 443075E–4498043N (ED50)
Map Sheet	458 II Fertilia (IGMI 1:25,000)
Elevation	20 m a.s.l.
Morphology	Flat to gently undulating alluvial terraces. Slope: <2%
Parent material	Pleistocene aeolian sand deposits strongly cemented by calcium carbonate over fine to medium Lower Pleistocene alluvial deposits
Rock outcrops	Absent
Surface stones	Very few boulders
External drainage	Good
Erosion	Absent
Land uses	Cereal crops and grazing

A	0–15 cm. Color (dry) 7.5YR 3/4 (strong brown). Common, very fine, rounded, weathered coarse fragments. Sandy clay loam. Medium, hard, not sticky and slightly plastic, moderate to strong subangular blocky. Common, fine pores. Calcareous. Well drained. Common fine to few coarse roots. Common biological activity. Abrupt and linear boundary.
Bkm	15 to 60–80 cm. Color (dry) 10YR 8/5 (very pale brown to yellow). Massive, very hard blocky. Medium sands, strongly cemented by calcium carbonate. Calcareous. Slow drainage. Pores, roots and biological activity absent. Abrupt, wavy to irregular boundary.
2Bkm	60–80 to 120 cm. Color (dry) 10YR 8/3 (very pale brown). Massive, very hard blocky. Fine sands, strongly cemented by calcium carbonate. Calcareous. Slow drainage. Pores, roots and biological activity absent. Abrupt, smooth boundary.
2Bkt	120–136 cm. Color (dry) 10YR 7/6 (yellow). Very few and very fine, coarse fragments. Medium, hard, strong subangular to massive blocky. Common, hard, very clear, elongate white carbonate nodules. Calcareous. Slow drainage. Pores, roots and biological activity absent. Abrupt, smooth boundary

(*continued*)

3Bkm	136–137 cm. Color (dry) 10YR 7/3 (very pale brown). Coarse fragments absent. Fine, moderate, slightly hard, subangular to massive blocky. Fine sands, platy, weakly cemented by carbonate. Calcareous. Slow drainage. Pores, roots and biological activity absent. Abrupt, smooth boundary.
3Btk	137–180 cm and more. Color (dry) 7.5 YR 5/6 (strong brown). Few, fine rounded, weathered coarse fragments. Sandy clay loam to clay loam. Many, distinct clay cutans on ped faces and voids. Medium, strong, hard angular blocky. Few, fine, hard, very clear, rounded white carbonate nodules. Calcareous. Slow drainage. Pores, roots and biological activity absent. Abrupt, smooth boundary.
Soil Classification (Soil Taxonomy USDA, 2006): Petrocalcic Palexeralfs (eroded phases)	

The third profile was opened in a vineyard plot, inside the farm. The plot was interested, because the removal of shallow Bkm horizon (about 25–30 cm thick) during the nineties, followed by the reconstruction of the horizons sequence and by mechanical surface leveling. The parent materials of this profile are limy and clayey Plio-Pleistocenic alluvial deposits.

The site is characterized by a relatively shallow groundwater table, which comes up close to the surface in winter, especially in a nearby depression. The genesis of the now removed petrocalcic horizon and the present *gley* horizons is related to the evaporation processes of freatic fringe waters. The Bkm horizon development and the observed increase with depth of the calcium carbonate in the Bg horizons[3] may be related to the variation in time of either the groundwater table level or the content of calcium bicarbonate in soil solution.

Profile 3	
Location	*Sella e Mosca* Farm, Alghero (Italy)
Site coordinates	UTM32T 443721E–4498804N (ED 50)
Map Sheet	458 IV Santa Maria La Palma (IGMI 1:25,000)
Elevation	20 m a.s.l.
Morphology	Flat to gently undulating alluvial terraces. Slope: <2%
Parent material	Lower Pleistocene, fine alluvial deposits weakly cemented by calcium carbonate
Rock outcrops	Absent
Surface stones	Absent
External drainage	Good
Erosion	Absent
Land uses	Vineyard

[3]The maximum calcium carbonate content, about 30%, was observed in the present evaporation fringe.

Ap	0 to 30–65 cm. Color (moist) 10YR 3/3 (dark brown). Few, fine, rounded, weathered coarse fragments. Medium, firm, sticky and plastic, strong subangular blocky. Common, fine pores. Calcareous. Well to moderately drained. Common, fine and medium roots. Common biological activity. Gradual and linear boundary.
Btg	30–65 to 100 cm. Color (moist) 10YR 5/6 (yellowish brown). Few, fine and very few medium, rounded, weathered coarse fragments. Medium to coarse, firm, sticky and plastic and strong subangular to angular blocky. Common clay cutans. Very few, soft irregular, white, fine carbonate nodules. Many, very fine and fine, distinct, yellowish brown mottles. Few fine pores. Slightly calcareous. Slow drainage. Very few, fine roots. Scarce biological activity. Gradual and linear boundary.
Bgk	100–110 cm. Color (moist) 10YR 6/6 (yellowish brown). Common, fine to coarse, rounded, weathered coarse fragments. Medium to coarse, firm, sticky and slightly plastic, strong angular blocky. Common clay cutans. Common soft, irregular to rounded, white, fine carbonate nodules. Common, fine, distinct, brownish yellow mottles. Few fine and very fine pores. Strongly calcareous. Very slow to extremely slow drainage. Roots and biological activity absents.
W	110 cm and more. Groundwater table.
Soil Classification (Soil Taxonomy USDA, 2006): Calcic (Aquic) Palexeralfs	

4 Conclusions

The soil reclamation interventions based on the removal of calcrete horizon were carried out extensively in the study area, in particular during the last decade. From the economic point of view, the outcome of the interventions is considered satisfactory by the farms involved.

The works described deeply modified the landscape, by making possible the gradual substitution of the traditional rainfed crops, especially olive groves, with vineyards and other irrigated crops. However, the strongest impacts were made on soils and on pedogenetic processes. First of all, the horizons sequence was changed, leading to the transformation of the natural Petrocalcic Palexeralfs into Anthrosols or Xerarents. Secondly, the deeper penetration of roots and water in soil completely changed the soil environment and started up a new line of pedogenetic processes. These are acting with different modes and intensities under different irrigated crops (e.g., the *Sella e Mosca* vineyards), or under rainfed conditions such as the olive groves and the short cycle crops (cereals, forage, pastures). The comparison between these different evolutionary lines could allow for the creation of a first framework for interpreting and describing the recent evolution of the newly created Anthrosols. The scientific interest of such a study, only preliminarily dealt with here, would be enhanced by the high natural value of the soils of the area. Actually, the calcrete horizons observed could have a basic role in the study of palaeo-climatic conditions under which they have evolved, contributing to the research on the evolution of the environmental conditions during the Quaternary.

Acknowledgments The authors thank Monidri project and Dr. Rosario Napoli for making available their data on the soils of the *Sella e Mosca* farm.

References

Baldaccini P, Previtali F, Madrau S, Deroma MA (2002) I suoli del bacino del rio d'Astimini – Fiume Santo (Sardegna nord-occidentale). Note e Carta alla scala 1.25000.

Gile LH, Peterson FF, Grossman RB (1966) Morphological and genetic sequence of carbonate accumulation in desert soils. Soil Science, 101(5): 347–360.

IUSS Working group WRB (2007) World reference base for soil resources 2006, first update. World Soil Resources Reports, 103, FAO, Rome.

Kapur S, Zdruli P, Akça E, Arnoldussen AH, Kapur B (2004) Anthroscapes of Turkey: Sites of Historic Sustainable Land Management (SLM). In: van Asselen S, Boix-Fayos C, Imeson A (Eds) 'Briefing papers' of the second SCAPE workshop in Cinque Terre (Italy), 13–15 April 2004.

Kubilay N, Saydam AC, Yemenicioğlu S, Kelling G, Kapur S, Karaman C, Akça E (1997) Seasonal chemical and mineralogical variability of atmospheric particles in the coastal region of the Northeastern Mediterranean. Catena, 28(3–4): 313–328.

Machette MN (1985) Caliche soil of the southwestern United States. In: Weide DL, Faber ML (Eds) Soils and quaternary geology of the southwestern United States. Geological Society of America Special Papers 203: 1–21.

Madrau S (2004) The soils as palaeoclimatic indicators. A sequence of paleosols in the Fiume Santo Area (Sassari, Sardinia). Preliminary observations. Geografia Fisica e Dinamica Quaternaria, 27: 47–53.

Madrau S, Deroma MA, Pittalis D, Putzu G, Zucca C (2007) Il modello di valutazione del rischio di desertificazione ESAs. Il caso della Sardegna nord-occidentale (comuni di Alghero, Portotorres, Sassari e Stintino). Nucleo Ricerca sulla Desertificazione, Dipartimento di Ingegneria del Territorio, Università di Sassari. Gallizzi ed., Sassari.

Servizio Geologico d'Italia (1961) Carta Geologica d'Italia. Foglio 179 Porto Torres. L.A.C., Firenze.

Sommer M, Schlichting E (1997) Archetypes of catenas in respect to matter. A concept for structuring and grouping catenas. Geoderma, 76: 1–33.

Stokes M, Nash DJ, Harvey AM (2006) Calcrete fossilization of alluvial fans in SE Spain: The roles of groundwater, pedogenic processes and fan dynamics in calcrete development. Geomorphology, 85(1–2): 83–64.

Wilking LP, Owen P, Woodruff CM, Jr (2001) Study of Caliche Soil as a filter medium of treatment and disposal of wastewater. Texas A&M University, College Station, TX; Consulting Geologist, Austin, Texas. www.towtrc.state.tx.us/common/studyofCalicheSoil.pdf.

Yaalon DH (1971) Soil-forming processes in time and space. In: Yalon DH (ed) Paleopedology. Origin, nature and dating of Paleosols. Israel University Press, Jerusalem, 29–40

Stimulated Soil Formation in a Degraded Anthroscape: A Case Study in Southeast Spain

J. M. Arocena, J. M. van Mourik, and Á. Faz Cano

Abstract Extensive extractions of natural resources such as metals led to degraded landscapes oftentimes inhospitable to plants and other organisms. Conversion of such unhealthy environments to productive ecosystem requires the formation of soils. This paper reports on our efforts to accelerate the development of soils in mine wastes deposits in southeast Spain. We investigated the structural, chemical and mineralogical properties of mine tailings. These materials have low organic C (<0.5%) and N (<0.05%), high acidity (pH as low as 2.0), high electrical conductivity (up to 20 dS m^{-1}) and high levels of metals such as Zn (up to 22,000 mg Zn kg^{-1}) and Pb (up to 7,000 mg Pb kg^{-1}). Our results showed that organic matter (OM) from sewage sludge and pig manure when added together with calcite from marble cuttings resulted to accumulations of OM as cappings on calcite particles. These OM caps hold enough water and nutrients to support the establishment of seedling in amended tailing materials. It is clear from micromorphological examinations that granular structure starts to form in the interface of OM caps and root zone of seedlings. Our results suggest that OM can be effectively enriched in mine wastes deposits through simultaneous additions of OM and calcite. Our findings might be useful to accelerate the establishment of functional ecosystem characterized by healthy soils with granular structure in degraded anthroscapes.

Keywords Microstructure · Organic matter caps · Soil development · Electrical resistivity

J.M. Arocena (✉)
Canada Research Chair in Soil and Environmental Sciences, University of Northern British Columbia, Prince George, BC V2N4Z9, Canada
e-mail: arocenaj@unbc.ca

J.M. van Mourik
IBED-Earth Surface Processes and Materials, Universiteit van Amsterdam, Nieuwe Achtergracht 166, 1018 WV Amsterdam, The Netherlands

Á. Faz Cano
Departamento de Ciencia y Technología Agraria, Universidad Politécnica de Cartagena, Paseo Alfonso XIII, 52, 30203 Cartagena, Murcia, Spain

S. Kapur et al. (eds.), *Sustainable Land Management*,
DOI 10.1007/978-3-642-14782-1_8,

1 Introduction

Degraded anthroscapes abound southeast Spain due to >2,500 years of extensive metal extractions in the region. The major ore minerals exploited in the area were sphalerite [(Zn, Fe)S], galena (PbS) and pyrite [pyrite ($Fe^{2+}S_2$), pyrrhotite ($Fe_{0.95}{}^{2+}S$), and marcasite ($Fe^{2+}S_2$)] (Oen et al. 1975). Mining activities came to an end in early 1990s leaving behind large volumes (i.e., usually 700,000–900,000 m^3 per deposit) of mine wastes stored in many tailing ponds (or balsa). These waste materials are quite inhospitable to plant growth because of low organic matter, high acidity and electrical conductivity, elevated contents of Pb, Zn and other metals. Furthermore, high evaporation rates during the summer months results in the seasonal formation of thick crusts (~2–3 cm) of salt efflorescence on the surface of tailing ponds (Carmona et al. 2009). Hammarstrom et al. (2005) reported that secondary sulfate minerals can incorporate minor and trace metals, such as Co and Cd, into their structure composed mainly of iron, Al, Mg, Cu, Mn, Zn, and sulfate.

Recently, the use of plants to stabilize mine tailing ponds is proven to be one of the best reclamation techniques available to date. Vegetation cover provides physical protection to prevent wind erosion and surface runoff (Norland and Veith 1995), and plants contribute soil organic matter (SOM). Soils in properly-functioning ecosystem have good quality SOM to supply essential nutrients such as N, P, S, K, Ca, Na, and Mg, increase soil water holding capacity and cation exchange capacity, and promote favorable aggregated soil structure (Stevenson 1994; Sollins et al. 2006). Soil aggregation is one of the best indicators for fertile soils because it bridges the physics and biochemistry of soil systems (Young and Crawford 2004).

The main objective of this paper was to highlight the results of our efforts to transform unproductive anthroscapes left behind by extensive metal mining activities in southeast Spain into functional and productive ecosystems. Specifically, we selectively report the chemical and mineralogical properties of the mine wastes and tested the utility of organic and calcium carbonate-amendments to initiate the formation of soils in these biologically-unhealthy substrates in degraded anthroscapes. Our paper demonstrates the utility of understanding the fundamental principles of soil formation to accelerate the establishment of sustainable ecosystems in degraded anthropogenic landscapes.

2 Materials and Methods

2.1 Study Area and Addition of Amendments

The Sierra de Cartagena study area is situated in the province of Murcia (southeast Spain) (Fig. 1). The climate is typical Mediterranean with average monthly temperature ranging from 9.3°C in January to 24.4°C in July and mean annual precipitation of 275 mm, mostly in autumn and spring. The potential evapo-transpiration reaches

Fig. 1 Relative location of mine tailing deposits in southeast Spain

900 mm year^{-1}. We selected several mine tailing ponds selected to understand the key limiting properties of the deposits towards the establishment of functional ecosystems.

Although we had sampled many tailing ponds in the study area, we will limit the results to La Union, Belleza, El Lirio, Brunita, Gorguel and Mazarron. La Union, Belleza, El Lirio, Brunita and Gorguel were formed from tailings discharged during metal extraction from Cu-, Pb- and Zn-containing sulfate minerals (Conesa et al. 2007a, b; Martinez-Pagan et al. 2009). Mazarron site is typical of low-elevated sites where salt efflorescence accumulates in periods of high evapo-transpiration rates (Carmona et al. 2009).

Selected sites at Brunita tailing ponds were subjected to additions of organic (i.e., pig manure and sewage sludge) and industrial waste (i.e., marble cuttings) to increase the pH and organic matter contents (Zanuzzi 2007). The plots were left alone in the field and were sampled for micromorphological analyses 3 years after the application of amendments.

2.2 *In Situ Characterization of Mine Tailings by Electrical Resistivity Measurement*

We used a Syscal R1 resistivity meter (IRIS Instruments 2001) with multiconductor cable designed for up to 72 electrodes. Each steel electrode was 30-cm long and connected to the resistivity meter through take out clips. In El Lirio, nine 2D ERI profiles were determined with eight profiles located across the pond and one profile in lengthwise orientation (Martinez-Pagan et al. 2009).

2.3 Chemical and Mineral Characterization

We determined the routine physical and chemical properties of the soil and salt samples. Soil samples were dried for at least 48 h and passed through a 2-mm sieve prior to chemical analyses. The total amounts of C and N were measured using an Elemental Vario EL CNS analyzer (Hanau, Germany) while pH was determined in 1:1 water/soil ratio (Peech 1965). Electrical conductivity (EC) of the soil was measured in soil solution extracted from a mixture of 1:5 water/soil ratio (Bower and Wilcox 1965). Total content of metals in soil and salt samples was determined from digests obtained using the nitric/perchloric acid digestion (Risser and Baker 1990). We used a UNICAM 969 atomic absorption spectrometer to measure Cd, Cu, Pb and Zn.

We used a Bruker D8 diffractometer with a General Area Detector Diffraction System (GADDS) to determine the mineral composition of salt efflorescence. X-ray diffraction of salt minerals was conducted on ground samples and randomly mounted on glass slide using mineral oil. The diffractometer was operated at 40 kV and 20 mA to generate a Co-Kα radiation collimated to 850 μm. Identification and semi-quantitative estimation of the minerals were achieved by comparison of the XRD pattern with the JCPDF Powder Diffraction Database 2 using EVA™ software (Socabim 2001). Morphology of salt minerals was observed using a Philips XLS 30 scanning electron microscope (SEM) equipped with EDAX™ energy dispersive system (EDS). The SEM was operated at 20 and 1.94 A filament current and the semi-quantitative ($\pm 5\%$) elemental composition of minerals was estimated using EDAX-4 software (EDAX 1995).

2.4 Micromorphological Analysis

Thin sections from undisturbed soil samples collected using Kubien boxes were prepared according to the technique described by Jongerius and Heintzberger (1975). Thin sections (5×7.5 cm) were described under an Olympus BH-2 polarizing microscope following the the concepts and terminology in Brewer and Pawluk (1975) for microfabric, Brewer (1976) for plasma fabric, types of void and pedofeatures Bullock et al. (1985) for microstructure.

3 Results

3.1 In Situ Characterization of Materials in Tailing Deposits

Electrical resistivity (ER) had inverse relationship with EC (Fig. 2). ER values generally increased with depth from surface to the bedrock, while very low to intermediate values were found to be associated with mine tailings. Mine wastes

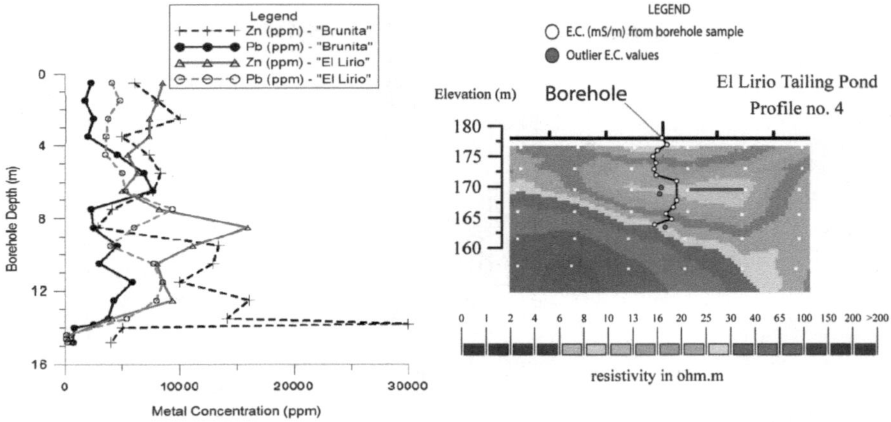

Fig. 2 Variations in electrical resistivity (ER), electrical conductivity (EC) and total Pb and Zn with depths at mine tailing deposits in El Lirio and Brunita sites [modified from Martínez-Pagán et al. (2009)]

stored in El Lirio tailing pond had ER <8 Ωm and were much lower than the ER for bedrock (>150 Ωm). Mine tailings with ER <8 Ωm occupied the center of tailing ponds with a thickness of ~20 m and were composed of fine-textured (<0.4 mm in ø) materials. Materials in the ponds with intermediate ER values (8–150 Ωm) were located at the perimeter with coarse (0.1–1.0 mm in ø) particle size.

3.2 *Chemical and Salt Mineral Characterization*

Selected chemical properties of mine tailings materials show very low pH, high EC and metals (Table 1). The lowest pH (2.3) was measured in Brunita while La Union tailings had the highest pH at 7.6. Belleza tailings had the highest EC at 18 dS m^{-1}. Both C and N contents were low in all tailings and ranges from 0.3 to 0.5 and nil-1.02%, respectively. Tailing materials had higher Zinc and Pb contents compared to Cd and Cu. La Union tailings had the highest observed Zn contents at 22,000 mg Zn kg^{-1} tailings while Belleza materials had the most Pb (7,000 mg Pb kg^{-1} tailings). Martínez-Pagán et al. (2009) reported that the contents of Pb and Zn varied with depth at El Lirio and Brunita tailing ponds (Fig. 2).

The minerals in salt efflorescence in Mazarron were dominated by well-crystalline hydrated sulfate minerals of Al, Fe, Mg, Mn, and Zn (or halotrichites) (Carmona et al. 2009). For example, euhedral crystals of copiapite had tabular {010} habits of ~3 μm long and <0.5 μm thick and stacked in layers consisting of 5–10 plates (Fig. 3). These sulfate salt minerals contained metals such as Cd, Zn and Cu (Carmona et al. 2009).

Table 1 pH, EC (dS m^{-1}), total organic C, N (%), Pb, Zn, Cu and Cd (mg kg^{-1}) contents in selected mine tailing deposits in southeast Spain

	pH	EC	OC	N	Pb	Zn	Cu	Cd	Reference(s)
La Union	2.9–7.6	3–8	–	–	3,000–6,400	9,000–22,000	110–240	–	Conesa et al. (2007a)
Belleza	3.0	18	0.5	0.04	7,000	5,400	380	–	Conesa et al. (2007b)
El Gorguel	7.2	3–8	0.13–0.4	0.02–1.02	5,200	9,100	84	–	Conesa et al. (2007a), Ottenhof et al. (2007)
El Lirio	5.7–6.4	4–14	0.4	0.00–0.01	3,100–4,900	6,800–12,800	90–120	20–40	Conesa et al. (2007a), Zanuzzi (2007)
Brunita	2.3–3.0	2.0–20	0.3	0.00–0.03	670–1,800	980–2,000	60–200	1–4	Conesa et al. (2007a), Zanuzzi (2007)

Fig. 3 Copiapite ($Fe_{0.69}Fe_4(SO_4)_6(OH)_2(H_2O)_{20}$) crystals observed in salt efflorescence collected from mine tailing pond in Mazarron, southeast Spain

3.3 Micromorphology of Amended Mine Tailing Deposits

Tailing materials in mine waste deposits were dominated by laminar type of microstructure (Fig. 4a). Addition of organic amendments in combination with calcite from marble cuttings resulted to the accumulation of organic matter cappings on primary calcite particles (Fig. 4b).

We observed that natural seedling establishments were primarily taking place on organic matter cappings. Three years after treatment, initial development of porous granular soil microstructure was observed in places where seedlings had established (Fig. 4c, d).

4 Discussion

4.1 Non-destructive Characterization of Mine Tailing Deposits

Our data suggest that electromagnetic techniques such as electrical resistivity (ER) imaging (ERI) when combined with soil chemical analyses can be used to determine the structural and chemical composition of mine tailing ponds to assess efficient measures of environmental protection. The inverse relationship between EC and ER are proven useful to map the spatial changes in the concentrations of Pb and Zn in El Lirio and Brunita sites. Generally, materials with ER $<8\ \Omega m$ and $>150\ \Omega m$ represent mine tailings and bed rock, respectively (Martínez-Pagán et al. 2009). The ERI can also be used to estimate the volume of mine tailings stored in a deposit. Our results are consistent with Chouteau et al. (2005) who used ERI techniques to image the internal structure of waste rock pile in mining areas.

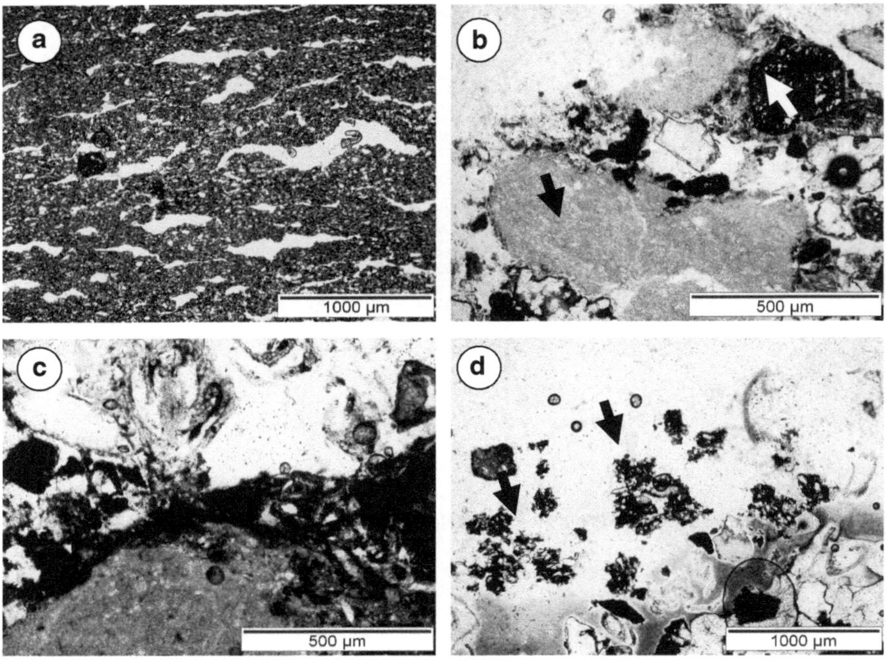

Fig. 4 Micrographs of mine tailings at various stages after organic (i.e., sewage sludge and pig manure) and industrial (i.e., marble wastes) amendments. (**a**) sedimentary microstructure of unamended tailings, (**b**) capping of organic materials (*white arrow*) on primary calcite (*black arrow*) from marble wastes amendment, (**c**) seedling establishment on organic matter caps, and (**d**) granular structure (*black arrows*) at the interface of organic matter caps and seedling root system

4.2 Plant Establishment in Mine Tailings

The anthroscapes left behind by more than 2,500 years of metal extractions in southeast Spain were characterized by materials that are inhospitable to plants and organisms. The very low organic matter contents (<0.5 %) is not sufficient to hold water and essential nutrients to support vigorous plant growth. A typical productive prairie soils in north America may contain 5–6% soil organic matter in 0–15 cm surface layer (Stevenson 1994). The high acidity from the oxidation of pyritic materials mobilized metals such as Pb^{2+} and Zn^{2+} from ores (e.g., Heikkinen and Räisänen 2007) and caused high concentrations of these metals in the soil. Lead and Zn contents in mine wastes exceed the environmental regulations in Spain and other European countries (Conesa et al. 2007a, b; García et al. 2005). Zinc contents of the soils are more than the 300 mg Zn kg^{-1} soil threshold reported by Kabata-Pendias and Pendias (1992).

The low pH in many of these tailings immobilizes metals up to toxic levels and/or renders nutrients unavailable to plants and organisms. For example, phosphorus becomes unavailable at low pH (<5.0) due to the formation of stable

Al-phosphates (McBride 1994). Electrical conductivity in these tailing materials are considered high because EC as low as 2.0 mS cm^{-1} has been reported to impair growth of sensitive plants through reduction in water potential, ion imbalance or disturbances in ion homeostasis and toxicity (US Salinity Laboratory Staff 1954; Parida and Das 2005).

In one of our earlier reports, we calculated that salt harvesting (or removal) of sulfate minerals during the summer months can extract up to 20 times more Zn than the most efficient plant species tested in phytoextraction of Zn in mined soils (Carmona et al. 2009). Removal of these metals through salt removal may alleviate the inhospitable conditions and may allow plants to establish in these mine waste deposits.

4.3 Simultaneous Additions of OM and Calcite Stimulate Soil Formation in Anthroscapes

Our results showed that sewage sludge or pig manure when added at same time with calcite may promote soil organic matter (SOM) accumulations in mine waste deposits. The build-up of SOM resulted to stable SOM-calcite complex in SOM cappings on calcite particles. These SOM cappings provide water and nutrient to support initial seedling establishment in mine tailings. The affinity of SOM to calcite is well established in the literature. For example, oxalic acid precipitates as calcium oxalate in calcareous soils (Ström et al. 2001) while lime addition increases hydrophobic and carboxylic groups in organic matter (e.g., Karlik 1995; Andersson et al. 2000) and precipitates SOM due to flocculation or adsorption by cation bridging in a high Ca^{2+} environment (Römkens and Dolfing 1998). This process is similar to the accumulation of pedogenic calcite in Canada (Landi et al. 2003). These SOM accumulations are conducive for seedling establishment due to increase in cation exchange and water holding capacities.

The initial accumulation of SOM in these materials may represent the early stage of soil formation. Our results show granular structure formation on the interface between SOM caps and root zone of seedlings. With time, we expect that soil organisms such as spring tails, millipedes and earthworms will colonize these mine wastes and lead to the formation of healthy soils. It is known that soil organisms, such as earthworm are the principal agents in the formation of granular microstructure in soils, i.e., the granular shape originating from the fecal materials (e.g., Pawluk 1987; Van Mourik 2003).

5 Conclusions

Anthroscape from many years of mining activities in southeast Spain resulted to degraded and inhospitable landscapes to plants and organisms. Mine wastes materials have low organic matter and pH, high EC and elevated levels of metals. Salt harvesting (or physical removal) can alleviate both high salinity and metal contents.

We suggest that environmental management of these anthroscapes should focus on techniques to stimulate the accumulation of organic matter. The build-up of soil organic matter in these materials represents the early stage of soil formation. Once SOM build-up is initiated, granular structure will form leading to the development of healthy soil where cation exchange, water holding capacities, porosity, pH and nutrient contents are sufficient to support growth of plants and organisms.

For efficient retention of organic matter, our results suggest that organic-rich materials must be added simultaneously with calcite to promote the formation of Ca-SOM complexes.

Acknowledgements We acknowledge the Fundación Séneca of Comunidad Autónoma Regiőn de Murcia (Spain) and Natural Sciences and Engineering Research Council and the Canada Research Chair program (Canada) for the financial support. Words of appreciation are also due to past and present GARSA graduate students and research assistants at the Polytechique University of Cartagena (Spain).

References

Andersson S, Nilsson SI, Saetre P (2000) Leaching of dissolved organic carbon (DOC) and dissolved organic nitrogen (DON) in mor humus as affected by temperature and pH. Soil Biol. Biochem. 32:1–10

Bower C, Wilcox L (1965) Soluble salts. In: C.A. Black (Ed.). Methods of Soil Analysis. Soil Science Society of America, Madison, WI, p. 933–951

Brewer R (1976) Fabric and mineral analysis of soils. 2nd ed. R.E. Krieger Publishing Co., Huntington

Brewer R, Pawluk S (1975) Investigation of some soils developed in hummocks of the Canadian sub-Arctic and southern Arctic regions. I. Morphology and micromorphology. Can. J. Soil Sci. 55:301–319

Bullock P, Fedoroff N, Jongerius N, Stoops G, Tursina T (1985) Handbook for soil thin section descriptions. Waine Research Publications, UK

Carmona DM, Faz Cano Á, Arocena JM (2009) Cadmium, copper, lead, and zinc in secondary sulfate minerals in soils of mined areas in southeast Spain. Geoderma. 150:150–157

Chouteau M, Poisson J, Aubertin M, Campos D (2005) Internal structure and preferential flow of a waste rock pile from geophysical surveys. In: Proceedings of the Symposium on the Application of Geophysics to Engineering and Environmental Problems, Atlanta, Georgia, p. 289–297

Conesa HM, Faz Cano A, Arnaldos R (2007a) Heavy metal accumulation and tolerance in plants from mine tailings of the semiarid Cartagena-La Union mining district (SE Spain). Sci. Total Environ. 366:1–11

Conesa HM, Faz Cano A, Arnaldos R (2007b) Initial studies for the phytostabilization of a mine tailing from the Cartagena-La Union Mining District (SE Spain). Chemosphere. 66:38–44

EDAX (1995) DX-4 Users Manual: Revision 5. EDAX International, Mahwah

García G, Zanuzzi AL, Faz Á (2005) Evaluation of heavy metal availability prior to an in situ soil phytoremediation program. Biodegradation. 16:187–194

Hammarstrom JM, Seal II RR, Meier AL, Kornfeld JM (2005) Secondary sulfate minerals associated with acid drainage in the eastern US: recycling of metals and acidity in surficial environments. Chem. Geol. 215:407–431

Heikkinen PM, Räisänen ML (2007) Mineralogical and geochemical alteration of Hitura sulphide mine tailings with emphasis on nickel mobility and retention. J. Geochem. Explor. doi:10.1016/j.gexplo.2007.09.001

IRIS Instruments (2001) Syscal switch v. 11.4: User's manual, Orleans. France. 66 pp

Jongerius A, Heintzberger G (1975) Methods in soil micromorphology. Soil Survey Papers 10, Wageningen, 48 pp

Kabata-Pendias A, Pendias H (1992) Trace elements in soils and plants. 2nd ed. CRC Press, Boca Raton

Karlik B (1995) Liming effect on dissolved organic matter leaching. Water Air Soil Pollut. 85: 949–954

Landi A, Mermut AR, Anderson DW (2003) Origin and rate of pedogenic carbonate accumulation in Saskatchewan soils, Canada. Geoderma. 117:143–156

Martínez-Pagán P, Faz Cano A, Aracil E, Arocena JM (2009) Electrical resistivity imaging (ERI) revealed the spatial chemical properties of mine tailings ponds in the Sierra Minera (SE Spain). J. Environ. Eng. Geophys. 14:63–76

McBride MB (1994) Environmental chemistry of soils. Oxford University Press, New York

Norland MR, Veith DL (1995) Revegetation of coarse taconite iron ore tailing using municipal waste compost. J. Hazard Mater. 41:123–134

Oen IS, Fernandez JC, Manteca JI (1975) The lead-zinc and associated ores of La Union, Sierra de Cartagena, Spain. Econ. Geol. 70:1259–1278

Ottenhof CJM, Faz Cano Á, Arocena JM, Nierop GJ, Verstraten JM, Van Mourik KJM (2007) Soil organic matter from pioneer species and its implications to phytostabilization of mined sites in the Sierra de Cartagena (Spain), Chemosphere. 69:1341–1350

Parida AK, Das AB (2005) Salt tolerance and salinity effects on plants: a review Ecotox. Environ. Safety. 60:324–349

Pawluk S (1987) Faunal micromorphological features in moder humus of some Western Canadian soils. Geoderma. 40:3–16

Peech M (1965) Hydrogen-ion activity. In: C.A. Black (Ed.). Methods for Soil Analysis. American Society of Agronomy, Madison, WI, p. 914–916

Risser J, Baker D (1990) Testing soils for toxic metals. In: R. Westerman (Ed.). Soil Testing and Plant Analysis. 3rd ed. Soil Sci. Soc. Amer. Spec. Publ. 3, Madison, pp 275–298

Römkens PFAM, Dolfing J (1998) Effect of Ca on the solubility and molecular size distribution of DOC and Cu binding in soil solution samples. Environ. Sci. Technol. 32:363–369

Socabim (2001) EVA application version 7

Sollins P, Swanston C, Kleberc M, Filley T, Kramere M, Crowa S, Caldwella BA, Lajthaa K, Bowden R (2006) Organic C and N stabilization in a forest soil: evidence from sequential density fractionation. Soil Biol. Biochem. 38:3313–3324

Stevenson FJ (1994) Humus chemistry: genesis, composition, reactions. 2nd ed. Wiley, New York

Ström L, Owen AG, Godbold DL, Jones DL (2001) Organic acid behaviour in a calcareous soil: sorption reactions and biodegradation rates. Soil Biol. Biochem. 33:2125–2133

United States Soil Salinity Laboratory Staff (1954) Diagnosis and improvement of saline and alkali soils. USDA Handbook No. 60, United States Department of Agriculture, Washington, DC, USA

Van Mourik JM (2003) Life cycle of pollen grains in mormoder humus forms of young acid forest soils: a micromorphological approach. Catena. 54:651–663

Young IM, Crawford JW (2004) Interactions and self-organization in the soil-microbe complex. Science. 304:1634–1637

Zanuzzi A (2007) Reducción de impactos ambientales asociados a suelos contaminados por metales pesados mediante el uso de residuos antropogénicos. PhD Thesis, Technical University of Cartagena, Spain. 468 pp

Asian Anthroscapes: China and Taiwan

Zeng-Yei Hseu, Chen-Chi Tsai, Heng Tsai, Zueng-Sang Chen, and Hari Eswaran

Abstract Anthroscapes are important in the assessment of human-induced land degradation. In case of soils affected by land use changes in large-scale farming, the anthropogenic processes have been created in China for over a few thousands years. The Chinese started to use soil for crop production as early as in the Stone Age. Soil properties in China and Taiwan have been greatly influenced by various anthropogenic processes to form anthroscapes. Paddy soils or rice-growing soils are more than 30 million hectares, distributed in China and Taiwan, which are one of the world's most important soil resources for food production. According to the Chinese Soil Taxonomy [ISS/CAS (Institute of Soil Science, Chinese Academy of Sciences) (ed.) (2001) Chinese Soil Taxonomy. Science Press, Beijing, 203p], Anthrosols meet the requirements of the combination of horizons, including anthrostagnic, fimic, siltigic and cumulic epipedons and the hydragic horizon. The changes of soil characteristics after planting rice are degradation of soil organic matter (SOM), redistribution of exchangeable bases, translocation and segregation of iron (Fe) and manganese (Mn) by the cycling of reduction and oxidation processes, as well as decomposition and synthesis of clay minerals. The genetic horizons of paddy soils in China and Taiwan include a cultivated gray horizon, a plowpan, a percogenic horizon, and a plinthitic horizon or illuvial horizon with various redoximorphic features. The landscape position is the major factor to

Z.-Y. Hseu
Department of Environmental Science and Engineering, National Pingtung University of Science and Technology, Pingtung 91201, Taiwan, ROC

C.-C. Tsai
Department of Natural Resources, National I-Lan University, I-Lan 26501, Taiwan, ROC

H. Tsai
Department of Geography, National Changhua University of Education, Changhua 50018, Taiwan, ROC

Z.-S. Chen (✉)
Department of Agricultural Chemistry, National Taiwan University, Taipei 10617, Taiwan, ROC
e-mail: soilchen@ntu.edu.tw

H. Eswaran
USDA, Natural Resources Conservation Service, PO Box 2890, Washington, DC 20013, USA

S. Kapur et al. (eds.), *Sustainable Land Management*,
DOI 10.1007/978-3-642-14782-1_9, © Springer-Verlag Berlin Heidelberg 2011

control the duration of saturation and reduction in the paddy soils of Taiwan. An anthrogleyic epipedon in Taiwan was also common in the surface horizon in paddy soils. Irrigation has induced the enrichment of the soil organic carbon (SOC) stock in the paddy soils. The free Fe contents of paddy profiles tend to be increased as the length of cultivation history increases in China, but this trend is much less profound in non-paddy profiles.

Keywords Anthroscapes · Land degradation · Anthropogenic processes · Chinese Soil Taxonomy · Anthrosols · History of paddy soils · Anthrogleyic epipedon · Landscape position

1 Introduction

The Anthropogenic process is recognized as a soil formation process which consists of a collection of geomorphic and pedological processes resulting from human activities. These human activities include deep tillage, intensive fertilization, addition of extraneous materials, irrigation with sediment-rich waters and wet cultivation (Kosse 1990). Additionally, anthroscapes are important in the assessment of human-induced land degradation. These may be sites where major negative changes lead to declines in land quality, and thus understanding the soil-landscape relationships on anthroscapes is a prerequisite for addressing global land changes in the sustainability of the environment. The anthropogenic soil-forming processes, relating to long-term agricultural use, affect soil properties over hundreds or thousands of years to form anthroscape soils (Dazzi et al. 2009).

In the case of soils affected by land-use changes in large-scale farming, the anthropogenic processes functioning in China for over a few thousand years are particularly based on numerous agricultural and livestock activities undertaken on soils, such as the puddling of surface soils (mechanically stirring and mixing surface soil with water and making it into a muddy paste) for aquatic rice production and dumping detritus on grassland. With the increasing population and the discovery of mechanical power, the area of paddy soils or rice-growing soils expanded, particularly since the middle of the twentieth century in Asia, especially in India, China, Indonesia, Philippines, Thailand, Japan, Korea and Taiwan.

In addition to non-paddy anthropogenic soils, paddy soils will be focused and illustrated in this chapter, because the aquatic rice-growing system is not only always strongly associated with the core of the Chinese cultural history but also are within the world's most important media for food production. Paddy lands are concentrated in Monsoon Asia, where they add up to slightly more than 90% of the worlds rice growing land. In the countries of Monsoon Asia, frequently more than half the total arable land is allocated to the aquatic rice cultivation system (Kyuma 2004). Rice cultivation had been a wide plantation in the basins of the Yellow and Yangtse Rivers in Mainland China by the third century AD (Li and Sun 1990). Paddy soils produce one-quarter of the grain for China's market and are

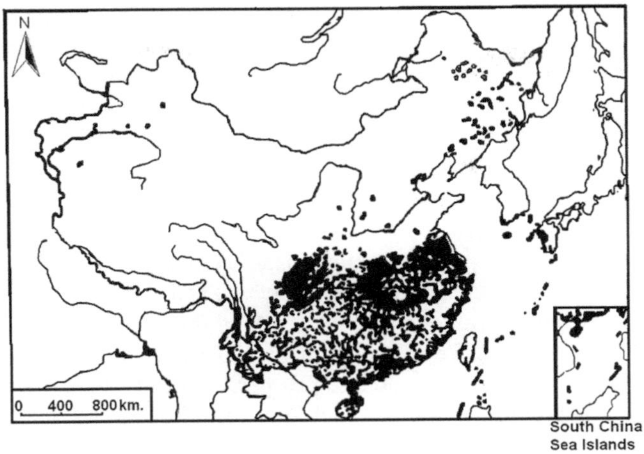

Fig. 1 Sketch map of the distribution of paddy soils in China and Taiwan [from Gong (1992)]

widespread all over China and Taiwan, but more than 90% are concentrated in the vast expanse of plains, hills, and mountain areas to the south of the line from the Qinling Mountain to the Yellow River as shown in Fig. 1. The total area of anthropogenic soils is around 397,200 km^2 in China (Shi et al. 2006). However, paddy soils are more than 30 million hectares in China and Taiwan (23% of the world's total irrigated lands) and occur mainly in the area south of the Yangtse River (Zhang and Gong 2003; Shi et al. 2006).

In this chapter, we will initially introduce soil science development in ancient China to emphasize the long-term history of the anthroscape creation from China – Country of ancient civilization. The anthropogenic soils (Anthrosols) will further be characterized by the soil classification rationale and land uses particularly for paddy soils. Moreover, the paddification processes play crucial roles in the formation of Anthrosols for aquatic rice production around Monsoon Asia, and thus the anthropic epipedon will be subsequently demonstrated by case studies in China and Taiwan.

2 Origin and Development of Soil Science in Ancient China

Based on many ancient records, archaeological discoveries, and current studies, the knowledge on soils, including their classification, distribution, and utilization was understood 2,000 years ago in China (Gong et al. 2003). Carbonized grain and paddy unearthed in Banpo village of Xi'an and Hemodu village of Zhejiang have ^{14}C dating of 5,600–6,080 year BP (Zhu 1964) and 6,890–7,040 year BP (Zhejiang 1978), respectively. Farmers in the Yellow River region began to plant millet, while the people along the Yangtse River started to plant aquatic rice about 6,000–7,000 year BP. Rice grain unearthed at Pengtoushan, Fengxian county, Hunan Province in

1988 revealed dates of 9,100 ± 120 and 8,200 ± 120 year bp (Pei 1989), which is new evidence of rice planting of about 2,000 years earlier than the previously known. According to ^{14}C dating in Taihu Lake region, Jiangsu province, rice and wheat pollen spores were buried in soil about 5,000–6,000 year BP, clearly indicating the long history of agriculture in China (Gong and Liu 2002). New evidence has revealed that people started to plant crops as early as in the Stone Age, which was followed by expansion of nationwide agricultural activities.

The scholar Xunshen, of the East Han Dynasty, considered soil as part of the land that provided nutrition for the living things. The book *Yingjing li* indicated that all food, grasses, and forests come from soil, and the soil was a basic concept linked to the whole world. The writings in Guanzhi during the Spring and Autumn Period and the Warring States Period pointed out that land was the basis for administration, while the soil was the basis for people to prosper. Thus, in ancient times, soil was considered as necessary for growing crops, and soil affected the rise or the fall of the administrations (Gong et al. 2003).

Soil classification in the Xia Dynasty in China, about 4,000 year BP, was recorded in the book *Yugong* that was thought to be written in the Warring States Period about 2,500 year BP. There is an old story about a man called Yu who fought against the big floods. He was away from home for more than 13 years until he found the effective ways of building dikes and dredging river courses to divert flood waters to the sea. Yu traveled throughout China and attained vast knowledge on soils and agriculture and included his classification on soils in the book *Yugong* (Gong et al. 2003).

Yu recognized and described the lands of the nine Provinces of the Shang Kingdom. He determined the high mountains and the great rivers as the two major geomorphic units with soil types along with the commodities and characteristic products of each province. Soil classification in *Yugong* was based on some soil characteristics such as soil fertility, color, texture, moisture, and vegetation. Soils of the Nine Provinces were classified into nine groups as *pale rang*, *black fen*, *red chi-fen*, *mud tu ni*, *blue li*, *yellow rang*, *white fen*, *lu*, and *zhi* according to their colors such as black, white, red, yellow, and blue as well as some terms related to soil texture (Table 1). We assume that the *rang was* roughly related to loess soils and their derivative alluvial silty soils, *fen* reflected humus-rich soils, *lu* stood for

Table 1 Soil description of the studied Provinces of China

Province	Soil description			Approximate terms in China soil taxonomy
	Dominate soil	Color	Texture	
Ji	Pale rang	White	Loam	Halosols, Aquic Cambosols
Jing	Tu ni	Grey	Clay	Gleysols, Ferrosols
Liang	Blue li	Dark	Clay	Aquic Cambosols, Gleysols
Qing	White fen	White	Clay	loam Agrosols, Halosols
Xu	Red chi-fen	Red	Clay	Udic Agrosols
Yan	Black fen	Black	Clay	loam Vertosols, Agrosols
Yang	Yellow rang	Grey	Clay	Gleysols, Ferrosols
Yong	Yellow rang	Yellow	Silty loam	Ustic Argrosols
Yu	Rang, fen, and lu	Yellow (black)	Loam	Aquic Cambosols, Vertosols

From Gong et al. (2003)

dark and compacted soils with an implication of clay pans and shajiang horizons, *zhi* stood for all sticky soils containing high amounts of clay, and that *chi* unquestionably referred to saline soils of the solonchak type. Each of these soil groups had taxa classes based on soil fertility levels (Gong et al. 2003).

3 Anthropogenic Soils in China

The term "anthropogenic soils" is not a taxonomic unit in many soil classification systems in the world. However, Anthrosols are defined by the World Reference Base (WRB) as "soils that have been modified profoundly through human activities of irrigation and cultivation and the addition of organic materials or household wastes." They can have a hortic, irragric, plaggic or terric horizon or an anthraquic horizon and an underlying hydragric horizon. These terms have been defined in the WRB system with diagnostic features developing along the long-term and continuous cultivation with their properties depending strictly on the soil management practices, such as intensive fertilization, irrigation, wet cultivation and continuous addition of materials (IUSS Working Group WRB 2006).

In *Soil Taxonomy* (Soil Survey Staff 2006), the original authors kept undisturbed soils and the cultivated or otherwise human-modified equivalents in the same taxa insofar as possible. Despite this goal, the system provides diagnostic horizons and taxa for soils together with soil morphologic properties induced by human intervention. Two epipedons, the anthropic and plaggen, and one diagnostic subsurface horizon, the agric horizon, were defined. In addition, the sulfuric horizon is commonly the human modification of soils containing sulfidic materials. Anthraquic conditions were added in Soil Taxonomy in 1992 (Soil Survey Staff 1992) for cultivated and irrigated soils formed under human induced flooding, such as aquatic rice and cranberry culture. The suborder Arents is defined using fragments of diagnostic horizons that are arranged in an order that cannot be attributed to soil forming processes that have been deeply mixed by humans (NSSC Staff 2002).

Human-induced processes that markedly change soil properties and consequently form diagnostic horizons are the anthropic processes. To be recognized as a soil forming process, both the cause and effect must be clearly established. In the Chinese Soil Taxonomy, anthropic surface horizons are diagnostic surface horizons that are the result of agricultural activities that have made major changes in soil processes at or near to the soil surface (e.g., submerged cultivation of paddy soils), or caused the formation of at least 50-cm thick layers on top of the natural soil surface, usually in combination with intensive agricultural practices, sometimes including (part of) the natural topsoil in the man-made layers. According to the *Chinese Soil Taxonomy* (ISS/CAS 2001), Anthrosols meet the requirements of the following combination of horizons having:

1. An anthrostagnic epipedon and hydragic horizon
2. A fimic epipedon and phosargic horizon
3. A siltigic epipedon or cumulic epipedon

3.1 Formation of Paddy Soils in China

Paddy soils are developed from various other soils. The recognition of paddy soils is mainly based on the obvious impacts of anthropogenic activities on the soils. Artificial and seasonal water saturation bring about enhanced reduction, illuviation and oxidation processes of iron (Fe) and manganese (Mn) in the soil profile, often leading to the formation of special layers characterized by Fe and Mn distribution (Yu 1985; Zhang and Gong 1993). Puddling is often a necessary practice for easy seedling transplantation (Fig. 2), which causes the development of a poor soil structure in the plow layer and a compacted plow pan, where the latter is useful in saving irrigation water provided it does not adversely affect the root growth. Therefore, a principal distinction of paddy soils from upland soils is the variation of their oxidation and reduction statuses.

Land leveling and terracing produces changes in the soil moisture regime (Fig. 3). The seasonal irrigation and drainage conditions cause the alternating oxidation and reduction processes and the build-up of the corresponding conditions for the translocation and transformation of the soil particles. The application of lime

Fig. 2 Puddling is a necessary practice for easy seedling transplanting of rice

Fig. 3 Landscape of land leveling and terracing for aquatic rice production in Yungnan, south-western China

materials and plant ashes in acidic soils or irrigation with calcareous groundwater leads to the accumulation of base cations. The river mud widely applied as manure in polder areas may raise the ground surface by 0.25–0.5 cm each year. The changes of the soils after planting rice are summarized as in the following:

3.2 *The Degradation of Soil Organic Matter*

Organic matter contents of paddy soils are generally higher than that of the corresponding non-paddy soils, but the ratio of humic to fulvic acids, the percent of the aromatic compounds and their molecular weights are decreased. The artificial soil moisture regime of the paddy soils is believed to be favorable for the accumulation of SOM, as their contents are statistically higher than that of the non-paddy soils (Pan et al. 2003; Cheng et al. 2009). Paddy soils evidently play a significant role in SOC storage with the consequent mitigation of climate change and enhancement of the potential of the world carbon pools.

3.3 Eluviation and Illuviation of Cations

A redistribution of exchangeable bases occur in the soils after planting rice as a consequence of the eluviation of cations in the saturated soils and the illuviation in the unsaturated soils. In general, the base saturation percentage (BSP) of red earth is only 20% or less, but it increase to over 50% after planting rice, with calcium ions dominating the bases. The accumulation process of cations begins from the cultivated horizon and gradually increases with depth.

Regarding the three cultivated soils derived from red earths on the aged alluvial terraces of the Taoyuan County, northwestern region in Taiwan (Fig. 4), the BSP values through Oxisol profiles were always lower than 10%; however, the BSP increased in a paddy Ultisol with about 40 years cultivation, and were further over 50% in a paddy Alfisol that integrated from an Ultisol cultivated for rice for about 100 years (Hseu and Chen 1996, 2001; Lin et al. 2005, 2007; Tsai et al. 2008; Jien et al. 2010). Additionally, some of the paddy soils on the Karst landscape and Zhujiang Delta not only are saturated with cations but also have an accumulation of calcium carbonate in the cultivated layers, even in the plowpan, due to the

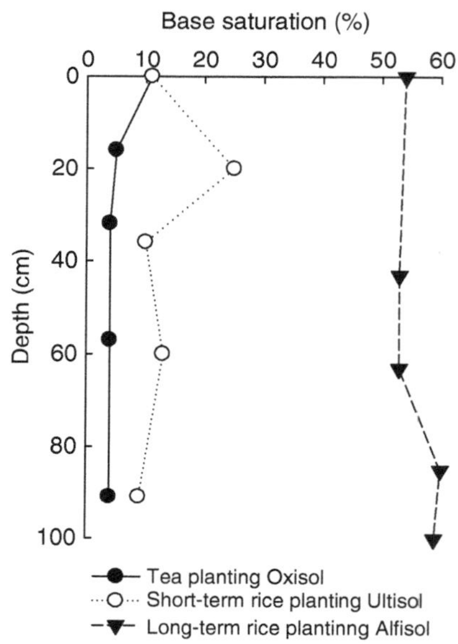

Fig. 4 The base saturation percentages of two paddy soils compared to a tea planting Oxisol developed from red earths on Taoyuan Terrace, northwestern Taiwan

irrigation with Ca-rich water or over liming. However, Brinkman (1970) indicated, that the changes on exchangeable cations, and cation exchange capacity (CEC) induced by ferrous-ferric alternation, and "ferrolysis" was found as a mechanism for acidification of paddy soils under strong leaching and less chemical fertilization conditions.

3.4 Eluviation and Illuviation of Iron and Manganese

Translocation and distribution of Fe and Mn caused by alternative reduction and oxidation is an important process affecting soil characteristics in paddy soils. Even in a strongly acidic soil solution with a pH value of 4–4.5, the concentrations of trivalent Fe and tetravalent Mn are negligible, where under reduction, Fe and Mn are highly activated. In addition, the concentrations of Fe and Mn in soil solution are increased due to the complexes formed by the metal ions with soluble organic substances. The eluviation of Fe and Mn by reduction is not obvious in the young history of paddy soils, but it is very apparent in relatively long cultivated soils (Yu 1985). In the overall profile, the Fe and Mn contents are lower in the cultivated surface horizon but relatively higher in the illuvial horizon and lowest in the gley horizon.

3.5 Decomposition and Synthesis of Clay Minerals

The composition of clay minerals of paddy soils depends commonly upon their initial soil characteristics, with the exception of paddy soils developed from the soils containing abundant potassium minerals, in which the composition of clay minerals was changed greatly due to the marked de-potassium process during soil formation (Gong 1992). In the case of paddy soils derived from calcareous soils with abundant Fe oxides, the content of hydrous mica was reduced but that of vermiculite was increased as compared with that of the initial soil (Gong 1992). As for the paddy soils developed on lateritic red earths with very low amounts of K minerals, the changes in the composition of clay minerals are slight compared with their initial conditions (Gong 1992), with a clear leaching loss of the free Fe ion.

3.6 Paddification Processes in Paddy Soils

Paddy soils with different water regimes vary under different redox statuses. In surface-water paddy soils, the cultivated horizon is in a state of reduction, and subsurface horizons are relatively oxidized in the profile during the rice growing season; following soil drying after the harvest period the whole profile is in a state of oxidation (Hseu and Chen 1996, 2001; Lin et al. 2005, 2007). In groundwater paddy soils, the profile is reduced through the year, particularly during the wet seasons. After irrigation in a paddy soil, the cultivated horizon and the upper part of the plowpan are saturated with water, causing the anaerobic decomposition of organic matter leading to the reduction of the cultivated horizon except a very thin layer at the surface. The subsoil remains unsaturated and in an oxidized condition, due to the stagnation of the irrigation water at the plow-pan.

The eluviation in the formation processes of paddy soil involves dissolution, reduction, and complexation of Fe and Mn. In the dissolution process, base cations are mostly leached. As a consequence of the complexation, Fe and Mn ions are complexed with dissolved organic matter and converted from solid phases into liquid phases and leached with depth. Compared with the irrigation water, the percolated water in upland red earths has a higher content of silicic acid, while that in paddy soils contains more Fe^{2+}, Mn^{2+}, and base cation ions. The migration of the elements in the red earth is in the following order: Ca > Na > Mg > K > SiO_2 > P, Mn, Fe, and Al, while the migration of the ions in paddy soils is Ca > Mn > Mg > Na > P > K > Fe > SiO_2 > Al. Therefore, the leaching loss of P, SiO_2, and base cations are significantly increased in paddy soils.

In addition to such elemental migration by the alternative redox process during paddification, the pedogenic horizons of paddy soils include a cultivated horizon, a plowpan, a percogenic horizon, and a plinthitic or illuvial horizon, in which the plinthitic horizon or illuvial horizon is the diagnostic horizon of paddy soils differing from other soils (Gong 1992; Hseu and Chen 1994, 1997; Lin et al. 2005, 2007; Jien et al. 2010).

3.7 Cultivated Horizon

During the submerging season, the cultivated horizon is reduced, except the surface soil which is in a state of oxidation, while after drainage during the season of growing upland crops (rotation cropping) it turns into two layers. The first layer is 5–7 cm thick with a surface soil of less than 1-cm in thickness constituted by relatively dispersed soil particles. Below the surface soil is the solum dominated by microaggregates with abundant rice roots. The second layer is dark colored with large clods and macropores, as well as Fe-Mn mottles or red coatings on the walls of the soil pores.

3.8 Plowpan

It is relatively compact with more clayey soil texture and more or less laminated in structure with some Fe-Mn mottles on the surface with an additional plowpan.

3.9 Percogenic Horizon

The percogenic horizon is often found beneath the plowpan and formed under conditions of seasonal percolation and eluviation induced by irrigation water (Gong 1992). There are two types of percogenic horizons, one is formed under weak eluviation of Fe and Mn segregated in situ with only a few rust streaks and spots; and the other is formed under strong eluviation with a lower content of clay, Fe and Mn and a color of light gray. The latter is generally known as the albic horizon.

3.10 Linthitic Horizon or Illuvial Horizon

This horizon contains more clay, organic carbon, base cations, Fe and Mn as well as some elements capable of apparently changing their valences within the illuviated Fe and Mn. Based on its redox intensity, this horizon may be further divided into the three subtypes of a plinthitic horizon as the brown, gray, and blue-gray.

3.11 Gley Horizon

The gley horizon is formed under strongly reduced conditions. Due to the infiltration of some dissolved oxygen, the mottles may occur in the upper part of gley horizon. Under the influence of the groundwater on Fe, the gley horizon may turn to grayish white. In addition, it is highly dispersable in some cases.

3.12 Anthropic Epipedon of Anthrosols

In the Chinese Soil Taxonomy (ISS/CAS 2001), soils with properties induced by subsequent to long-term soil management for agriculture or other land uses are considered as Anthrosols. They are equivalent to the Anthrosols defined in the WRB system, and partially correspond to Inceptisols, Mollisols, Alfisols, and Ultisols of the Soil Taxonomy (Soil Survey Staff 2006). Paddy soils are Hydragric Anthrosols in the Chinese Soil Taxonomy. Various agricultural engineering processes, such as terracing of sloping land and the reclamation of bog and marsh land, create the basic land features and an environment suitable for planting rice (Gong 1992). The changes of properties and overall quality of soils from different origins follow different pathways, and the length of time soils are used for rice production affects the magnitude of the changes. Suborders of Anthrosols in the Chinese Soil Taxonomy include Siltigic, Cumulic, and Fimic epipedons defined in the following sub-sections (ISS/CAS 2001).

3.13 Siltigic (or Irragric) Epipedon

The siltigic epipedon is an anthropic epipedon rich in silt size particles and irrigated since long with sediment-rich irrigation water responsible for gradual siltigation. Due to the long standing influence of fertilization and cultivation, oxygen is lost in the siltigic epipedon. The siltigic epipedon should meet the following requirements:

1. Thickness of 50 cm or more
2. Homogenous in soil color, texture, structure, consistence, and content of calcium carbonate; the neighboring subhorizons have similar soil texture classes
3. Weighted average of SOC content in the upper 50-cm layer is 4.5 g/kg or more and SOC content decreases with increasing soil depth, but it is at least 3 g/kg at the bottom of the epipedon
4. After the soil sample is submerged for 1 h and passed through an 80-mesh sieve in water, the dense, semi-rounded platy soil peds can be observed, and the silting lamination can be detected by a hand lens; or, the features of tillage disturbance by the evidence of semi-rounded, round fine granular clods and the relicts of silting lamination may be detected in the inner parts of the clods by polarized microscopy
5. Contains artifacts such as coal cinder, charcoal, bricks, tiles and porcelain remains. Siltigic evidence (or irragric evidence) prevails at a surface horizon with soil properties that meet all requirements except the thickness for a siltigic epipedon. The thickness is defined as 20–50 cm for siltigic evidence.

3.14 Cumulic Epipedon

The cumulic epipedon is an anthropic epipedon developing from the application of large amounts of earthy manure and compost, pond mud, or mellow by cultivation. It meets the following requirements:

1. Thickness of 50 cm or more
2. Homogenous in soil color, texture, structure, consistence, and the neighboring sub-horizons have similar soil texture classes
3. Weighted average of soil organic carbon content in the upper 50 cm depth is 4.5 g/kg or more
4. Has one of the following features besides particle composition similar to that of the original soils in vicinity, due to the influence of the source of cumulic materials
   (a) The features of hydromorphic or semi-hydromorphic soils
   (b) Some fragments of diagnostic horizons or diagnostic characteristics similar to that of automorphic soils in the vicinity
5. Has artifacts such as coal cinder, charcoal, bricks, tiles, and porcelain remains. Cumulic evidence is an evidence of the surface horizon which has the properties

that meet the requirements for the cumulic epipedon except that its thickness is between 20 and 50 cm.

3.15 Fimic Epipedon

The fimic epipedon is a highly mature anthropic epipedon, it forms by intense vegetable cultivation and the application of large amounts of human and animal excreta, manure, and organic garbage with frequent irrigation over a long period. It meets all the following requirements:

1. Has a thickness of 25 cm or more (including the upper highly fimic sub-horizon and the lower transitional sub-horizon)
2. Has a weighted average of SOC of 6 g/kg or more
3. Has a weighted average of 0.5 M $NaHCO_3$ extractable phosphorus (P) content of 35 mg/kg or more
4. Has abundant worm-casts and 50% or more worm channels that are less than 10 cm in length
5. Contains artifacts such as coal cinder, charcoal, bricks, tiles, and porcelain remains

A Fimic surface horizon is one, which has the properties that meet requirements for the fimic epipedon except one of the following:

1. The thickness is between 18 and 25 cm
2. The weighted average of 0.5 M $NaHCO_3$ extractable P content is between 18 and 35 mg/kg
3. The weighted average of the SOC content is between 4.5 and 6 g/kg

3.16 Anthrostagnic Epipedon

The anthrostagnic epipedon is an anthropic epipedon formed under submerged cultivation, including the cultivated horizon and the plow pan of a paddy soil. It meets the following requirements:

1. Has a thickness of 18 cm or more
2. Has anthrostagnic soil moisture regime for at least 3 months of the year in most years when the soil temperature is more than 5°C
3. The upper part of the epipedon (cultivated horizon) is pudding due to the disturbance of tillage with water for at least half a month of the year in most years when soil temperature is more than 5°C
4. Has a soil color value, moist, of 4 or less, chroma, moist of 2 or less, hue of 7.5YR or more yellow, even GY, B or BG, when submerged
5. Has abundant red mottles after drained

6. The ratio of the bulk density of the lower part (plow pan layer) to that of the upper part (cultivated horizon) is 1.10 or more, when drained

Anthrostagnic features of a surface horizon develop due to the weak influence of submerged cultivation. The horizon with anthrostagnic evidence lacks a plow pan; or has a weakly developed plow pan, but the ratio of the soil bulk density of the plow pan to the cultivated horizon is less than 1.10.

3.17 Hydragric Horizon

The hydragric horizon is a subsurface horizon affected by rice cultivation. The formation of the hydragric horizon is due to the reduction of the Fe and Mn oxides and consequent leaching from the anthrostagnic epipedon or to a subsoil layer with gleyic features. The hydrargic horizon is further oxidized to accumulate in the subsoil and meets the following requirements:

1. Has a thickness of 20 cm or more, and its upper boundary is at the bottom of the anthrostagnic epipedon
2. Has one or more of the redoximorphic features described in the U.S. Keys to Soil Taxonomy (Soil Survey Staff 2006)
3. The content of DCB-extractable iron in the horizon is at least 1.5 times of that in the cultivated horizon
4. Has gray humus-silt-clay films in a thickness of 0.5 mm or more on ped surfaces and walls of channel voids
5. Has well developed prismatic structure and/or angular blocky structure

Hydragric evidence is an evidence of a subsurface horizon which has the features meeting the requirements of a hydragric horizon but only has a thickness of 5–20 cm.

3.18 Carbon Stocks in Paddy Soils of China

The possibility of estimating the global carbon (C) pools on the earth has attracted scientists for centuries. The greenhouse effect, resulting in the global consequences of climate alterations, has led scientists to study the global carbon cycle (Eswaran et al. 1993; Lal 1999, 2004; Lal et al. 2004). Additionally, global concerns have increased over greenhouse gas emissions and their potential impact on climate change over the past two decades. As a result, an international agreement, the Kyoto Protocol, was signed to mitigate greenhouse gas concentrations in the atmosphere by improving the terrestrial carbon sinks.

Soils are regarded as the most important sink to sequester the atmospheric carbon dioxide (CO_2) and to reduce the trace gases, e.g. CO_2, methane (CH_4), and nitrous oxide (N_2O), which are active in producing these gases and enhance the greenhouse effect (Batjes 1996). Estimates of global C content in soils have been made. Globally,

1,576 Pg (Petagram = 10^{15} g) of C is stored in soil systems globally, with about 506 Pg of this in soils of the tropics, where about 40% of the C in soils of the tropics is stored in the forest soils (Eswaran et al. 1993). Batjes (1996) calculated that the total soil C pools for the entire land area of the world, excluding C stored in the litter layer and charcoal, ranged from 2,157 to 2,293 Pg C in the upper part of 100-cm layer, where SOC is estimated to be 684–724 Pg C in the upper 30-cm layer, 1,462–1,548 Pg of C in the upper 100-cm layer and 2,376–2,456 Pg C in the upper 200 cm.

Carbon loss from agricultural soils has been extensively discussed in the literature (major refs). The information on SOC storage and sequestration in paddy and corresponding non-paddy soils is necessary for the designation of the anthroscapes continuum. Such a landscape continuum, i.e., the anthroscape, that is developed in chronosequence on fairly uniform parent materials under nearly identical climate conditions and landscapes around the Yangtze Estuary in eastern China, manifest significant increases in surface horizon carbon contents (Fig. 5) (Cheng et al. 2009).

Pan et al. (2003) obtained data selected from the second National Soil Survey Project which was completed in the early 1980s for C content calculation and C pool estimation of paddy soils in China. The second National Soil Survey in China also identified 525 soil series to be included as paddy soil pedons. Since the upper parts of a typical paddy soil is composed of a plow layer and a plowpan (Li 1992), they used the mean values of C in two layers to calculate the C pool in different depths at 30, 50, and 100 cm depth from the soil surfaces.

Pan et al. (2003) also indicated that about 60% of the soil series of the area of paddy soils and have SOC contents ranging from 12.5 to 22.5 g/kg, while 95% of the soil series and 99% of the area have SOC contents ranging from 7.5 to 30 g/kg. The means and area weighted means of SOC contents in the plowpan layer are lower than those of the plow layers by 4 g/kg. According to the frequency distribution of the mean topsoil C contents in terms of numbers of the paddy soil series calculated by Pan et al. (2003), 80% of soil series and 90% of all paddy soils have a SOC content of 20–60 ton/ha in the top soil, while 99% of the overall paddy soils ranged from 20 to 100 ton/ha. Ultimately, a wide variation of SOC storage was determined in the top soils of the paddy fields.

The SOC data allow a reliable estimation of SOC storage by different subgroups of paddy soils in terms of their water regimes. As shown in Table 2, significant variations of the SOC content and SOC storage were also found among the Soil Subgroups. The subgroups of gleyization and de-gleyization in paddy soils have a higher SOC bulk density due to their occurrence in lowlands to favor the SOC accumulation. In contrast, paddy soils with redox and percolation conditions, which are typically classified as hydragric paddy soils with well established diagnostic horizons and higher contents of ferric hydroxides in the soil profile (Li 1992) have a medium level of 30 ton/ha of SOC. The lower SOC contents of the other Soil Subgroups can be attributed to the soil constraints, such as higher salinity, continuous reduction conditions, and the lack of SOC protective materials of clay or hydroxides. However, two thirds of the total C stock was found in the two major subgroups of redox and percolating paddy soils (Gong 1992). These two subgroups

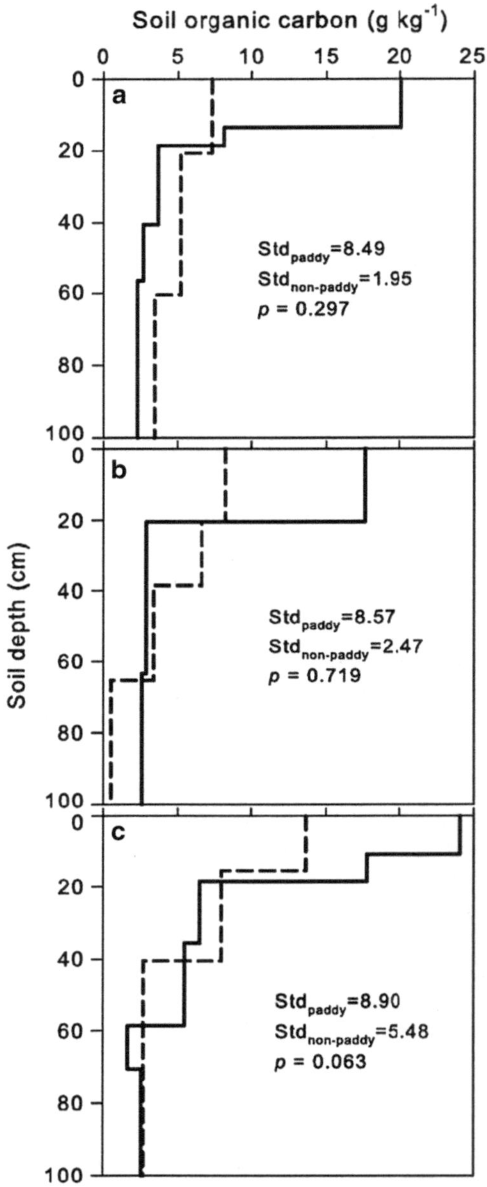

Fig. 5 Vertical distribution of soil organic carbon in paddy (*solid line*) and non-paddy (*dash line*) profiles over 0–100 cm depth with 50 year (**a**), 100 year (**b**), and 300 year (**c**) of cultivation history [from Cheng et al. (2009)]

have been considered as the typical paddy soils in China that have been formed under long term hydragric pedogenic processes, and were classified as Typic Stagnic Anthrosols (Gong 1992).

The statistical data of the SOC contents of the different regions of China are appropriate to utilize for the estimation of SOC storage enhancement induced by

Table 2 Distribution of topsoil carbon storage among the subgroups of paddy soils in China

Subgroup	Area (ha)	Sample no.	Soil organic carbon of plow layer (g/kg)	Soil organic carbon of plowpan (g/kg)	Carbon density (ton/ha)	Total carbon pool (Tg)
Redoxing[a]	1,421.9	85,285	16.42	12.38	44.6	634.6
Waterlogged[b]	3,803.4	15,388	13.40	9.96	33.3	145.6
Percolating[c]	5,543.4	17,091	16.53	12.47	44.9	248.8
Gleying[d]	2,645.6	14,662	18.62	14.14	49.1	129.8
De-gleying[e]	1,052.6	10,922	22.27	17.07	56.0	58.9
Bleached[f]	710.1	5,890	11.72	8.61	34.5	24.5
Saline[g]	353.3	1,256	9.92	7.16	30.4	10.7
Salt and acid[h]	82.5	49	16.36	12.33	44.5	3.7
Total	28,702.7	150,543	15.65	11.76	44.2	1,256.7

Tg (Teragram = 10^{12} g)
[a]Mostly Fe-leaching-Stagnic-Hydro-agric Anthrosols
[b]Mostly Hap-Stagnic Anthrosols
[c]Fe-accumulating Stagnic Anthrosols
[d]Gleyic Stagnic Anthrosols
[e]Paleo-Gleyic Stagnic Anthrosols
[f]Albic Stagnic Anthrosols
[g]Mostly sodic fluvents
[h]Mostly sulfaquents

aquatic rice cultivation (Pan et al. 2003). The results of this calculation are summarized in Table 3. It is estimated that the total enhanced topsoil SOC storage in paddy soils of China is up to 0.28 Pg under irrigated conditions. The highest C stock is preserved in Southern and Eastern China, where the irrigation is frequently available for rice production due to the higher annual precipitation and also the long standing agricultural history of higher rice production under precision farming practices. Ultimately, the well management of paddy soil systems can play an important role in the enhancement of the regional SOC stock in China.

The estimated total SOC pool in China's paddy top soils is 1.3 Pg, which is 2% of China's total storage in 1 m depth and is 4% of China's total storage in the total topsoil (the plow layer and the plowpan), at an area of 3.4% in the whole of the country. In this C pool, 0.85 Pg is in the plow layer, which is formed by agricultural practices, and 0.45 Pg is in the plowpan layer. The SOC contents of the paddy topsoil is higher than that of the corresponding soils in the dry crop land, although somewhat lower than that of the world mean value (Pan et al. 2003).

3.19 Pedogenic Iron and Manganese Characteristics of Paddy Soil Chronosequences in China

Zhang and Gong (1993) selected three sets of soil chronosequences from the alluvial plains, a bog area and well drained terraces. The age of rice cultivation ranged from 0 to 600 years, estimated by thee local farmers or from the records of old paddy areas.

Table 3 Change in soil organic carbon content and Carbon storage increase due to irrigated rice production in China's agricultural soils

Region	Soil organic carbon content (g/kg)			Increased carbon density (ton/ha)		Increased C storage (Tg)		
	Un-irrigated topsoil	Paddy plow layer	Paddy plowpan	Paddy plow layer	Paddy plowpan	Paddy plow layer	Paddy plowpan	Total
East China	8.79	14.26	10.56	9.27	2.33	136.89	34.40	171.3
South China	10.69	16.86	12.8	12.06	4.83	80.40	32.20	112.6
North China	6.81	9.65	6.15	5.97	−1.50	4.47	−1.13	3.3
Northwest China	14.00	14.70	12.04	1.49	−3.06	0.32	−0.66	−0.3
Northeast China	16.21	17.90	11.82	4.05	−9.17	3.64	−8.22	−4.6
Total/mean	9.60	16.26	12.25	9.69	2.43	225.72	56.59	282.3

Tg (Teragram = 10^{12} g)

Soil characteristics are different on soil redox potential (Eh), clay content, dithionite-citrate-bicarbonate (DCB) extractable Fe and total Fe contents over ages of rice cultivation. These changes could not be exclusively assigned as the consequences of rice cultivation only, because of the unaccounted changes brought about by the upland cropping system. The vertical movement of Fe in the soil profile is generally restricted due to the low solubility of Fe except in reduced forms or in organic complexes. The solubility of Mn is also controlled by the reduction and formation of organic complexes; however, Mn is reduced under a higher redox potential than Fe and, thus, Mn is more readily soluble in paddy soils (Collins and Buol 1970). Therefore, the illuviation of Mn is usually lower than that of Fe in the paddy soil profile.

A series of paddy and non-paddy soil profiles that are developed in chronosequences from approximately 50–1,000 years of exposure periods, on fairly uniform marine deposits, under nearly identical landscape and climate conditions, were selected from Cixi, Zhejiang Province, China, where aquatic rice has been grown under water submergence (Cheng et al. 2009). The non-paddy soil profiles showed more uniform distribution of free Fe oxyhydrates than those of the paddy soil profiles. The average of the free Fe hydroxides from the surface to 100 cm depth of all paddy profiles were 13.1 g/kg, in contrast to the 10.4 g/kg in non-paddy profiles. The free Fe contents of paddy profiles tend to be significantly increased as the length of cultivation history increases, but this trend is less profound in non-paddy soil profiles (Fig. 6).

The management of paddy soils significantly increased the SOC, clay and total Fe contents, in the surface soil horizons as compared with non-paddy profiles, but decreased the total Mn contents (Cheng et al. 2009). The soil development is obviously more advanced in paddy profiles in terms of distribution or redistribution patterns of the total Fe and free Fe oxyhydrates, which are not evident in non-paddy profiles. A longer history of rice cultivation tends to produce more diversified horizons of paddy soil profiles, whereas the longer history of non-paddy cultivation does not have any effect on horizon development (Cheng et al. 2009).

3.20 *Fimic Characterization of Anthrosols in China*

Vegetable-cultivated soils are significantly influenced by the long-term application of human and animal wastes and other organic residues, as well as by intensive cultivation, frequent irrigation and other human activities. In the course of time, the surface horizon becomes soft and fluffy, dark colored, and contains high amounts of organic matter and available P, and qualifies as a Fimic epipedon. The characteristics of fimic epipedons were significantly different from those of the original parent soils (Gong et al. 1999). These distinct anthropedogenic processes have been recognized as soil ripening or fimic ripening processes (Gong 1992). The soils with a fimic epipedon, which formed under fimic ripening processes, were classified as Fimic Anthrosols in the Chinese Soil Taxonomy. Each of these soils

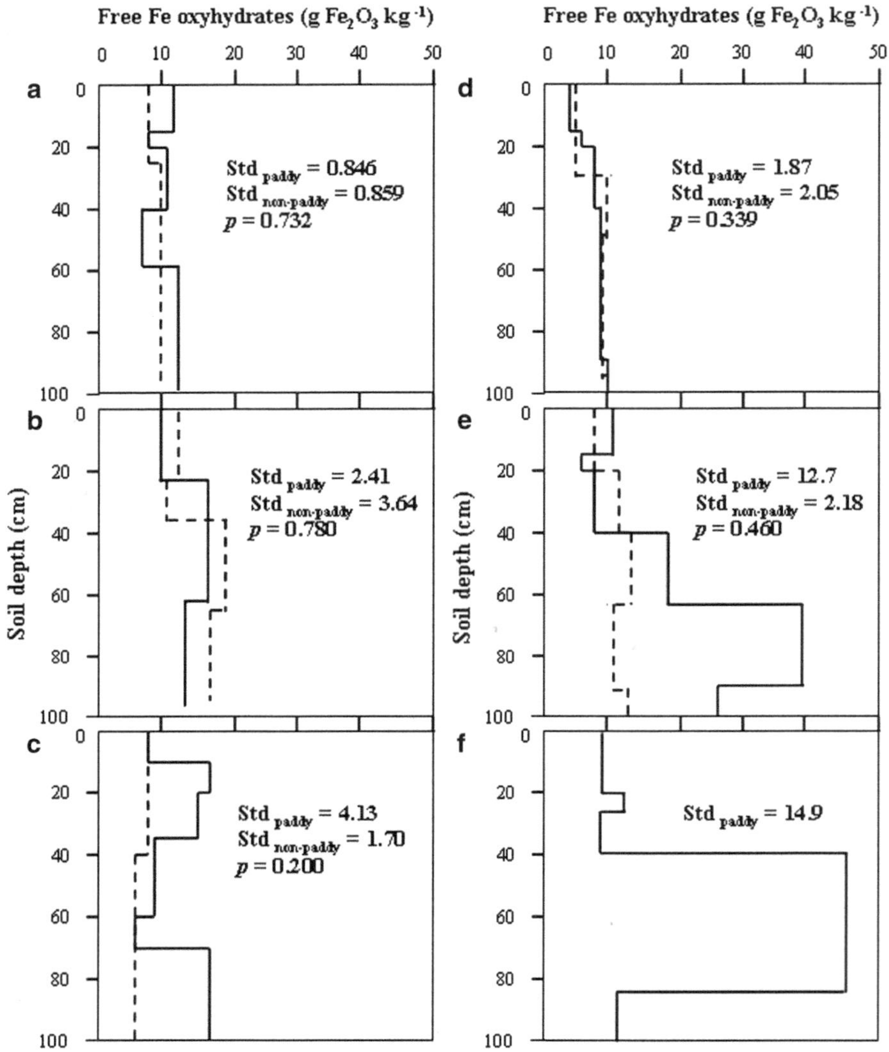

Fig. 6 Vertical distribution of free iron oxyhydrates in paddy (*solid line*) and non-paddy (*dash line*) profiles over 0–100 cm depth with 50 year (**a**), 100 year (**b**), 300 year (**c**), 500 year (**d**), 700 year (**e**), and 1,000 year (**f**) of cultivation history [from Cheng et al. (2009)]

has been so transformed by anthropedogenic processes that the original soil is no longer recognizable or survives only as a buried soil.

Nutrient enrichment in the Fimic Anthrosols under long-term vegetable cultivation was responsible for the accumulation of high organic matter contents, intensive enrichment of available P and enhancement of soil bioactivities (Gong 1992). As a result, most of the fimic surface horizons are well supplied with nutrients and have favorable physical properties (with good structure, high porosity, high water retention capacity). However, there is also a distinct and increasing tendency of the total and available contents of potentially toxic metals (Gong et al. 1999) sufficient to pose toxicity to crops, animals and humans.

Thirty-seven soil profiles from various original parent materials were collected by Zhang et al. (2003) from Changsha, Nanjing, Taian, Jinan, Beijing, Shanghai, Suzhou, Wuxi and other locations to explore the pedogenic characteristics of Fimic Anthrosols and the relationships between Fimic Anthrosols, different vegetable-cultivated soils and parent soils in China. They indicated that the most important pedogenic characteristics of Fimic Anthrosols were the accumulation of organic matter and enrichment of available P in the surface, subsurface and transitional horizons. Therefore, the thickness of the fimic epipedon, including fimic surface and subsurface horizons, the contents of available P and organic matter are regarded as the diagnostic criteria for the classification of Fimic Anthrosols in the Chinese Soil Taxonomy (Gong et al. 1999). These criteria clearly reflect the pedogenic characteristics and the soil-forming processes that the vegetable-cultivated soils gradually undergo as they develop into Fimic Anthrosols.

3.21 Anthroscapes in Taiwan

Paddy soils, the predominant anthroscapes in Taiwan, occupy about 50% (450,000 ha) of the agricultural lands on the island (Chen 1984). These anthroscapes are located on the alluvial soils and terraces, which are distributed on the western and eastern alluvial plains and also on the Quaternary terraces with different histories of paddification, parent materials, and topography. The pedogenic processes of paddy soils in Taiwan are mainly affected by irrigation water, groundwater table, and dissolved oxygen concentrations in the water. The paddy soils are characterized by the gleyization of the surface soil, formation of a compact subsurface layer (plow pan), accumulation of organic matter in the surface soil, formation of redoximorphic features such as Fe/Mn nodules and clay/Fe depletions, and redistribution of base cations by irrigation water (Chen 1984, 1992; Chen and Chang 1985; Lin et al. 2005, 2007).

3.22 Gleyization Process of Paddy Soils in Taiwan

Artificial flooding on the soil surface produces a perch water table and induces the reducing conditions. This process forms the Anthrogleyic Epipedon (Lynn 1988) and is characterized by massive structures, greater than 10 cm of soil thickness and soil color with hue of 2.5Y or yellower and chroma of 2 or less in the Ap horizon than that of the underlying horizon. These soil morphological characteristics are special common phenomena in the surface horizon of red soils for more than 30 years in paddy soil cultivation. The soil characteristics of the Ap horizon coincide with the definition of the Anthrogleyic Epipedon proposed by Lynn (1988).

3.23 Redoximorphic Features (Clay Accumulation and Segregation of Iron and Manganese)

Redoximorphic features are formed by the processes of reduction, translocation, and oxidation of Fe and Mn oxides in paddy soils. The amount of Fe depletions increased while the amount of Fe concentrations decreased as soils are saturated and reduced for longer periods (Vepraskas and Guertal 1992). Some examples of interpretations of soil redoximorphic features were illustrated by Vepraskas (1992).

Accumulation and depletion of Fe, Mn and clay in Btg or Bg horizons are the significant pedogenic processes in the paddy soils of Taiwan, especially in the upper part (40–60 cm) of the pedons of the red earths and alluvial soils, which cause the formation of the red-brown colored mottles. These processes are quite different from the illuviation of clay, Fe, and Mn in the lower parts (60–120 cm or more) of the pedons before paddification (Chen 1984, 1992).

The paddy agric horizon was proposed in the Chinese Soil Taxonomy for the translocation and accumulation of clay, silt, Fe, and Mn below the plow pan layer (Yang et al. 1988). Chen and Chang (1985) indicated that the changes of different forms of free oxides (Fe, Al, or Mn) can be explained as an important index of pedogenic processes of paddy soils formed on the surface of red earths or soils formed in the alluvial plain for a longer time in Taiwan. Chen and Chang (1985) also proposed the schematic representation of different depth functions of pedogenic oxides, clay, and organic carbon of diagnostic horizons in the well drained paddy soils in northern Taiwan (Fig. 7).

The downward movement of the free Fe oxides and clay had the same tendency in the Btg (or Bg) horizons and the order of the rate of the downward movement was as Mn $>$ Fe $>$ Al. A larger activity ratio (AR) of iron (Fe_o/Fe_d, the ratio of oxalate extractable Fe to dithionite-citrate extractable Fe) and smaller crystallization index (CI) of iron [$(Fe_d - Fe_o)/Fe_t$] were found in the waterlogged plow layer than that of the lower horizon of the pedon. Therefore, the iron activity ratio (Fe_o/Fe_d) in the Btg (or named as Bgir, or Bgmn) horizon was very low (range from 0.05 to 0.10). This ratio may be regarded as characteristic to the high illuvial and oxidized iron in the Btg horizon or in the plinthite of the Btv horizon of paddy soils in Taiwan (Chen 1992; Lin et al. 2005, 2007).

Gray mottles are abundantly found in the alluvial paddy soils of Taiwan, especially on the slate, sandstone, and shale-derived soils on the western part of Taiwan. The gleyed horizon (hue 10Y or bluer, matrix chroma 1 or less), with strong reducing conditions, formed by the reduction and movement of Fe and Mn during long periods of water saturation in poor drainage conditions. The soil characteristics of the gleyed epipedon were also proposed in the Chinese Soil Taxonomy (Yang et al. 1988).

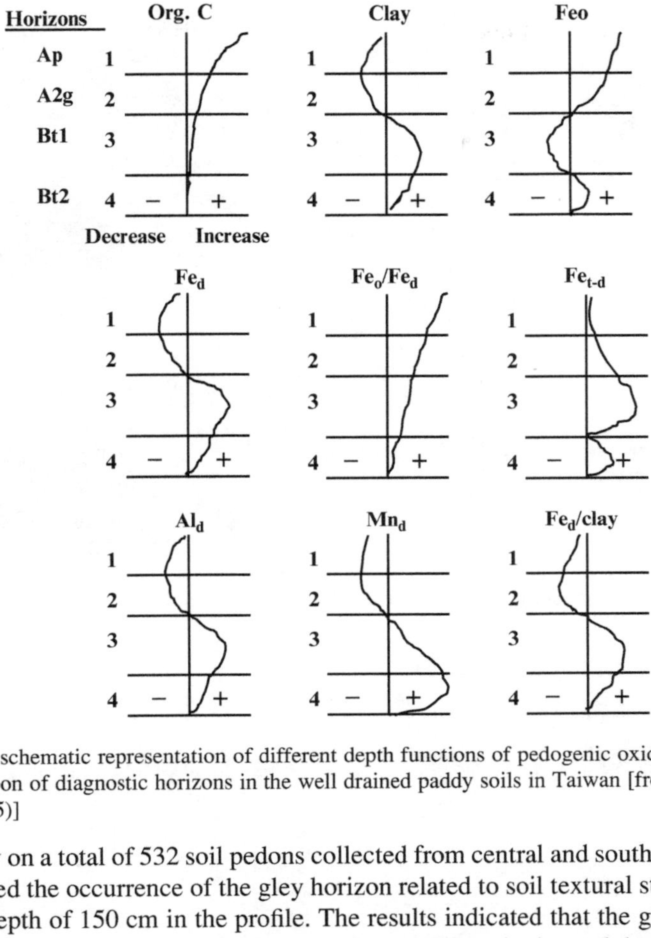

Fig. 7 The schematic representation of different depth functions of pedogenic oxides, clay, and organic carbon of diagnostic horizons in the well drained paddy soils in Taiwan [from Chen and Chang (1985)]

A study on a total of 532 soil pedons collected from central and southern Taiwan has revealed the occurrence of the gley horizon related to soil textural stratification within a depth of 150 cm in the profile. The results indicated that the gley horizon was usually formed at the interface of the textural discontinuity and the occurrence of the soil mottles seemed more dependent on the uniformity of texture in paddy soils in Taiwan (Sun and Yang 1989).

3.24 *Paddy Soils on the Red Alluvial Soils in Northern Taiwan*

Terraces are the dominant geomorphic features of the Taoyuan area (about 60% of the total landforms) in northwestern Taiwan (Fig. 8). In the study of Hseu and Chen (1995), three experimental plots (3 × 5 m) were selected for monitoring the redox potential and saturation conditions along the Taoyuan Terrace over a distance of 4 km on the Siaoli soil (Typic Plinthaqualf), Potu soil (Plinthaquic Paleudalf) and Luchu soil (Plinthic Paleudalf). When the studied soils were reduced, The reduced soluble Fe and Mn diffused from the voids to the lower horizon and then

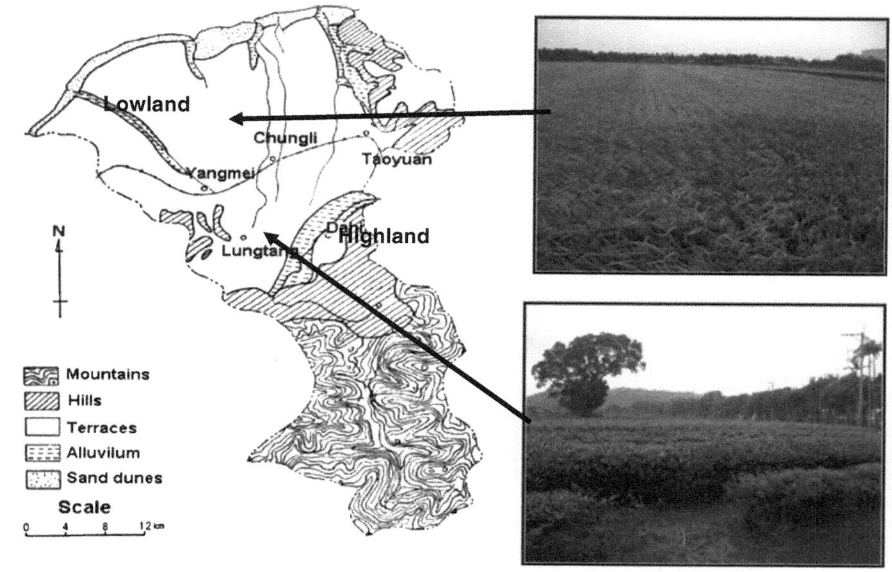

Fig. 8 Landscape of paddy soils distributed at Taoyuan area, northern Taiwan

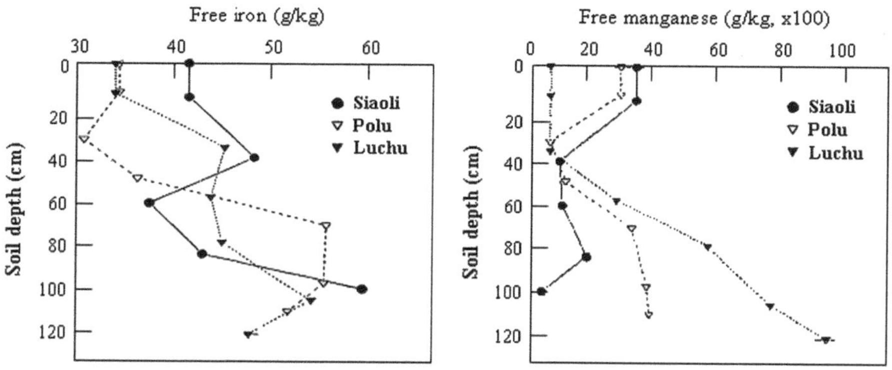

Fig. 9 The depth distribution of (**a**) DCB????-extractable iron and (**b**) DCB????-extractable manganese selected soils in Taiwan [from Hseu and Chen (1995)]

precipitated when the soils were reoxidized. Figure 9 shows that, the maximum concentrations of DCB-extractable Fe (Fe_d) and Mn (Mn_d) were found in the lower Bt horizons associated with higher redox potentials. This trend of free Mn (Mn_d) was more evident than that of the free Fe (Fe_d) because Mn oxides became soluble prior to Fe oxides when the soil was reduced.

The depth redistributions of Fe and Mn were promoted by paddification of rice-growing soils (Chen 1984). In this study, soil saturation, i.e., the anthropic soil saturation, was the most important factor, whether the soil was reduced or not. The surface water irrigation and the shallow groundwater influenced the morphology of these soils simultaneously. Results indicated that the effect of the irrigation water on the pedogenic process with respect to saturation, reduction, and redoximorphic features was greater than that of the shallow groundwater in Siaoli and Luchu soils. On the contrary, the shallow groundwater was the main contributor to saturation below 50 cm depth in the Potu soil.

An Ultisol hydrosequence including three soils was selected to monitor the water table and redox potential (Eh) in 1996, and to study the micromorphology of the pedons located on the Chungli Terrace of Taoyuan County, northern Taiwan (Hseu and Chen 1997). The results revealed, the increasing depth of the plinthite with the increasing drainage conditions in the soils studied, as in the following: The Houhu soil (Typic Plinthaquults) < Hsinwu soil (Typic Plinthudults) < Lungchung soil (Plinthaquic Paleudults). The Houhu pedon was reduced and saturated almost throughout the year with a chroma of 1, with a strong reduction of the pedon disturbing the orientation of the illuvial clay (Fig. 10). In the Hsinwu pedon, elongated clay coatings with strong orientation developed due to the optimum alternate oxidation and reduction cycles at a 75 cm depth in a period of 25% of the year. Less than 2 months of saturation conditions and only 12% of the year under reduction conditions were found at a 75 cm depth of the Lungchung pedon. Soil solutions sampled by the tension free method also indicated the differences in composition and concentration of ions between the soil profile and water chemistry (Hseu et al. 2000). The monitoring data indicated that the detectable Mn was more easily correlated with the contemporary concentration in the soil solution and redox regimes than Fe. The low Fe activity ratio (Fe_o/Fe_d) in the Btv horizons, with the redoximorphic features in these aged Ultisols, and the minor dilution effect of the irrigation water was one of the major factors responsible for the low concentration of Fe in the soil solution.

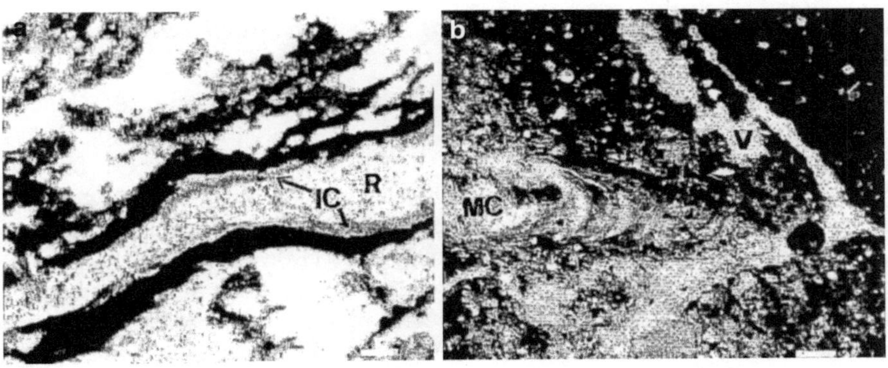

Fig. 10 Representative micrographs of the study soils with magnification ×40. The lengths of bars are 0.25 mm: (**a**) Houhu Btvl horizon, 66–82 cm, *pale* coloured illuvial clays (IC) along root channel (R); plane-polarized light (PPL); (**b**) Houhu Btv3 horizon, 102–122 cm, microlaminated clay infills (MC) of chamber in the angular blocky microstructure surrounded by planar voids (V), PPL [from Hseu and Chen (1997)]

The formation of redoximorphic features can be explained hypothetically as follows (Hseu and Chen 2001): (1) When the parent material of the red soils was deposited on the terrace in the Quaternary Period, the soil horizons developed under humid subtropical conditions that promoted the illuviation of clay and Fe in the soil profile, and (2) the redox concentrations occurred as soft masses and concrete nodules associated with the fluctuations of seasonal high water tables, especially in the region close to sea level. Finally, after the soils were converted to paddy fields, a plow layer formed and the subsequent development of the groundwater by precipitation and human activities led to more diverse redoximorphic features. Therefore, when the soil was used for rice production, tonguing plinthites including soft masses, Fe-Mn concretions, and mangan distributions increased in the Btv horizons.

Many studies of anthraquic conditions have also documented that redoximorphic features can form very quickly, in just a few weeks in some cases (Mitsuchi 1992; Somasiri and Deturck 1992; Hseu and Chen 1996). The root system of rice formed varying void types in the soils causing the development of Fe-Mn concretions with different shapes in the upper parts of the Btv horizons. Iron oxides were originally accumulated in these voids, such as root channels, to form concrete nodules, which were also identified by Silva and Ojea (1991).

The landscape position is one of the factors controlling the duration of saturation and reduction in the soil (Jien et al. 2004, 2010; Lin et al. 2005, 2007). About 10% Fe-Mn concretions were correlated to more than 80% of saturation and reduction in the Houhu soil located on the toeslope position. About 20% Fe-Mn concretions were correlated to 50% of saturation and 25% reduction in the Hsinwu soil located in the footslope position. About 15% Fe-Mn concretions were correlated to 40% of saturation and 10% of reduction in the Lungchung soil located in the lower backslope position.

3.25 Paddy Soils of Slate Alluvial Soils in Central Taiwan

According to the soil survey undertaken by the Taiwan Agricultural Research Institute of the Council of Agriculture of Taiwan, there are five major Soil Orders including Entisols, Inceptisols, Alfisols, Ultisols, and Oxisols, that can form in west-central Taiwan. These soils cover the surfaces belonging to three geologic terrains namely the Pleistocene anticlines, Holocene coastal plain and modern alluvial fan (Lin 1957; Ho 1986; CGS 2004) (Fig. 11a). The Chang-Hua coastal plain, with altitude less than 20 m above sea level, is located at the western part of the anticline terrain along the west-coast region of Taiwan. The surface is covered with alluvium of clay, silt, sand and boulders by the Tadu and Choshui rivers flowing from the Central Ridge at an altitude of more than 3,000 m above sea level. The high yield of sediments from the Choshui River, the longest river in Taiwan, developed a large modern alluvial fan (known as Choshui alluvial fan) adjacent to the south of the Changhua coastal plain as the youngest geologic terrain of the area (Lin 1957; Ho 1986).

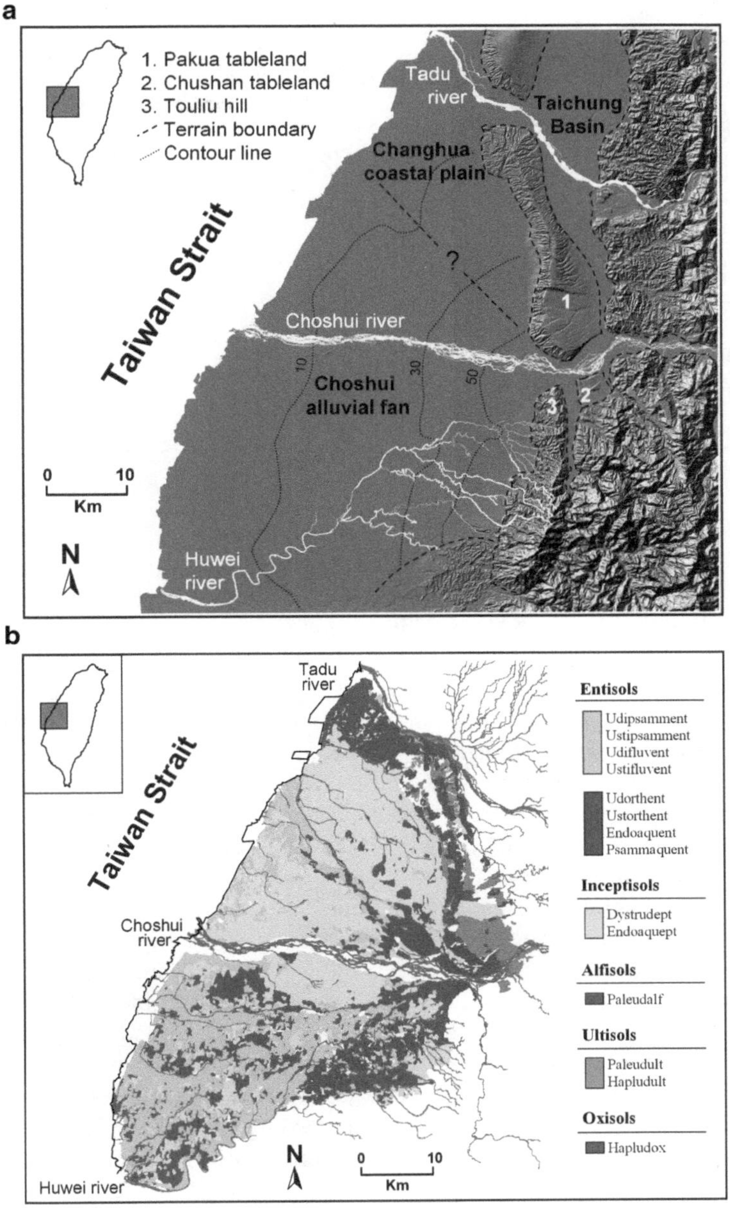

Fig. 11 (**a**) The shaded relief map of the study area in Central Taiwan; (**b**) the soil map of the study area in Central Taiwan

Different soils are distributed on the surfaces of the Changhua coastal plain. The Choshui alluvial fan covers three major Soil Taxonomic Orders, including the Entisols, Inceptisols and Alfisols (Fig. 11b). Entisols cover most of the Choshui

Table 4 Basic properties of the soil varieties occurred in the study area of Taiwan soil types[a]

	pH	Organic carbon (%)	Cation exchange capacity (cmol(+)/kg)	$CaCO_3$ (%)	Fe_d (%)
Entisols					
Udipsamment	7–8	0.1–1.3	2.8–17.1	0.1–3.6	0.9–1.9
Ustipsamment	7–9	0.3–2.1	5.6–16.7	0.3–3.7	0.7–1.9
Udifluvent	7–8	0.6–1.0	10.5–19.7	na	1.4–2.5
Ustifluvent	7–9	0.4–2.0	6.2–18.3	0.5–4.7	0.5–2.1
Udorthent	7–8	1.3–2.4	1.6–14.7	0.1–3.6	0.9–1.9
Ustorthent	7–9	1.0–3.4	17.0–22.5	0.1–0.5	na
Endoaquent	7–8	0.3–2.9	3.8–20.0	0.1–3.6	0.8–3.1
Psammaquent	na[b]	na	na	na	na
Inceptisols					
Dystrudept	5–8	0.3–3.5	4.5–22.7	0.1–3.6	0.6–3.1
Endoaquept	6–8	2.2–3.5	15.5–19.4	3.1–7.6	na
Alfisols					
Paleudalf	5–8	0.3–2.3	7.5–29.1	0.1–0.4	0.7–3.3

[a]Based on keys to soil taxonomy (Soil Survey Staff 2006)
[b]na: data not available

alluvial fan, with patches of Inceptisols scattered on the surface. The Changhua coastal plain, on the contrary, is covered mainly by Inceptisols, with zonal distribution of Entisols about 2–4 km in width extending northwestward from the apex of the Choshui alluvial fan. Alfisols are the least abundant soils, being irregularly distributed in a confined area of the southeast region. Table 4 lists the different Great Groups of the soils identified in west-central Taiwan. The soil characteristics of the seven Great Groups of Entisols are neutral to strongly alkaline, with mostly SOC $< 2\%$, CEC < 20 cmol(+)/kg, 0.1–4.7% of $CaCO_3$ content, and DCB-extractable Fe (Fe_d) content $<3\%$. Inceptisols and Alfisols have similar soil characteristics, such as soil pH, CEC and Fe_d contents. But Alfisols have less SOC and $CaCO_3$ contents than those of Inceptisols.

A landform with Inceptisols on the surface is expected to be older than that of Entisols formed on the surface during the geomorphic evolution. The boundary between Inceptisols and Entisols observed in this study area distinguishes the geologic terrains between the Changhua coastal plain and the Choshui alluvial fan, which support the idea that the alluvial fan was built on and above the coastal plain. The sediments discharged by the Choshui River covered most of the coastal area. However, this area may have been exposed by the tides until its uplift since the middle Holocene. The coastal area is further developed by the massive deposition of alluvium from the Choshui River, which constitutes a large alluvial fan on the coastal plain.

The effective time interval of soil formation for each soil order is critical, which involves information on soil ages. For example, the soil ages of $\geq$1,300 and 19,000 years for the Inceptisols distributed on the Changhua coastal plain and the Pakua tableland suggest the required time for soil development toward an Inceptisols. A longer period of time is required for Inceptisols intergrading to Ultisols, where the minimum age of the Ultisols lies between 19,000 and

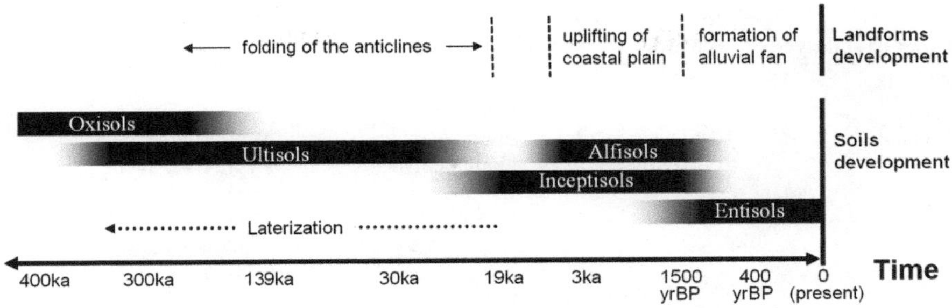

Fig. 12 The pathway of the progressive pedogenic development in taxonomic class of Soil Order in Taiwan. Note that for easy illustration

30,000 ka with a maximum time constraint of more than 300,000 year determined for Oxisol ages. Ultimately, the developmental sequence of soil pedogenesis with time constraints is established for soils graded from Entisols, Inceptisols, Ultisols, to Oxisols (Fig. 12).

3.26 *Paddy Soils of Sandstone and Shale Alluvial Soils in Southern Taiwan*

The Cha-Nan alluvial plain, located at the southern-central Taiwan, is the biggest plain (about 2,730 km^2) in the country. The length of the Cha-Nan plain from south to the north is about 110 km, including Yunlin, Chiayi, and Tanan counties. The mean annual precipitation of the Cha-Nan alluvial plain is about 1,200–2,600 mm, increased from west to east and concentrated from May to September with higher than 100 mm precipitation per month. The soil moisture regime is ustic in the west area along the coast and udic in interior alluvial plain. Unlike the Changhua coastal plain, the dominant parent material is the sandstone and shale alluvium from the eastern high mountain range, which covers more than 64% of the total area in the Cha-Nan plain, and the minor is slate, sandstone and shale mixed alluvium (18%). The history of the man-made irrigated paddy soils in the Cha-Nan alluvial plain begins in 1930, where at the same time the construction of the Wushantou Reservoir was completed for increasing the production of rice. Corn, sorghum and other crops are also common today in the Cha-Nan alluvial plain.

According to the rural soil survey database in three counties (Guo et al. 2005), 144 soil series were classified into four Soil Orders. Table 5 shows the results of the soil classification based on Soil Taxonomy (Soil Survey Staff 2006). Entisols occupied 59% of the soil series, followed by Inceptisols (33%) and Alfisols and Ultisols (8%). The average content of DCB-extractable Fe (Fe_d) in 150 cm depth is >2% in Alfisols, <1.7% in Inceptisols and <1.5% in Entisols (Table 6). These Fe_d contents are less than those of the Ultisols located at the Taoyuan paddy soils. We concluded that the relative lower Fe_d in the Cha-Nan alluvial plain could be

Table 5 Soil classifications of 144 soil series in Cha-Nan alluvial plain of Taiwan[a]

Order of soil	No.	Suborder	No.	Great group	No.	Subgroup	No.
Alfisols	6	Ustalf	3	Paleustalf	3	Typic Paleustalf	3
		Udalf	3	Paleudalf	3	Typic Paleudalf	3
Entisols	85	Aquent	2	Endoaquent	1	Typic Endoaquent	1
		Fluvent	11	Udifluvent	2	Typic Udifluvent	2
				Ustifluvent	9	Typic Ustifluvent	9
		Orthent	51	Udorthent	50	Lithic Udorthent	6
						Aquic Udorthent	5
						Typic Udorthent	39
				Ustorthent	1	Udic Ustorthent	1
		Psamment	21	Udipsamment	15	Aquic Udipsamment	1
						Typic Udipsamment	14
				Ustipsamment	6	Typic Ustipsamment	6
Inceptisols	48	Aquept	3	Endoaquept	3	Typic Undoaquept	3
		Udept	30	Dystrudept	30	Aquic Dystrudept	12
						Typic Dystrudept	18
		Ustept	15	Dystrustept	15	Typic Dystrustept	15
Ultisols	5	Udult	5	Hapludult	1	Typic Hapludult	1
				Paleudult	4	Typic Paleudult	4

[a]Based on keys to soil taxonomy (Soil Survey Staff 2006)

Table 6 Dithionite carbonate bicarbonate-extractable Fe (Fe_d) content (%) of Alfisols, Inceptisols, and Entisols in Cha-Nan plain of Taiwan

Depth (cm)	Alfisols (n = 5)		Inceptisols (n = 6)		Entisols (n = 6)	
	Mean	Standard dev.	Mean	Standard dev.	Mean	Standard dev.
0–20	2.22	0.11	1.27	0.19	1.00	0.46
20–40	2.48	0.12	1.56	0.27	1.12	0.49
40–60	2.59	0.14	1.71	0.18	1.13	0.63
60–80	2.50	0.12	1.62	0.10	1.30	0.81
80–100	2.42	0.14	1.56	0.13	1.46	0.86
100–120	2.41	0.39	1.60	0.15	1.35	0.72
120–150	2.13	0.47	1.60	0.11	1.25	0.68

attributed to the lower annual precipitation and higher atmospheric temperature in turn causing higher evaporation at the soil surface than the water infiltrating into the soil. Lower infiltration rates suggest lower precipitation with lower leaching and soil weathering.

3.27 *Classification of Paddy Soils in Taiwan*

Paddy soils in Taiwan are mainly classified as Entisols, Inceptisols, Alfisols and Ultisols, with minor Mollisols and Oxisols based on Soil Taxonomy (Soil Survey Staff 2006). Chen (1984) and Sheh and Wang (1989) established the classification system of the paddy soils in Taiwan with diagnostic subsurface horizons, typical for

Table 7 Approximate area of major soil groups in paddy soils of Taiwan

Soil order	Suborder	Great group	Area (km^2)	Percentage of total surveyed area (%)
Entisols	Aquents	Haplaquents	20	
		Fluvaquents	150	
	Fluvents	Udifluvents	530	
			(Total: 700)	5
Inceptisols	Aquepts	Haplaquepts	1,500	
	Ochrepts	Dystrochrepts	1,500	
		Eutrochrepts	1,100	
	Umbrepts	Haplumbrepts	300	
			(Total: 4,400)	30
Alfisols	Aqualfs	Ochraqualfs	1,000	
	Udalfs	Hapludalfs	1,600	
	Ustalfs	Haplustalfs	1,000	
			(Total: 3,066)	25
Ultisols	Aquults	Paleaquults	100	
		Plinthaquults	100	
	Udults	Hapludults	500	
	Humults	Haplhumults	30	
			(Total: 730)	5
Mollisols	Udolls	Hapludolls	110	
	Ustolls	Haplustolls	15	
			(Total: 125)	1
Oxisols	Peroxes	Kandiperoxes	50	
	Udoxes	Hapludoxes	50	
			(Total: 100)	<1
Other soils and upland soils		(5,000)	35%	
Total surveyed area in Taiwan	14.600	100%		

Taken from Chen (1992)

the paddy soils of Taiwan including, (1) An Ap horizon, (2) A plow-sole (or plow pan Bg) horizon with relatively high bulk density, (3) An argillic horizon (Bt, Btg or Btv) or a cambic horizon (Bw or Bg) with various redoximorphic features and with significant translocations of Fe and Mn below the plow-sole horizon, and (4) A gleyed horizon with seasonal fluctuations of a high water table.

Aquents, Fluvents and Psamments are the major Soil Suborders in Entisols with no pedogenic features except the Ap horizon, which were mostly distributed on the young alluvial fans near the coastal areas in the western and eastern plains. The above Soil Suborders are different in soil moisture, amount and vertical distribution of organic carbon, and soil texture. The total area with Entisols of the paddy soils in Taiwan is about 700 km^2 (5% of total survey area) (Table 7). Generally, Aquents are much more waterlogged through the profile than Fluvents and Psamments, and thus they are different in soil hydrological conditions such as saturation and reduction durations of the year at 50-cm depth.

Inceptisols are generally the soils derived from the recent alluvium of sandstone, shale and slate in western Taiwan covering a total area of 4,400 km^2. The Soil

Suborders are Aquepts, Ochrepts and Umbrepts with Bwg or Bw horizons. Furthermore, Haplaquepts, Dystrochrepts, Eutrochrepts and Haplumbrepts are the main Great Groups of the alluvial soils of Taiwan.

Alfisols are mainly on the older alluvial plain derived from sandstone, shale, and slate in the western part of Taiwan. Argillic horizons (Bt or Btg) are formed in the lower parts of the pedons (Chen 1992). Aqualfs, Udalfs and Ustalfs are the most important Soil Suborders and have different soil moisture regimes. Ustalfs are locally on the southern part of Taiwan with higher evaporation. Ochraqualfs, Hapludalfs and Haplustalfs are the main Great Groups of the alluvial soils.

Ultisols are mainly found in the paddy soils of red earth and old alluvium from the red earth on Quaternary terraces in northwestern Taiwan (Hseu and Chen 1996) and are used for lowland rice production for over 50 years, with significant morphological characteristics of color change in the epipedon due to flooding by irrigation water. The diagnostic subsurface horizon is the argillic horizon with plinthites (Btv) that are Fe-rich and humus poor mixtures of clay with quartz and other minerals (Soil Survey Staff 2006), commonly occurring as dark red redox concentrations that usually form platy, polygonal or reticulate patterns. These patterns change irreversibly to irregular Fe and Mn concretion or nodules on exposure to the wet and dry cycles. The Great Groups of Aquults are Paleaquults and Plinthaquults, respectively. On the other hand, Udults are the dominant Soil Suborders in the paddy soils of red earth with Hapludults and Paleudults as the main Great Groups.

Mollisols for lowland rice production can be only found in eastern Taiwan, particularly on igneous rock and mudstone alluvial soils or andesite-derived soils. The diagnostic horizon is a mollic epipedon, without the subsurface diagnostic horizon of the Mollisols. The soil depth is generally shallow (<60 cm). However, the high organic matter and low color value and chromas of the surface soils are the dominant characteristics of these paddy soils. Udolls are the main Soil Suborder, and Hapludolls are the main Great Group.

Oxisols are mainly on the older terrace surfaces and have an Oxic horizon. They have undergone a 50-year paddification with Kandiperoxes and Hapludoxes as the main Great Groups in Taiwan.

Although the soil classification in the rural and hilly lands of Taiwan is the system proposed by the USDA Soil Taxonomy, there has been a consistent desire to relate it to the World Reference Base (WRB) system for easy translation. As there are differences in the definitions and the structure of the two soil classification systems, a direct equivalence in terminology may not always be relevant. Table 8 attempts to provide a broad equivalence with the subunit of the WRB system.

There are direct equivalents for some categories, while others may have more than one. Endoaqualfs in Soil Taxonomy approximately correspond to Endostagnic Luvisols; however, Paleudalfs and Paleustalfs generally correspond to Haplic Luvisols. Moreover, Entisols with different soil moisture regimes and various textures provide the units of Regosols, Fluvisols, and Arenosols in the WRB system. Haplohemists easily correspond to Hemic Histosols. Inceptisols with aquic conditions are equivalent to Anthroaquic or Umbric Gleysols, which mostly occur in paddy soils. Additionally, Dystrudepts are directly corresponding to Haplic Cambisols. Mollisols in

Table 8 Correlation of the Taiwan soil series between soil taxonomy (Soil Survey Staff 2006) and the WRB system (IUSS Working Group WRB 2006)

Great group in soil taxonomy	WRB soil group
Endoaqualfs	Endostagnic Luvisols
Paleudalfs	Haplic Luvisols
Paleustalfs	Haplic Luvisols
Endoaquents	Endostagnic Regosols
Fluvaquents	Stagnic Fluvisols
Psammaquents	Haplic or Protic Arenosols
Sulfaquents	Haplic Regosols
Udifluvents	Haplic Fluvisols
Ustifluvents	Haplic Fluvisols
Udorthents	Haplic Regosols
Ustorhents	Haplic Regosols
Udipsamments	Haplic or Protic Arenosols
Ustipsamments	Haplic or Protic Arenosols
Haplohemists	Hemic Histosols
Endoaquepts	Anthroaquic Gleysols
Humaquepts	Umbric Gleysols
Dystrudepts	Haplic Cambisols
Hapludolls	Haplic Phaeozems
Hapludoxes	Geric Ferralsols
Kandiudoxes	Vetic or Ferralic Nitisols
Endoaquults	Endostagnic Acrisols
Paleaquults	Satgnic Acrisols
Plinthaquults	Satgnic Plinthosols
Haplohumults	Umbric Acrisols
Hapludults	Haplic Acrisols
Paleudults	Haplic Acrisols
Plinthudults	Acric Plinthosols
Dystruderts	Haplic Vertisols

Taiwan are almost classified as Hapludolls, and thus approximately correspond to Haplic Phaeozems. Oxisols with oxic horizons and kandic horizons appear to be Ferralsols and Nitisols, respectively. All Ultisols are corresponding to Acrisols, with the exception of Ultisols with the Plinthic Great Group to Plinthosols. Eventually, Vertisols are additional examples where Soil Taxonomy and WRB systems have equivalent definitions in the rural and hilly lands of Taiwan.

4 Conclusion

Soil properties in China and Taiwan have been greatly influenced by various anthropogenic processes to form anthroscapes, particularly by long-term agricultural use over hundreds or thousands of years. The total area of anthropogenic soils is around 397,200 km^2 in China. Paddy soils are more than 30 million hectares in China and Taiwan (23% of the world's total irrigated lands), occurring mainly in the area of the Yangtse River. Farmers along the Yangtse River have planted aquatic rice for about 6,000–7,000 years, i.e., as early as in the Stone Age, which was followed by expansion of agricultural activities nationwide.

According to the Keys of the Chinese Soil Taxonomy, Anthrosols meet the requirements of the combination of horizons including anthrostagnic, fimic, siltigic, and cumulic epipedons and hydragic and phos-argic horizons, respectively. The changes taking place in the soils after planting rice are degradation of SOM, redistribution of exchangeable bases, translocation and segregation of Fe and Mn by reduction and oxidation, and decomposition and synthesis of clay minerals. Additionally, the genetic horizons of paddy soils in China include cultivated horizons, plowpans, percogenic horizons, and plinthitic or illuvial horizons, in which the plinthitic horizon or illuvial horizon is the diagnostic horizon of paddy soils different from other soils.

Typical horizons in the paddy soils of Taiwan includes the Ap horizon, plowpan, argillic horizon or cambic horizon with various redoximorphic features and with significant translocation of Fe and Mn below the plow pan, along with a gleyed horizon with seasonal high water table. The Anthrogleyic Epipedon in Taiwan was mainly characterized by a thick horizon (10 cm thick or more) with soil color changes to 2.5Y or yellower and chroma of 2 or less in the Ap horizon than that of the underlying horizon. These soil morphological characteristics are common in the surface horizon in paddy use. Gray mottles are very common in paddy soils on the alluvium of Taiwan. The gleyed horizon with strongly reducing conditions was also proposed in the Chinese soil Taxonomy.

The SOC contents of the paddy surface soils are higher than those of the non-paddy soils. Irrigation has induced the enrichment of the SOC stock in paddy soils. The free Fe contents of the paddy profiles tend to increase as the length of cultivation history increases, but this trend is less profound in non-paddy profiles. However, the main factor controlling the duration of saturation and reduction of the paddy soils of Taiwan is the landscape.

References

Batjes NH (1996) Total carbon and nitrogen in the soils of the world. Eur. J. Soil Sci. 47, 151–163.

Brinkman R (1970) Ferrolysis, a hydromorphic soil forming process. Geoderma 3, 199–206.

CGS (2004) Geologic map of Changhua. Central Geological Survey of Taiwan, scale 1:50,000, 1 sheet.

Chen ZS (1984) A model for the nowadays soil survey and classification of paddy soils in Taiwan: the study on the formation, genesis and classification of the seashore district paddy soils in Taoyuan, Taiwan. PhD Dissertation. Graduate Institute of Agricultural Chemistry, National Taiwan University, Taipei, Taiwan.

Chen ZS (1992) Morphological characteristics, pedogenic processes, and classification of wet soils in Taiwan. In: Kimble JM (ed.), Proceedings of the Eighth International Soil Classification Workshop, October 6 to 21, 1990, Baton Rouge, Houston, and San Antonio, pp. 53–59.

Chen ZS, Chang JM (1985) An investigation on the composition, distribution, and their pedogenetic characteristics of free oxides in well drained paddy soils. Memoirs of the College of Agriculture of the National Taiwan University, 25, 22–39.

Cheng YQ, Yang LZ, Cao ZH, Ci E, Yin S (2009) Chronosequential changes of selected pedogenic properties in paddy soils as compared with non-paddy soils. Geoderma 151, 31–41

Collins JF, Buol SW (1970) Effects of fluctuations in the Eh-pH environment on iron/or manganese equilibria. Soil Sci. 110, 111–118.

Dazzi C, Lo Papa G, Palermo V (2009) Proposal for a new diagnostic horizon for WRB Anthrosols. Geoderma 151, 16–21

Eswaran H, Van der Berg E, Reich P (1993) Organic carbon in soils of the world. Soil Sci. Soc. Am. J. 57, 192–194

Gong ZT (1992) The formation of paddy soils. In: Li QK (ed.), Paddy Soils of China. Science Press, Beijing, pp. 3–29 (in Chinese).

Gong ZT, Liu L (2002) Holocene palesols in Kunshan, south Jiangsu Province. Acta Pedol. Sin. 39, 618–627 (in Chinese, with English abstract).

Gong ZT, et al. (ed.) (1999) Chinese Soil Taxonomy: Theories, Methods and Practices. Science Press, Beijing (in Chinese).

Gong ZT, Zhang X, Chen J, Zhang G (2003) Origin and development of soil science in ancient China. Geoderma 115, 3–13

Guo HY, Liu CH, Chu CL, Chiang CH, Yei MC (2005) Taiwan soil information system. Taiwan Agricultural Research Institute, Council of Agriculture, Taichung, Taiwan (http://www.tari.gov.tw/index_in.htm).

Ho CS (1986) An introduction to the geology of Taiwan: explanatory text of the geologic map of Taiwan, 2nd edition. Central Geological Survey of Taiwan, Taiwan, 165pp.

Hseu ZY, Chen ZS (1994) Micromorphology of rice-growing Alfisols with different wetness conditions in Taiwan. J. Chinese Agri. Chem. Soc. 32, 647–656

Hseu ZY, Chen ZS (1995) Redox processes of rice-growing Alfisols with different wet conditions. J. Chinese Agri. Chem. Soc. 33, 333–344

Hseu ZY, Chen ZS (1996) Saturation, reduction, and redox morphology of seasonally flooded Alfisols in Taiwan. Soil Sci. Soc. Am. J. 60, 941–949

Hseu ZY, Chen ZS (1997) Soil hydrology and micromorphology of illuvial clay in an Ultisol hydrosequence. J. Chinese Agri. Chem. Soc. 35, 503–512

Hseu ZY, Chen ZS (2001) Quantifying soil hydromorphology of a rice-growing Ultisol toposequence in Taiwan. Soil Sci. Soc. Am. J. 65, 270–278.

Hseu ZY, Chen ZS, Leu IY (2000) Soil solution somposition, water tables, and redox potentials of anthraquic Ultisols in a toposequence. Soil Sci. 165, 869–880

ISS/CAS (Institute of Soil Science, Chinese Academy of Sciences) (ed.) (2001) Chinese Soil Taxonomy. Science Press, Beijing, 203p.

IUSS Working Group WRB (2006) World Reference Base for Soil Resources 2006, World Soil Resources Report No. 103, 2nd edition. FAO, Rome.

Jien SH, Hseu ZY, Chen ZS (2004) Relations between morphological color index and soil wetness condition of anthraquic soils in Taiwan. Soil Sci. 169, 871–882

Jien SH, Hseu ZY, Chen ZS (2010) Hydropedological Implications of Ferromanganiferous Nodules in Rice-Growing Plinthitic Ultisols under Different Moisture Regimes. Soil Sci. Soc. Am. J. 74:880–891.

Kosse AD (1990) Diagnostic horizons in Anthrosols. In: Rozanov BG (ed.), Soil Classification, Reports on the International Conference on Soil Classification, Centre for International Projects. USSR State Committee for Environmental Protection, Moscow, pp. 264–273.

Kyuma K (2004) Paddy Soil Science. Kyoto University Press, Kyoto, 280p.

Lal R (1999) Soil management and restoration for C sequestration to mitigate the acceleration greenhouse effect. Prog. Environ. Sci. 1, 307–326.

Lal R (2004) Soil carbon sequestration impact on global climate change and food security. Science 304, 1623–1627.

Lal R, Griffin M, Apt J, Lave L, Morgan GM (2004) Managing soil carbon. Science 304, 393.

Li Q (1992) Paddy Soils of China. Science Press, Beijing, pp. 232–248 (in Chinese).

Li C, Sun Q (1990) Soils of China. Institute of Soil Science, Chinese Academy of Sciences, Science Press, Beijing, 873p (in Chinese).

Lin CC (1957) Topography of Taiwan. Taiwan Provincial Documentary Committee Publication, Taiwan, 424p (in Chinese).
Lin YS, Chen YG, Chen ZS, Hsieh ML (2005) Soil morphological variations on the Taoyuan terrace, Northwestern Taiwan: role of topography and groundwater. Geomorphology 69, 138–151.
Lin YS, Lin YW, Wang W, Chen YG, Hsu ML, Chiang SH, Chen ZS (2007) Relationships between topography and spatial variations in groundwater and soil morphology within the Taoyuan-Houko Tableland, Northwestern Taiwan. Geomorphology 90, 36–54
Lynn WC (1988) Discussion in the Fifth International Soil Management. Workshop: Classification and Management of Rice-Growing Soils. Wufeng, Taichung, Taipei, Taiwan, December 11–23, 1988, 3 volumes, Eswaran, H. and T.C. Juang (editors). Food and Fertilizer Technology Center for the Asian and Pacific Region (FFTC/ASPAC), Taiwan Agricultural Research Institute (TARI) and Soil Management Support Service (SMSS) of USAID.
Mitsuchi M (1992) Anthropically induced wet soils. In: Kimble JM (ed.), Characterization, Classification, and Utilization of Wet Soils. Proceedings of 8th International Soil Correlation Meeting. USDA-SCS Publication of National Soil Survey Center, Lincoln, pp. 179–184.
NSSC Staff (National Soil Survey Center Staff) (2002) Soil Taxonomy and Anthropogenic Soils Classification. Website of International Committee for Anthropogenic Soils (http://clic.cses.vt.edu/icomanth/) (Verified April 22, 2009).
Pan GX, Li LQ, Wu LS, Zhang XH (2003) Storage and sequestration potential of topsoil organic carbon in China's paddy soils. Glob. Chang. Biol. 10, 79–92.
Pei P (1989) Remains of rice planting culture in Pengtoushan and prehistoric rice planting agriculture in China. Agric. Archaeol. 2, 25–30 (in Chinese).
Sheh CS, Wang MK (1989) The soils of Taiwan. Soil Survey and Testing Center, Department of Soil and Environmental Sciences, National Chung Hsiung University, Taichung, Taiwan, 205p.
Shi XZ, Yu DS, Yang GX, Wang HJ, Sun WX, Du GH, Gong ZT (2006) Cross-reference benchmarks for translating the genetic soil classification of China into the Chinese Soil Taxonomy. Pedosphere 16, 147–153
Silva MTB, Ojea FG (1991) Iron oxide accumulations in tertiary sediments of the Roupar basin, Galicia, NW Span. Catena 18, 31–43
Soil Survey Staff (1992) Keys to Soil Taxonomy, 6th edition. U.S. Department of Agriculture, Soil Conservation Service, Washington.
Soil Survey Staff (2006) Keys to Soil Taxonomy, 12th edition. U.S. Department of Agriculture, Soil Conservation Service, Washington.
Somasiri S, Deturck P (1992) Anthropic wet soils of Sri Lanka. In: Kimble JM (ed.), Characterization, Classification, and Utilization of Wet Soils. Proceedings of 8th International Soil Correlation Meeting. USDA-SCS Publication of National Soil Survey Center, Lincoln, pp. 240–247.
Sun TL, Yang TC (1989) The occurrence of poor drainage conditions in alluvial soils of Taiwan. J. Chinese Agri. Chem. Soc. 27, 38–45
Tsai H, Maejima Y, Hseu ZY (2008) Meteoric ^{10}Be dating of highly weathered soils from fluvial terraces in Taiwan. Quart. Int. 188, 185–196.
Vepraskas MJ (1992) Redoximorphic features for identifying aquic conditions. Technical Bulletin 301. North Carolina Agricultural Research Service, North Carolina State University, Raleigh.
Vepraskas MJ, Guertal WR (1992) Morphological indicators of soil wetness. In: Kimble JM (ed.), Characterization, Classification, and Utilization of Wet Soils. Proceedings of the 8th International Soil Correlation Meet. USDA-SCS Publication of National Soil Survey Center, Lincoln, pp. 307–312.
Yang X, Effiand WR, Fenton TE (1988) Some diagnostic subsurface horizons in the proposed China Soil Classification System. Soil Surv. Horiz. 29, 88–94.
Yu TR (1985) Physical Chemistry of Paddy Soils. Science Press, Beijing, 217pp.

Zhang GL, Gong ZT (1993) Geochemical features of element migration under artificial submergence. Acta Paedol. Sin. 30, 355–365

Zhang GL, Gong ZT (2003) Pedogenic evolution of paddy soils in different soil landscapes. Geoderma 115, 15–29

Zhang M, Ma L, Li W, Chen B, Jia J (2003) Genetic characteristics and taxonomic classification of fimic Anthrosols in China. Geoderma 115, 31–44.

Zhejiang M (1978) Archaeological report on Hemudu first excavation. Acta Archaeol. Sin. 1, 39 (in Chinese, with English abstract).

Zhu X (1964) Lou Soils. Chinese Agricultural Press, Beijing, pp. 1–22 (in Chinese).

An Anthroscape from Morocco: Degraded Rangeland Systems and Introduction of Exotic Plant Material and Technology

C. Zucca, F. Previtali, S. Madrau, E. Akça, and S. Kapur

Abstract Fodder shrub plantations (*Atriplex nummularia*) were extensively introduced to rehabilitate degraded rangeland and to mitigate desertification in the Rural Municipality of Ouled Dlim (Marrakech Province, Morocco). The original rural landscape, characterised by sparse or absent natural vegetation cover due to a long history of intense grazing activities, has been deeply modified during the last decade by these agroforestry interventions. The purpose of this contribution is to describe the occurred environmental changes under the perspective of the Anthroscape concept, by giving emphasis to the description of the Anthroscape units affected by the plantations and to the geo-pedological processes that influenced the technical success obtained by the interventions. Some critical aspects and possible scenarios of the future evolution of the anthroscapes are also discussed.

Keywords Atriplex nummularia · Oldman saltbush · Rangeland degradation · Land restoration

C. Zucca (✉) and S. Madrau
DIT (Department of Territorial Engineering), University of Sassari, Via de Nicola, 9, 07100 Sassari, Italy
and
NRD (Nucleo Ricerca Desertificazione), University of Sassari, Sassari, Italy
e-mail: clzucca@uniss.it

F. Previtali
DISAT, Department of Environmental Sciences, University of Milano-Bicocca, Milano, Italy

E. Akça
University of Adiyaman, Department of Technical Programs, Adiyaman, Turkey

S. Kapur
University of Çukurova, Departments of Soil Science and Archaeometry, 01330 Adana, Turkey

S. Kapur et al. (eds.), *Sustainable Land Management*,
DOI 10.1007/978-3-642-14782-1_10,

1 Introduction

The anthroscape described by the present work is an example of rural landscape reshaped in the course of centuries by the traditional land use and further modified during the last decade by agroforestry interventions aimed at supporting rural development and mitigating desertification processes. More specifically, fodder shrub plantations were introduced to rehabilitate the productivity of severely degraded rangeland and to mitigate the environmental degradation processes, especially soil erosion by water and wind. These interventions, carried out with species well adapted to aridity conditions, became a common practice at least a couple of decades ago and, according to Le Houérou et al. (1991), were in many cases successful for the isoclimatic Mediterranean arid zones. The same authors describe these plantations as a "new man-made agro-silvo-pastoral production system", thus anticipating some of the concepts discussed below.

The area studied by the present work is located in the Rural Municipality of Ouled Dlim, in the Marrakech Region of Morocco. Since the early 1990s participatory projects based on these interventions, were carried out by the Moroccan government in the frame of the national strategies for rangeland rehabilitation. The shrub species used for the rehabilitation of the land was *Atriplex nummularia* ("Oldman saltbush"). Such species, native to Australia, proved to be well adapted to local conditions and was introduced for its resistance to aridity and grazing (Mulas and Mulas 2009; Le Houérou et al. 1991; Osmond et al. 1980). It can grow in dry conditions and on marginal soils, being characterized by a high efficiency in the use of water (Correal et al. 1990). Furthermore it is a halophyte being among the most salt-tolerant plants. As *Atriplex nummularia* can survive in conditions that prohibit the growth of most of other plants, it has a considerable potential for providing forage and fuel-wood in the arid and semiarid regions of North Africa. It is palatable for the livestock and can provide a high fodder production: according to Le Houérou (1995), up to 30 tons of dry matter per ha per year if irrigated and with EC conditions of soil around 15–20 dS/m. Furthermore it provides a year-round grazing green biomass and a feed availability into dry periods of foliage-wise high crude protein contents.

More recently (Zucca et al. 2005), in the same zone, this approach was extended and given wide visibility by a Euro-Mediterranean Cooperation Project, funded by MEDA-SMAP (Short and Medium Term Priority Environmental Action Programme), that was coordinated by the NRD of the University of Sassari. The Project ("Demonstration Project on Strategies to Combat Desertification in Arid Lands with Direct Involvement of Local Agropastoral Communities in North Africa", 2002–2007) aimed the application of the approach at a demonstrative level and on an extensive scale (about 2,000 ha). The project has been implemented through established groups of farmers (breeders Cooperatives) who signed partnership agreements. Local communities have been involved in project planning, placed their land at disposal and contributed to cover plantation and management costs. A high level of participation was also assured by the creation of a community nursery where all the plants used by the project were grown.

Several thousands of hectares have been planted so far in the region, consequently modifying the land use spatial patterns and practices and affecting the economic productivity along with the ecological functions of the land. However, the local communities conserved their life style based on grazing and sheep breeding.

The objective of the present chapter is to describe the landscape and the changes that have occurred from the perspective of the anthroscape concept.

An Anthroscape can be defined as a landscape clearly and prevalently bearing visible features imprinted by human activities (Previtali (2010)). The concept should be specifically applied to situations where marked differences from the natural landscapes are observed (Kapur et al. 2004), in such cases it would offer a model to understand the underlying processes which created the particular landscapes.

Emphasis is given here to the description of the anthroscape units affected by the plantations and to the on-going geo-pedological processes that influenced the clear technical success obtained by the interventions. Moreover, some critical aspects and possible scenarios of the future evolution of the anthroscapes are also briefly discussed.

2 The Study Area

The study area covered by the SMAP project and considered in this study is located in the Rural Municipality of Ouled Dlim, in Morocco, about 30 km North-West of Marrakech (31°45′N to 32°N and 8°15′W to 8°45′W; Fig. 1). The region is characterized by a semi-arid Mediterranean type of climate, with warm and dry summers and rainfalls concentrated in the winter season, with a high inter-annual and intra-annual variability. In particular, in the considered area, the average annual rainfall is 212 mm, the mean annual temperature 19, 9°C (Ouled Dlim station, years 1978–1997), with a main rainy period at the end of autumn and during winter and a dry season between April and October. The study area is located in the central-west part of the Palaeozoic Jebilet, the oldest relief of Morocco (Huvelin 1970) and it corresponds to two main geological units, at the west and the east respectively of a fault system. The west side (Upper Carboniferous) is characterized by schist and carbonatic schist formations, where similarly the east side (Lower and Middle Carboniferous, Visean and Namurian respectively) is dominated by the metamorphic formations of the "Sarhlef" schists in particular.

The landscape is characterised by rounded relieves, moulded by water erosion processes acting on the highly weathered and erodible rock formations (Fig. 2). A great part of the study area is nowadays characterized by moderate water erosion indicated in some places also by on-going processes of gully development. Erosion and accumulation pediments (*glacis*) are the most typical local landforms. At present the soils are highly degraded because of overgrazing and subsistence

Fig. 1 Left : the study area, included in the Rural Muncipality of Ouled Dlim, Marrakech region, Morocco. Right: the ideal "anthroscape transect" drawn on a Landsat TM map. The transect extends from the highest part of Jebilet relieves (Jebel Bou Gader) down to the Tensift river and, crossing the southern bank of this one, to a higher terrace surface. *Black crosses* correspond to soil profiles (the four bigger ones are the reference profiles mentioned in the text)

Fig. 2 A panoramic view of the rounded relieves and landforms of the study area. Photo by Anna Grignaschi

cropping (nutrient yielding by rainfed cereals). Consequently, the natural vegetation cover is sparse or absent and the soil is subjected to water and wind erosion.

In this regard, the area is a representative site of the more general and well known situation of rangeland degradation in northern Africa and in Morocco (Benbrahim et al. 2000). Unfortunately, the arid and semi-arid lands of the region possess a fragile natural resource base and offer limited alternatives for sustainable increases in agricultural productivity under purely rainfed conditions (ICARDA 1995). Agricultural activity in such areas is frequently dominated by range dependent small ruminant production systems (FAO 1995). However, during the past 2 or 3 decades, increases in livestock populations and people sedentarization have raised fears of irreversible environmental degradation and increased poverty for the inhabitants of such areas (IFAD 1996). Agricultural activities are often carried out by women because men migrate to look for jobs in the urban centres, due to the limited and irregular forage production in the region. These tendencies are widespread in the study area due to the proximity of the fast developing Marrakech hinterland.

3 The Anthroscape Before the Interventions

As already mentioned, all the area has been deeply modelled by ancient and present erosion processes, accelerated by land use practices. The area has been traditionally used as extensive pasture land. It is actually inhabited by a formerly nomadic breeder's community native of the south (present Mauritania and former Western Sahara): they left their native region during the sixteenth century and settled in Ouled Dlim, but only fairly recently they got fully sedentary. Sheep grazing is their most important economic activity.

Centuries of grazing activities led to the degradation of the natural shrub and tree vegetation cover in rangeland. However, sparse individuals of *Ziziphus lotus* (locally "*Jujubier*") have remained, due to their function as sheep shelters and fuel wood sources, along with commonly scattered species typical of degraded pasturelands, such as *Peganum harmala*. Despite the positive functions attained by these and other indigenous cover plants since millennia, rainfed crops contribute to soil erosion as well as to the degradation of the natural vegetation cover. Cereals, especially barley and wheat, are grown as fodder for livestock. Although seeding is done only in particularly rainy winters, and, as an average, every 4–5 years, in most places soil is ploughed almost every year. This appears to be in relation with the uncertain, collective land tenure, leading the farmers to leave on the ground visible signs of their "informal ownership". Land tenure and breeders organisation are further discussed later on.

Although these anthropic factors created a sort of geomorphic uniformity on the landscape, the land and soil units conserved features that document an ancient and complex evolution. These features, that have developed, along with the

Anthroscape reshaping process, had an influence on the success of the plantation interventions undertaken in the area.

The main land and soil features of the local anthroscape are described along a representative topo-sequential transect schematically drawn in Fig. 1. The related topographic profile and anthroscape units are graphically represented in Fig. 3 and described in Table 1. This table makes reference to soil profiles described for the area during a survey carried out in the frame of the SMAP project (Zucca and Previtali 2009).

The sequence shown in Fig. 3 extends from the highest part of Jebilet relieves (Jebel Bou Gader) down to the Tensift river and, crossing the southern bank of its bed, to a higher terrace surface.

One of the most characteristic and dominant features in the study area is the presence of old calcrete[1] fragments and relict calcrete surfaces, especially in Unit 3 and in some parts of Unit 2. These can be observed both lying on the eroded *LUVISOLS* (Unit 3, Profile M1, by the project nursery indicated in Fig. 3) and directly encrusting the outcropping schists (Unit 3, Profile M27). Also, *petrocalcic* horizons were observed within some profiles close to the M27 site (Petric *CALCISOLS*). Features of the in situ *petrocalcic* horizons were better preserved from erosion at hilltop positions.

The other characteristic and widespread feature is salinity and in particular the relative abundance of the soluble salts. The accumulation of calcium carbonate decreases with elevation and is almost absent in profile M30, at 560 m a.s.l. Salt accumulation can be observed even at higher locations, as reported for another profile (M25, 580 m a.s.l., classified as SOLONETZ), not included in the transect discussed here, but not far from M30. The source of both carbonate and salt is obscure if compared to the acidic composition of the dominant bedrocks and the absence of salt-rich formations in the area. An in-depth and still on-going study of

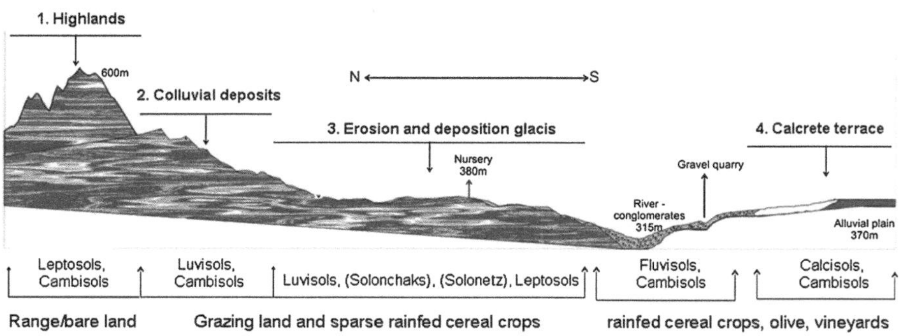

Fig. 3 The anthroscape units described along the studied transect. (1) Highlands; (2) Colluvial deposits; (3) Schist outcrops and pediment (glacis); (4) Calcrete terrace

[1]The definition of *calcretes* and a description of their main features is given by Madrau and Zucca (in this book).

Table 1 Anthroscape units of a representative transect in the study area (the soil profiles mentioned were described during a previous survey) (Zucca and Previtali 2007)

Anthroscape unit	Landforms and typical soils (FAO/ ISRIC/ISSS 2006)	Land use and potentialities
Highlands	Schist bedrock, moderate to steep slope and dissected morphology. Different Palaeozoic formations, in particular from Upper Carboniferous, showing relevant tectonic structures and displacements Soil profile shallow and weakly developed (LEPTOSOLS) On colluvium: CAMBISOLS Profile M30, *Kat Laazara* site: Haplic CAMBISOL (Colluvic)	Vegetation cover almost absent. The area is used as extensive rangeland. According to local land planning this domain is to be allocated to "forestry" or to rehabilitation and conservation of soils and environment Some areas on colluvia were extensively planted with fodder shrubs (Fig. 5a). Here soils are relatively deeper (up to 40 cm) and lighter. They are free of carbonates and show abundant biological activity, but in most of the situations the slope is too steep to allow for their agricultural use
Colluvial deposits	Pediment (*glacis*) surfaces, gently undulating, mainly developed on colluvial deposits Slope angle gradually decreases downslope. Accelerated water erosion has created moderate to deep incisions on the steepest slopes Profile M29, *Douar Boushab* site: Calcic Cutanic LUVISOL (Chromic)	Sparse shrub cover; herbaceous and degraded pasture land; rainfed cereal crops Soils are stony and calcareous; part of the coaese fraction is made of calcrete fragments. M29 shows a subangular prismatic structure in Btk horizons, inherited from ancient pedological processes. The A horizon was almost completely eroded. This soil is about 40 cm deep and relatively suited to conservative rainfed herbaceous crops Fodder shrub plantations were made in several locations, including near profile M29
Schist outcrops and pediment	Pediment erosive surfaces (*glacis d'érosion*), gently undulating, mainly developed on bedrock (chloritoschists of the "Sarhlef" formation) and on isolated colluvial-alluvial deposits To the south, a wide outcrop of black chloritoschists overlaid by the relicts of an old calcrete surface and characterised by very shallow eroded soils (M27) Profile M1, *Abda Skarna* "nursery" site: Calcic Epileptic LUVISOL (Sodic) Profile M27, *Kedadra* site: Lithic LEPTOSOL (Eutric, Aridic)	Herbaceous degraded pasture land; rainfed cereal crops Soils are shallow (around 30 cm) and stony and the bedrock is often outcropping. They are calcareous and rich in salts In profile M1, the shallow Btk horizon, relict of an ancient pedological evolution, is sandy clay loam and rich in calcrete fragments. In profile M27 soils are very shallow (around 10 cm), with poor fertility and biological activity and do not show the evidence of a long evolution. Fodder shrub plantations are widespread over this unit
Calcrete terrace	Surface very gently sloping and characterized by sparse large	Herbaceous degraded pasture land; rainfed cereal crops; olives and

(*continued*)

Table 1 (continued)

Anthroscape unit	Landforms and typical soils (FAO/ ISRIC/ISSS 2006)	Land use and potentialities
	calcrete blocks, relicts of an old calcrete of likely lacustrine origin, with thickness up to half a meter or more	vineyards. Soils not surveyed (outside project area)

the soils of the study area, conducted via micromorphology and isotope analyses, will contribute to the interpretation of the source of the carbonates and indirectly of the salt, indicating multiple sources and suggesting a long and complex climatic and geomorphologic evolution.

As an example, the calcrete terrace observed in Unit 4, on the hydrographic left of the Tensift river, appears to be very ancient and of lacustrine origin, as well as part of the fragments coming from Unit 3. This explains in part the origin of the carbonates observed in the soils. Concerning the almost ubiquitous abundance of salts, the aeolian Saharan input seems to be, the most likely, responsible. As it is well known, the Saharan dust is transported by winds at long distances. The depositions are particularly relevant on the surroundings of the Mediterranean areas and are known as one of the major sources of clay minerals (such as palygorskite and kaolinite) and salts (Kubilay et al. 1997). The study area is quite often affected by Eastern Saharan winds and dust fall and more vigorously was affected in the past geological periods (Yaalon and Ganor 1979).

4 The Newly Created Anthroscape

Most of the land of the area is managed collectively and subjected to the legal property status of land locally called *Guich*.[2] Collective land management rules are defined and enforced by the community but remain unwritten. During the last decades the local government promoted the creation of cooperatives of breeders, especially aimed at promoting a more formal organisation of farming activities in view of the rational management of collective resources. As a consequence local groups got formally entitled to adhere and to participate to state programs and to projects such as the SMAP. This task was essential for the project and was undertaken by the adhering cooperatives playing an active role from the beginning. The farmers were also involved in training and sensitisation activities, starting from the first plantation interventions carried out by the local administration (DPA, Direction Provinciale de l'Agriculture) in 1996–1997, and continuing later by receiving

[2]*Guich* are lands that the State conceded in use to some tribes as a compensation for given services of military nature. These lands can be typically found around the former imperial cities: Meknès, Fès, Marrakech, Rabat.

constant technical advice. Finally, awareness was raised, concerning the positive changes they were introducing to the landscape.

The newly created Anthroscape is strongly characterized by the presence of *Atriplex* plantations from both the aesthetic and the ecological points of view. Figure 4 documents, for an area located in another sector of the project zone (not touched by the present transect), the striking landscape contrast introduced by a plantation in about 6 years.

Actually, the general technical success obtained by the interventions is widely recognised in the area, where plant development is most often good even in very unfertile sites, proving the high degree of adaptability of the species to the local conditions. Figure 5 shows examples of plantations located in transect areas corresponding to the main anthroscape units, with evident landscape changes. Apart from the aspects related to the specific land suitability and the effectiveness of the seedlings management and plantation techniques, etc., the positive technical outcome was principally made possible by the community management. So far, the community-based management ensured a sustainable exploitation of the parcels. These are opened to grazing only at least 3 years after planting and, although *Atriplex* edible biomass is green all the year round, it is preferably grazed after summer, when there is no natural forage left outside the perimeter of the planted area.

The assessment of the impacts and benefits of such kind of interventions would be complex (Zucca et al. 2006) and beyond the scope of this paper. An analysis of the economic, ecological and hydrological impacts of the interventions is not available so far. They will be dealt with in the next 3 years by a new European project bearing the acronym "PRACTICE" (DG Research-FP7), that just started.

However, considering the gains of plant production and landscape conservation in the area attained by this project work, the concrete positive impacts generated by the introduced fodder shrub can be given as below

- Increase of the standing green fodder biomass, especially as a reserve for the dry season.

Fig. 4 Pictures taken about 6 months after seedling plantation (*left*; October 2000) and about 6 years later (*right*; January 2007), from the same place. The average height of the plants in the first line in the 2007 picture is about 1 m. Downslope it reaches 2 m. Photos by Claudio Zucca

Fig. 5 Examples of plantations located in transect areas corresponding to the main anthroscape units. (**a**) very recent plantation in highland (unit 1, by profile M30; photo May 2007); (**b**) about 10 years old plantation, medium developed, after heavy grazing (unit 3, by the nursery; photo January 2007); (**c**) about 5 years old plantation, before grazing (unit 3, comparable landscape position as in the previous picture; photo January 2007); (**d**) about 4 years old plantation (unit 3, southern sub-unit, by profile M27; photo January 2005); (**e**) Same location as for picture (5d), showing the broken, shallow chloritoschists in the furrow. Photos by Claudio Zucca

Measurements of the rangeland biomass production before the intervention ranged around 40 FU/ha, while in mature plantations, the production of edible biomass increased considerably, according to Ferchichi et al. (2007), up 167 FU/ha for Atriplex foliage, and up 718 FU/ha for the underlying herbaceous layer (in favourable parcels, in the phase of maximum development).

- Interannual "buffer" feeding reserve of standing fodder for facing severe and prolonged droughts.
  Atriplex biomass is mostly complementary to sheep diet. In the years of appreciable climatic conditions, during exploitation, the natural vegetation developed between the plant lines was primarily grazed disregarding defoliation. In the years of drought, the biomass attained constitutes a strategic buffer reserve for the breeders.
- Establishment of a dense vegetation cover and mitigation of soil erosion and of decline of soil quality.
  Plant density (generally according to a regular 3 × 3 m grid, or about 800 plants per ha) and canopy development ensure a dense vegetation cover. This plays a role in terms of shadowing and microclimate, wind breaking, soil organic matter enrichment, biological activity, recovery of autochthonous herbaceous species etc., strongly favouring land rehabilitation and desertification control.
- Water effectiveness of the intervention; irrigation is needed the first year only and just for a couple of times: during plantation and after a few months. So the approach helps to save the water sources and provides abundant biomass with very low water consumption.
- Wildlife shelter.
  The oldest plantations are now increasingly becoming habitats for wild birds and small mammals that were almost disappeared in the area.
- Fuel wood production.
  Some years (8–10) after the plantation, the plants become senescent and the ligneous biomass should be pruned to promote re-sprouting. This allows for a recovery of the green biomass productivity during the following years and can provide significant amounts of fuel wood.
- Last, but not least, the consolidation of the organisation and cohesion of the local communities, along with the investments made for the establishment of the nursery, which is still functioning and producing even after the termination of the project, are of high merit.

5 Discussion and Conclusions

The basic technical outcome of the intervention (plant development) is not homogeneous. In the study area there is a remarkable correspondence between the success of the plantations and the distribution of the saline and carbonate rich soils. Other factors, such as clay contents and especially soil depth would be effective as well. The soil and land survey carried out, and especially the ongoing work on the palaeoenvironmental evolution are providing essential contribution to the understanding of the spatial pattern of the relevant pedological features in the area.

So far, these aspects have been assessed in a semi-quantitative way only. Based on the field experience, some empirical land suitability classes may be described for the studied region as in Table 2.

Table 2 Empirical land suitability classes related to the *Atriplex nummularia* plantation in the study area

		Land suitability class			
		Good	Medium	Poor	Not suited
Soil and site					
Mechanisation potential					
Slope (%)		0–10	10–20	20–35	35–50
Hard rock outcrop (%)		0–1	1–4	4–10	> 10
Depth available to roots (cm)		> 30	15–30	15–5	< 5
Texture[a]	USDA Class	Silt, silt loam, silty clay loam	Clay loam, silty clay, sandy clay loam, sandy loam	Clay[b], sandy clay[b]	
Erosion	Field observation	Absent/weak	Moderate	Intense	–

[a]Only texture classes observed on the ground were considered
[b]In case of compacted and/or eroded soil

The table highlights in fact the very limited requirements of the plants. When the plantation interventions began in the area more than a decade ago, both these aspects and the characteristics of soils of the area were unknown. So, the relative abundance of the saline soils was not really considered.

Even more recently, the choice of the parcels was mostly linked to social factors and based on a community decision. This was only in part related to the availability of water wells, and essentially based on the negotiation between the cooperative members. Although the choice may have been, sometimes, not optimal from the technical point of view, the outcome was in general satisfying, where the social and management constraints proved to be the determinant factors in the study area.

However, the knowledge generated by the studies carried out in the frame of the SMAP project (Bellavite et al. 2007) would most certainly support in the future a more appropriate development of the system introduced. Nevertheless, a high degree of social acceptance of the system was and is the first condition for the observed success. This has been due to the fact that the intervention did not change the way of life of the people. Only, a different kind of grazing system based on community management and control was introduced, stimulating a slight cultural change. From this point of view, the interventions had a societal impact as well, reinforcing the cooperative participation and the decision making processes, finally strengthening social cohesion. The nursery established by the project and managed by the local cooperative also contributed to improve these dynamics (Fig. 6).

As a consequence, the probable scenarios for the evolution of the "Atriplex Anthroscape" strongly depend on management factors. Following the termination of the external investments made by projects, and the decreasing support from local extension services that implemented the projects, the communities would inevitably have to explore the probabilities to take over the responsibility to valorise and preserve the preceding investments. As an example, the "coppicing" of the senescent plants requires man power and some limited equipment as the maintenance of the management and control system requires cohesion and common interest. Moreover, some local dynamics such as the land abandonment by the young people and

Fig. 6 The nursery realized by the project and managed by the local cooperative. Photo by Claudio Zucca

the attraction of the developing urban centres (Marrakech) along with the migrations abroad threaten the sustainability of the new production system.

If these are the challenges in a short term perspective, in the long term new visions will be needed. The long-term dynamics (and their ecological implications) of the newly established agro-forestry systems have not been fully considered so far. This is of particular importance because the introduced species are not autochthonous, although they are not invasive. It is clear that the present landscape should be seen as an intermediate one, allowing for the rehabilitation of a degraded land and for the improvement of rural livelihood. But it should be substituted by new systems based on a new equilibrium, possibly through local species or "re-naturalisation". Furthermore, alternative, complementary options should be proposed to favour a balanced rural development. If it is true that the local population developed a high degree of acceptance and ownership with regard to the implemented approach, it has to be reminded that such a solution came from the government and no other option was put on the table.

If in the short term the local communities will find the conditions to correctly manage and renew the plantations and if the needed long term vision will be developed by scientists and institutions, then the future Anthroscape will reflect this and stand as a mosaic of an evolving agro-forestry systems.

If the opposite would happen, most of the plantations, starting from the parcels located in relatively isolated locations, could be abandoned to free grazing and very quickly degrade back to the semi-desert conditions.

The land restoration approach described in this chapter could be an option for many arid production systems in northern Africa. There, a complex understanding of land use, ascending from political, economic and environmental factors, has had subsequent results in inducing the degradation of rangelands, creating a higher level of dependency for the producers on linkages between these extensive production systems and often-unstable imported or domestic feed sources from less arid or irrigated agricultural areas. There is an urgent need to move these production systems to a more sustainable basis, but caution in the design and implementation of development projects is required in order to take all the implications into account in the short and long terms.

References

Bellavite D, Zucca C, Belkheiri O, Saidi H (Eds) (2007) Etudes techniques et scientifiques à l'appui de l'implémentation du projet démonstratif SMAP de Lutte Contre la Désertification. NRD, Université des Etudes de Sassari, Italie.

Benbrahim KF, Ismaili M, Benbrahim SF, Tribak A (2000) Land degradation by desertification and deforestation in Morocco. Science et changements planétaires/Sécheresse 15(4): 307–320.

Correal E, Otal J, Sotomayor JA (1990) Utilization of sheep on oldman saltbush (Atriplex nummularia): palatability, browse efficiency, voluntary intake and chemical composition. Proceedings of 6th meeting of FAO European sub-network on Mediterranean pastures and fodder crop. October 17–19, Bari (Italy): pp.148–152.

FAO (1995) Sustainable Range-Dependent Small Ruminant Production Systems in the Near East Region FAO Regional Office for the Near East Cairo, Egypt.

FAO/ISRIC/ISSS (2006) World reference base for soil resources 2006. A framework for international classification, correlation and communication. World Soil Resources Reports, 103.

Ferchichi A, Yessef M, Belckheiri O (2007) Evaluation pastorale des plantations dans les zones du projet SMAP. In: Bellavite D, Zucca C, Belkheiri O, Saidi H (Eds) Etudes techniques et scientifiques à l'appui de l'implémentation du projet démonstratif SMAP de Lutte Contre la Désertification. NRD, Université des Etudes de Sassari, Italie.

Huvelin P (1970) Carte géologique et gîtologique des Jebilet 1:200000. Service Géologique du Maroc.

ICARDA (1995) International Centre for Agricultural Research in the Dry Areas. Annual Report 1995, Aleppo, Syria.

IFAD (1996) Combating Environmental Degradation. International Fund for Agricultural Development, Rome.

Kapur S, Zdruli P, Akça E, Arnoldussen AH, Kapur B (2004) Anthroscapes of Turkey: sites of historic sustainable land management (SLM). In: van Asselen S, Boix-Fayos C, Imeson A (Eds) 'Briefing papers' of the second SCAPE workshop in Cinque Terre (Italy), 13–15 April 2004.

Kubilay NN, Saydam AC, Yemenicioğlu S, Kelling G, Kapur S, Karaman C, Akça E (1997) Seasonal chemical and mineralogical variability of atmospheric particles in the coastal region of the Northeastern Mediterranean. Catena 28(3–4): 313–328, Elsevier, Holland.

Le Houerou HN (1995) Considérations biogéographiques sur les steppes arides du nord de l'Afrique. La Sécheresse 6: 167–182.

Le Houérou HN, Correal E, Lailahacar S (1991) New man-made agro-sylvo-pastoral production systems for the isoclimatic Mediterranean arid zone. Proceedings of IV Congrès International des Terres de Parcours, Montpellier, France, 1991.

Mulas M, Mulas G (2007) Potentialités d'utilisation stratégique des plantes des genres Atriplex et Opuntia dans la Lutte Contre la Désertification. In: Bellavite D, Zucca C, Belkheiri O, Saidi H (Eds) Etudes techniques et scientifiques à l'appui de l'implémentation du projet démonstratif SMAP de Lutte Contre la Désertification. NRD, Université des Etudes de Sassari, Italie.

Osmond CB, Björkman O, Anderson DJ (1980) Physiological Processes in Plant Ecology. Towards a Synthesis with Atriplex. Ecological Studies 36. Springer, Berlin, 468 pp. ISBN 3-540-10060-1.

Previtali F (2010) Mountain Anthroscapes: the case of the Italian Alps. In: Kapur S (Ed) Sustainable land management. Springer, Heidelberg.

Yaalon D, Ganor E (1979) East Mediterranean trajectories of dust-carrying storms from the Sahara and Sinai. In: Morales C (Eds) Saharan dust. J & W and Sons, New York, Ch. 9, pp. 187–193.
Zucca C, Bautista S, Previtali F (2006) Desertification prevention and reversal through mitigation and restoration programs. Encyclopedia of Soil Science. Taylor & Francis Group, New York. http://www.informaworld.com/smpp/title~db=all~content=g740169536~tab=toc.
Zucca C, Lubino M, Previtali F, Enne G (2005) The Euro-Mediterranean partnership: a participatory demonstration project to fight desertification in Morocco and Tunisia. In: Zdruli P, Trisorio Liuzzi G (Eds) Determining an income-product generating approach for soil conservation management. IAM Bari, Italy, pp. 317–325.
Zucca C, Previtali F (2007) Sols et Paysages de la zone d'étude du projet SMAP dans la Commune Rurale d'Ouled Dlim, Marrakech (Maroc). In: Bellavite D, Zucca C, Belkheiri O, Saidi H (Eds) Etudes techniques et scientifiques à l'appui de l'implémentation du projet démonstratif SMAP de Lutte Contre la Désertification. NRD, Université des Etudes de Sassari, Italie.

The Historical Anthroscape of Adana and the Fertile Lands

Nurettin Çelmeoğlu

Abstract Adana the capital city of ancient Cilicia, is embraced by a land of fertile soils, traditional crops and unique knowledge, which has been transferred from the earliest to the latest people of the area since millenia. Some of the export and industrial major crops of the country are part of the local socio-economic structure since the Early Byzantine period, whereas some have been cultivated by the renovations performed during the Ottoman Dynasty. Cotton cultivation is the primary example practiced since Justinians period creating the cotton Anthroscape (the human reshaped basin-wide scale land unit and/or parts of it, explained in the introductory paper of this book) i.e., the lower part of the Seyhan (Saros) Basin (the lower part of the Anthroscape) with particular soil inherent properties and use of the traditional water resources, that have developed via the long-standing historical good management practices.

Keywords Historical anthroscape · Adana · Fertile lands · Cotton · Floods · Keli

1 Introduction

Adana, capital of "*Cilicia*", which occupies the south-eastern corner of Anatolia, overlooking the Gulf of İskenderun, and falls into two sharply contrasted regions; the fertile coastal plain of Adana and the hill-country inland to the North-east of it, where the lines of Taurus are broken mainly by the upper courses of the Rivers Seyhan (Saros) and Ceyhan (Pyramos). People in Cilicia, contemporary "*Çukurova*" (the Lower Plain), settled from the Neolithic period onwards. The dates of the ancient settlements of the region from the Neolithic to the Bronze Age is as follows:

- Aceramic/Neolithic: 8th and 7th millennia BC
- Early Chalcolithic: 5800 BC

N. Çelmeoğlu
Greater Municipality of Adana, Adviser to the Mayor, Atatürk Bulv. 67, Kat 8, Adana, Turkey
e-mail: celmeoglun@gmail.com

S. Kapur et al. (eds.), *Sustainable Land Management*,
DOI 10.1007/978-3-642-14782-1_11,

- Middle Chalcolithic ca. 5400–4500 BC
- Late Chalcolithic: 4500-ca. 3400 BC
- Early Bronze Age: 3400–2700 (or 3400–2400 BC)

The area was known as the Lands of Kizzuwatna in the earlier Hittite Era (2nd Millenium BC). The earliest finding of this is the seal of the *King Isputahshu*, son of Pariyawatri, dated ca. 1650 BC, which is now displayed in the Adana Museum (Langlois 1861; Ünal 1999; Göney 1976). In the Assyrian inscriptions, the Area is called "*Khilakque*" or "*Khilikku*". In the early part of the first millenium BC, the Cilicians were one of the four chief powers throughout Western Asia. During the thirteenth Century BC, a major population shift occured as the "Sea Peoples" named by Egyptians who overran the area as part of the Palestinian, Cicilian, Tyrrehenian, Etruscan and Sardinian lands. The Hurrians that had settled in Cilicia deserted the area and moved northeast towards the Taurus Mountains to settle in the area of Cappadocia (Ener 1986; Yurt Encyclopedia 1981). In the eighth century BC, Cilicia was unified under the rule of *Mopsos*, the founder of Mopsushestia (Mopsuestia), contemporary Misis on River Ceyhan, eventhough the capital was Adana (Ener 1986; Çelmeoğlu April 1998). Cilica's multicultural character has been reflected in the bilingual inscriptions of the ninth and eighth centuries crafted both in the Indo-European hieroghliphic Luwian and West Semitic Phoenician. In 830 BC, the Assyrians began to conquer the plain and soon extended their authority over certain areas of the Taurus Mountains, which they called the "*Tunni Mountains*". "Tunni" in Assyrian means silver, which was mined in the area until late eighteenth century AD (Langlois 1861; Lloyd 1989). Cilicia was conquered in 540 BC by *Cyrus the Great* and became a Vassal Satrapy, paying a yearly tribute of 360 white horses (one horse per day) and 500 *talents* of silver, as per the inscriptions of Herodotus. "Talent" was a unit of weight used among the Jews, Greeks and Romans but varying considerably in different countries and in different regions of the same country.

The common Attic Talent was equal, on the usual estimate, to about 25 kg (Ener 1986). There were several sanctuaries remaining more or less independent from the Persian rule such as Castabala, Mazaca and Mallus. The last vassal king of Cilicia became involved in the Civil War between *Artaxerxes II* and *Cyrus the Younger*, siding with Cyrus, who was defeated and the king was dethroned. Thus, Cilicia became an ordinary Satrapy. The second to last Satrap (the Governor) of the area was *the Babylonian Mazaeus*. Shortly afterwards, his successor was expelled by *Alexander the Great* during 333 BC (Langlois 1861; Lloyd 1989). As of 67 BC, the Romans conquered the area which later became a Byzantine province in 395. During this period, Cilicia was exporting goats-hair cloth in huge quantities, which was called "*Cilicium*", the material for the ancient tents. The Islamic Forces took-over the lands by mid-seventh century and held it until it was reoccupied following a 4-year battle in 965 by *Nicephorus Phocas*. In 1097, the *Crusaders* came over and charged the *Armenians* to administer Cilicia on their behalf. Then the area was under the control of the *moslem Mameluks* by 1370 and the *Ottoman Empire*, ruling as of 1517 (Çelmeoğlu Sep. 1996; Langlois 1861).

During the third millennium B.C., the Semites became more crowded in the Mesopotamian Valley and controlled the area during the *Old Akkadian* period, from

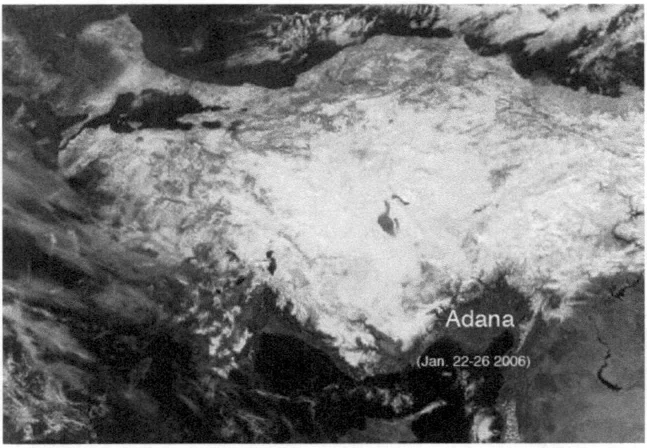

Fig. 1 Map of Asia Minor and Africa from Abraham Ortelius's Theatrum Orbis Terrarum, Antwerp 1595. It shows Alexander the Great's journey with a detail of the oracle of Ammon-Ra he visited in 331 B.C. The Greeks equated Ammon Ra with Zeus, the leading god of their pantheon

2360 to 2180. *Sargon* of Akkad, operating from his capital in the South of modern Baghdad, was the guiding light to the empire at the beginning of the period and was the first empire builder in the region (Ener 1986). King Sargon states, in his cuniferous tablets, that he was the only witness of the Cilician fertile lands on the spot (Fig. 1).

This very short historical brief shows that the Adana Province has been a meeting point for a great number of nations for some 8,000 years, tribes and civilizations not only because of its geographical position but also for both its fertile lands and mild climate (Fig. 2).

Between today's Mersin and Toprakkale, there are more than 250 historical sites. Some of these date back to the Neolithic times. For thousands of years, people have been benefiting from the fertile plains of the Taurus foreland. The legacy of the "rivers of Paradise" as the Arabian geographers called the Saros and Pyramos, were actually the contemporary rivers of Seyhan and Ceyhan.

2 Components of the Ancient Anthroscape

2.1 Irrigation

The contributions of nature were so generous in the area that and one could say that the geography of the area was made to facilitate the activities of man. Although many of the water sources disappeared during the last century, the number of springs in Adana is quite high, feeding hundreds of streams spilling into the rivers.

Fig. 2 Satellite photo of the Anatolian Penninsula taken during the 22nd to 24th January 2006. Adana outcrops as the main land with no involvement of the snow, signifying the mild weather

The water table over the majority of the plain has been very close to the surface. Hence, irrigation, the primary factor effecting cultivation after the climate, has been relatively easy to implement with increased yields.

2.1.1 Irrigation Versus the People of Adana

Effective irrigation methods were introduced by the *Assyrians* in 800–700 BC. The system consisted of a bifurcated inlet excavated upstreams feeding a canal constructed at the river bank. It consisted of a gentle slope that caused the water level in the canal to gradually become higher than the river. Hence, after some distance, the canal was high enough to spill on to the cultivated area, which was overlooking the river. With the same technique of water diversion, a great number of grain mills were operated by hydro-power (Fig. 3).

2.1.2 Horse Driven Wells

Until the early 1950s, irrigation in Adana was practised by means of irrigation canals and water wheels using both rivers and wells. The waterwheel system is

Fig. 3 Water elevated through a trench dug about 400 m upstreams, feeding the penstock of a water mill

known also as the traditional *Miao Irrigation System*, that has been the most attractive and efficient irrigation instrument. Such wheels were made of wood planks, and were driven by water power when they were constructed in the river and animal power in case the water source was a well (Abu Nasser 1960). By the mid-1950s, in the city of Adana itself, there were more than 30 horse driven water wheels.

Until the late 1950s, the author of this chapter witnessed many of the horse driven water wheels in and around the city of Adana. The wells used to reflect a rectangular cross section of about 80–100 cm width and 6–8 m length, and the entire faces protected with cut-stone masonry. The wells were dug upto 4–5 m deep but the water level was not lower than 2 m from the surface. Hence, the water wheels used to have a diameter of about 5–6 m, only a few of them were driven by buffalos. A simple gear system was located underground at the center of a circle of about 5–6 m, at the periphery of which the animal walked to pull a wooden rod connected to a vertical shaft that actuated the underground gear system. The wheel was turned by a shaft laid in an underground stone channel stretching between the gear system and the well.

The woods used for the planks and the transmission system were mostly cut from the cedar trees reputed for durability and strength. Some operators preferred to apply a kind of reinforcement by laths of forged iron.

Be it water force driven in the rivers (Fig. 4) or animal power driven at the wells, the wheels were equipped with any kind of water container, i.e., large tin cans, wooden buckets, clay pots, etc... These containers, filled with water when submerged into the water, were pouring the same into an intake of a gutter positioned just at the right point where containers were turning down (Çelmeoğlu Oct. 1997).

Fig. 4 Water Wheel lifting irrigation water at the river side in Adana (Ca. 1955)

Immediately after the inauguration of the *Hydropower Plant* in Adana on the 8th of April 1956, the electric power operated pumps started to be used in field and orchid irrigation, causing the rapid growth of the plantation area as well as the quantity of the crops per unit area, despite the on going research on the quality-wise properties and the taste related mineral and aromatic content determinations undertaken by the relevant institutions. Numerous doubts are still pending on the quality of the fruits and vegetables consumed in the area, which are also in need of further research on the positive and negative contributions of the agricultural chemicals and excess water use on human health.

Yet, the demand was ever increasing because of the radical steps conceived in the construction of roads connecting Adana to the major cities of the Country. Hence, Adana appeared to be developing in agriculture and agriculture based industry, namely *cotton textiles, cotton seed oil, vegetable ghee, margarine and in the manufacture of farming equipment.* By the mid-1960s, the majority of the plain had been cultivated, what gave the way to an efficient irrigation network consisting of large, medium and small canals and destroying the major part of the delta ecosystem (the lower Adana Anthroscape).

2.2 Diversion Canals Served After a 1,000-Year Period

The River Seyhan has at all the times kept its vital importance for Adana as the sources of water, mild weather, food and transport. Some engravings and the notes

Fig. 5 River Seyhan crossing Adana, looking from North to South. Currently it feeds several dams and power plants

of several well reputed travellers show that medium size boats were calling at Adana Port on the river in the nineteenth century. The river is fed mainly by the tributaries of Göksu, Eğlence, Körkün and Çakıt and some other smaller ones (Fig. 5).

One of or perhaps the only very old bridge of the world, which is still under use, is the Stone Bridge, which was constructed upon the instructions of the Roman Emperor *Hadrian* (117–138 AD). After several important amendments and good maintenance, *Justinian* (527–565) of the Byzantine Empire instructed for its enlargement and reinforcement so as to cope with the over all importance of the city of Adana, highlighting its significance.

To do so, it had almost to be rebuilt and therefore, a large construction site had to be drained. Judging from several historical sources, it seems that two diversion channels had to be dug at the left bank, starting from about 2 km upstreams of the bridge, where a cofferdam was spanning throughout the river bed. Although neither information regarding the length nor the exact position of the diversion channel are available, it is estimated that it was longer than 10–12 km, ending at the river immediate South of a downstream cofferdam.

After about a millenia, *Piri Mehmet Pasha* became the Beg (the Governor) of Cilicia (1518–1548), who seems to be the most contributing leader of the *Ramazanoglu Dynasty* (early fourteenth century – 1608; a vassal authority under Mameluks till 1517 and Ottoman Rule till 1608). Piri Mehmet Pasha conducted highly considerable innovations and improvements in terms of the over all agricultural traditions but his most important contributions were the improvement of the irrigation facilities. Apparently, he decided to use Justinian's diversion canals for both the irrigation and operation of the mills. During his reign, Adana possessed the largest number of ordinary mills but some of them were serving as sugar cane juice extractors to produce sugar, which became an important export item. The canal was also feeding the subsidiaries which were irrigating four large rice plantation sites sprawling over the southern lands of the left bank (Ener 1986, Çelmeoğlu July 1997). The map of *Piri Reis*, a well known Turkish Naval Commander and geographer, shows that a tributary of the river was protruding from Adana and spilling into a lake, which was the heritage of the Lagoon. It seems that this tributary was that of Justinian, what was later re-used by Piri Mehmet Pasha (Fig. 6).

Speaking about the irrigation Works conducted by Piri Mehmet Pasha, it should also be noted that another source of water from the river had been a secondary channel which was fed from the second and the third Eastern arches of the Stone Bridge, as per the Charter of the Piri Mehmet Pasha Foundation (the Wakf-"Vakıf" in contemporary Turkish) covering the period 1538–1554. This secondary channel was to serve the lands close to the city.

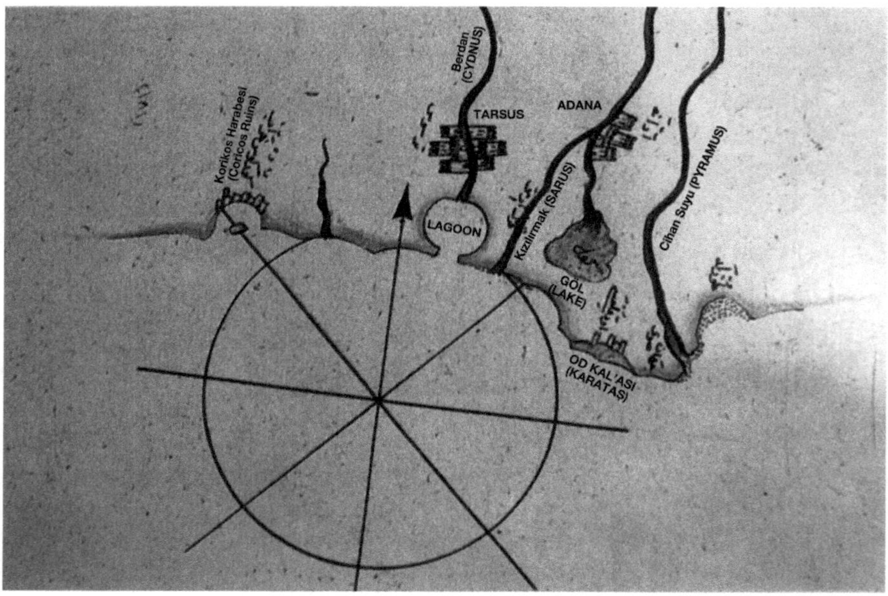

Fig. 6 The Map of Piri Reis; note the canal outcropping at Adana and ending at the Göl, the Lake. No doubt this lake was a part of the Lagoon which was once much larger than what is shown at the map. The lake is now dry and its area is called "Tuzla", what means the salt query. This could be explained by the capillary pressure as the result of evaporation what ascended the salt level apt o the surface. Some decades ago, both the State Agriculture and Hydraulic departments co-operated fort he rehabilitation and only after some remedial Works this area is now an arable land

2.3 Floods Reinforcing Fertility

Considerable floods were witnessed twice a year, in the historical anthroscape of Adana, called the "*Winter Floods*" and the "*Late Spring Floods*", both caused by the streams and rivers, submerging the entire lands where agriculture was practiced. Large amount of suspended organic matter in the flood water covered the surface with a rather thick layer of sediments making Çukurova a very fertile land.

Sequential winter floods were caused by the constant rains, while the late spring floods were created by the melting of the snow at the Taurus mountains. The Adana Plain experienced these floods almost every year until the Seyhan Dam and Hydroelectric Project was accomplished in April 1956.

Some years after the prevention of the floods, the elderlies have unfortunately noticed that both the taste and the flavour of the vegetables, especially tomatoes, cucumbers and chickpeas were gradually lost. This has been interpreted as the lack of the fresh sediments which were transported by the floods, i.e., by the natural sediment-laden water supply cycles of the earlier Adana Anthroscape that were and still are reigned by the wall of the Seyhan Dam (Göney 1976; Çelmeoğlu Nov. 1997).

2.4 Benefiting from Fauna and Agriculture

Common historical indications reflect that cultivation had been initiated conscientiously at the upper plains of the Taurus, Amanos, and the Lebanese Mountains by the available public awareness to conserve and sustainably exploit the natural resources. It is predicted that at these zones the climate was quite favourable and some domesticated animals were available for people, as it is believed that the sheep and goats of today have originated from Iran and Anatolia.

Taking into consideration the various historical facts, it can be affirmed that the cultivation in and around Adana has been a general practice at least since the reign of the *Summerians*, i.e., about 4000–5000 years BP.

It is a fascinating fact that the lands of Adana were overlying a geographical belt that was bordered by the Mediterranean Sea in the South and Taurus Mountains in the North, possessing plenty of water brought by two large rivers, *Saros* and *Pyramos*, *Cydnos* (*Berdan*) and many other rivers and streams. Certainly, the entire mountainous area, including the immediate gradients, were densely covered with huge forests, granting a natural regulation on the precipitations. On top of these, the rivers have generously been offering "fertility" by transporting and distributing alluvial material evenly all over the plain by means of floods. Floods, most of the time mentioned as "disaster"s have been extremely useful instruments to enrich the soil in the Adana Province (Çelmeoğlu Oct. 1996).

Adana was reputed also with its horses; as a clear evidence the stables of *King Soleman* were stocked with horses from Adana (Ener 1986). *Herodotus*, as earlier mentioned, informed that the Persians occupied the city in 540 BC and appointed a vassal Satrap, granting the autonomy against an annual tribute of 360 horses and a certain quantity of silver.

2.5 *Satsneferru, the Egyptian Nurse*

In 1862, an American Missionery in Adana, referred to as *Mr. Montgomery*, decided to construct a building at the heights of the *Tepebağ Mound* right at the heart of the city. When he started to construct the foundation, he discovered in the excavation some antique valuables and ordered the foundations to be dug deeper than usual practice. At a certain depth, an Egyptian statuette made of black granite was unearthed. This masterpiece remained in place in a building in Adana and later smuggled out of the country in potato sacks loaded in a yacht bearing the English flag. Finally, the statuette was brought to the *United States* where it remained in the possession of the Montgomery family before it was acquired by the *New York Metropolitan Museum of Art* (Çelmeoğlu 2010-unpublished book). At the end, it was understood that the person whose name the statuette bears was "*The Nurse Satsneferu*" or "*Sit Snefru*", the earliest nurse in the history (Fig. 7).

She was probably the person serving high level Egyptian officers resident in Adana. She ordered a statuette of herself declaring she could not leave the people without medical care and claiming that if she died in Adana and got buried without a statuette of herself, her soul would not be able to return back to Egypt. According to the inscriptions, the nurse was serving the colony during the 12th Dynasty, in early second millennium BC. The late suggestions indicate that this masterpiece is an evidence for the presence of an important residential Egyptian delegation in Adana, meaning that the Adana Province was very important to the Egyptians Empire. This statuette is also referred to in the book of Barber (1994) (Fig. 8) and attention is drawn on the linen clothing of the nurse.

Fig. 7 Egyptian Nurse of Adana

Fig. 8 Left: Stone statue of Sit Snefru, an Egyptian nurse of the Twelfth Dynasty (early second millennium B.C.), who accompanied the family she served all the way to Adana, at the northeastern corner of the Mediterranean. There she evidently adopted the local Near Eastern style of clothing; compare the typical wrapped tunic of the Mesoporamian woman on the right (this statue came from Tell Asmar, mid-third millennium B.C.), and the contrast of the sewn-up jumpers normally worn by Egyptian women

It is known that about 2,000 years ago, a linen institute was established in the *Anazarbos* (contempoary Anavarza) district of Adana. It was one of the outstanding capitals cities of the Adana Region. Without any doubt, the weaving industry was playing an important role in the economy of entire Cilicia. For instance, *Saint Paul* was also a well reputed weaver. Nevertheless, until the late nineteenth Century, linen was widely cultivated in the Region. Also, hemp was another remarkable income source as the seamen needed the ropes spun by hemp fibre.

2.6 Adana, the Paradise for Spinnable Fibres

It is also known that Adana has been a cotton growing area for at least 2500 years. Cotton was one of the reasons why Emperor Justinian valued Cilicia. The Stone

Bridge at River Seyhan which was ordered by Hadrianus of Rome (117–138) to be constructed was nearly rebuilt by the order of Justinianus.

Despite the available insufficient evidence, it is believed that the Emperor had allocated 1,000,000 sheep for both increasing the quantity and upgrading the quality of the wool produced in Cilicia. Justinian's contributions resulted in multiple productions in both varieties and quantities.

Justinian, with the intervention of two *Persian monks*, (probably bribing two Chinese monks of close contact and smuggled the seeds out of China) succeeded in bringing the *silk worm seeds* in scrolls to Cilicia along with prescriptions as to the proceedings including worm care, cocoon process, spinning, weaving and dyeing, disclosing, silk processing in China that was kept as a secret for about 2500 years. Ultimately, Adana and Antakya became the second most important silk production and process center in the world after China in 550 AD (Çelmeoğlu Mar. 1997). Nevertheless, cotton, linen and silk were not the only commodities Justinian was interested in, he also had instructed the construction of aqueducts, irrigation channels and efficient grazing systems for better production of wool also. This remarkable period of development did not last long. The weakened administration led to *corruption and disefficiencies in management*, where all extended to the inevitable end.

Furthermore, the area was several times hit by a series of frequent and devastating earthquakes described in the following section. Finally, Adana's place in production and supply of fibre, thread and fabric decreased drastically but still survives today (Ener 1986; Çelmeoğlu 2010-unpublished book).

2.7 Lions Preventing Agriculture

It is evident that soon after the Emperor Justinian passed away (13–14 November 565), a period of turmoil and chaos adversely affected the administration all over the country, including the Adana Region, and Cilicia. On top of this, the Area had experienced devastating earthquakes (in 561, 587, 650 and 660 AD) claiming some tens of thousands of victims; the local people, believing that the God was punishing them because their ancestors had tortured many believers (Many of Saints were martyred in Cilicia, According to the Catholic Church) and left the area.

Indeed, when the *Islamic Forces* decided to settle here, by the last quarter of the eighth century, there were almost no people. However, they noticed that the population of lions over the plain was so high that it was practically impossible to cultivate the land. As a solution, *they decided to import 400 buffalos* from India for use in land cultivation thinking to intimidate and drive the lions back to the uninhabited areas (Çelmeoğlu Nov. 1996; Ener 1986).

In fact, the buffalos at the Taurus mountains were reputed with their size and power but as they were undomesticated the import from India has been necessary. Nevertheless, it should be noted that the Province of Adana has later successfully domesticated this animal. Unfortunately, almost no water-buffalo is to be accounted for in the Adana Region nowadays, whereas they were in high numbers 50–60 years ago. Evliya Çelebi,[1] in the third volume of "*Seyahatname*" (the Travelbook) states that buffalos of the Adana region were the world's strongest and most powerful animals.

He also quoted; "Recalling from the cannons, it is not difficult to predict that the water-buffalos were not serving only as the source of valuable milk but also and mainly as the tracking force. Quite heavy farm carts were driven by a couple of buffalos". Taking into consideration the fact that about a century ago, *Adana was full of swamps*, the water-buffalos were the best surviving animals in the swampy areas. The Adana diet comprises many milk-related meals and desserts bound to the earlier water-buffalo populations. In addition, their skin's value in shoe making was extremely high especially from the point of view of durability. As the swamps were drained and land was so-called reclaimed for irrigated agriculture in the lowlands of Adana, the vast populations of the water buffaloes (well adapted to the local conditions despite their exotic origin), of the earlier Anthroscape of Adana, had to decrease and ultimately vanish (Ünal 1999; Çelmeoğlu 2010-unpublished book).

2.8 The Turks in Çukurova

Arabs, while reconstructing the cities, settled *turkish tribes* in the area at the beginning of the 800's (AD), while they introduced perfect irrigation and cultivation techniques, and created beautiful and prosperous fruit orchards.

Istahri, an Arab traveller, in his book "*Mesalik el Memalik*" (Professions in Countries), describes Anavarza (*Anazarbos*) as a city of fruits and date orchards, adding that the acreage of both the fertile arable lands and pastures were considerably large (Çelmeoğlu Nov. 1997–May 1999). Anazarbos, (Fig. 9) one of the most important capital cities of Antique Cilicia, was besieged by Nikephorus Focas of Byzantium from 962 until 965 AD, the date he instructed cutting down of 40,000 date trees, which probably were the dominant food supply of the natives who picked the fruits at nights. This slaughter of the trees was indeed helpful in convincing the people to hand over the fortified castle to *Nikephorus Focas*.

Leon Diakonos, a Byzantine historian describes the attacks of Focas as follows:

"When the Emperor (Nicephorus Focas) came close to Tarsus, the fields around his headquarters were all arable lands, flowers and trees. In order to provide a better

[1]*Evliya Çelebi* The first and the greatest representative of the Turkish Travel Literature, who journeyed throughout the territories of the Ottoman Empire over a period of 40 years. He was born in İstanbul, in 1611. Coming from a wealthy family he received an excellent education and begun his travels in 1640. Evliya Çelebi's collection of notes from all of his travels formed a ten-volume work called the "Seyahatname" (the Book of Travels).He died sometime after 1682; it is unclear whether he was in İstanbul or Cairo at that time. Although many of his descriptions were phrased in an exaggerating form or as the 3rd-source misinterpretation, his notes are widely accepted as a useful source of information about the cultural aspects as well as the lifestyle of the Ottoman Empire in the seventeenth century.

Fig. 9 Overlooking the city area from the catsle at Anavarza (Anazarboz). This photograph was taken on April 5th, 1905

visibility and to avoid unexpected attacks from the barbarians(!) hidden inside the woods, the Emperor ordered the clearance of all the forested area, the rich pastures, plantations and all kinds of fruit trees, thus converting the beautiful environment, developed at the Islamic Period from Ca 800 to 962 AD, into a poor, bare land." It would be worthwile to add that during the same period the cultivation of cotton had increased as well.

A Byzantine Historian, *Theophanes* (Died in 817 AD), who witnessed this period, mentiones that the Caliph Hisham devoted great efforts to improve the cultivation and the orchards by means of efficient irrigation facilities.

2.9 The Cotton Anthroscape of Adana

Besides linen and wool, and later silk, cotton had been the most consumed fibre for spinning and weaving in the region. In time a solid tradition of cotton growing and weaving culture developed in the area worthy to meet the term standing for a human reshaped land unit, i.e., the "Cotton Anthroscape of Adana".

Producers of the Adana Region were very lucky to be able to grow both the ordinary white cotton and the *light Brown-yellowish cotton* which enabled them to

Fig. 10 A piece of hand woven multipurpose rug; the raw weft yarn is made of the natural white and natural light brown cotton. This rug is being used for more than 40 years and despite it was washed very frequently, the white has acquired a kind of bleaching whereas the coloured one keeps its first day shade

produce highly fashionable goods since they were able to place stripes in warp and weft. By mixed threading, a third colour had also been obtained in production.

Karaisali, a district of the Adana Greater Municipality today, has been the last place for cultivating the varieties of cotton besides manual spinning and weaving at power looms. Unfortunately, the dyeing techniques acquired during the recent decades vanished and this naturally dyed cotton disappeared from the marketplace.

The author of this Chapter has visited every single village of this area to trace the traditional yellow cotton (a significant earlier crop of the historical Adana Anthroscape) from 2006 to 2008 and sadly by talking to the farmers and the people it was concluded that its cultivation was reduced gradually and was completely erased from the region (Fig. 10). Unfortunately, power-looms and warping-benches are no more recognised by the members of the younger generation even though they are still being used in some houses as decoration. Some elderly ladies expressed that "the young ladies have become modern and easy life seekers; they refrain from home economy, even dislike regular home tools unless they are some new appliances as dishwashers and laundry machines" (Ener 1986).

2.10 İbrahim Pasha's Contributions

Modernization in agriculture and upgrading of cotton cultivation was first extensively introduced by *Ibrahim*, the Egyptian, who invaded Cilicia and its surroundings from 1833 to 1840. He was the son of *Mohammed Ali Pasha*, the Khidiv

(A highly powerful governor) of Egypt, who soon after his appointment by the Ottomans acted as an autonomous king.

Just before the arrival of Ibrahim Pasha who was called "*Mısırlı*" ("The Egyptian") by the locals, Adana was lacking state authority. The administration was not injust to poor people who were punished instead of the real culprits. Security was almost non-existent and no one except the bandits could walk on the streets in the dark. The invader, immediately after his arrival, reestablished the *security and justice* at an unimaginable level. He improved the conditions of field workers considerably by granting them the following: working only five and a half days in a week, getting rest breaks twice a day during work, and receiving a minimum wage. He called some experts from Egypt, Sudan, Algeria and Tunisia to improve the general cultivation and to enhance cotton and sugar cane production. The reign of Ibrahim Pasha, brought *peace and prosperity* to the area and the locals were happy of the general conditions the Egyptian established in time, where his popularity increased in the area and was encouraged to occupy some other regions of the Ottoman Empire and the *French* were the first to become interested in Cilicia (Ener 1986; Yurt Encyclopedia 1981; Çelmeoğlu 2010-unpublished book).

Being aware of the fertile lands of the area, *Britain and France were not happy* with the ever increasing popularity of İbrahim Pasha and they suggested that the Palace in İstanbul needed some collaboration to "defeat" him. Actually, all these events were occurring during the period in which the Ottoman Empire was in a sharp decline. This gave hope to Europeans to inherit certain areas of the Empire. The French became really interested in Cilicia. In 1840, İbrahim Pasha, leaving a peaceful and rich city behind him, was compelled to leave the area because of the British and French fleets approaching the Bay of İskenderun (Yurt Encyclopedia 1981).

2.11 The American Civil War and the Cotton of Adana

The Great Civil War in North America which erupted on April 12, 1861 highly affected Adana and Egypt since the war stimulated the production of cotton in these regions. As a matter of fact, the *slave-based cotton cultivation* in the United States was the best source of raw material for European textile producers, the UK being first in line. During the years of the American Civil War, the sharp decline in production affected the European Textile Industry. Consequently causing enhancement of the European attempts in convincing İstanbul, the capital of the Ottoman Turkey, to declare official incentives to increase both the cotton production in Adana and Egypt and the concessions in cotton-exports. Thus, *Adana and Egypt*, the two Ottoman States, allocated more land for cotton plantation in order to meet the European demands (Tables 1–3).

While the Ottomans were allocating more agricultural land to the farmers, the British, French, German, Dutch, Italian, Austrian experts and traders as well as bankers poured into Adana, changing the over-all trade and lifestyle conditions of

Table 1 Cotton production in Adana by years (in bales of about 200–220 kg)

1896	2000
1905	45000
1906	50000
1907	56000
1908	46,000
1909	74,000
1910	65,000
1911	80,000
1912	115,000
1913	120,000
1914	70,000
1915	59,200
1916	86,000
1917	20,000
1918	50,000
1919	100,000 (French invasion; kept on till the end of 1921)

Table 2 Lint cotton production versus sawn lands in Adana

Years	Sawn acreage	Lint production (in 1,000 kg)
1940	175,566	30.299
1945	172,800	23.141
1950	211,600	47.400
1955	251,386	40.014
1960	265,250	65.037
1965	281,716	102.717
1967	280,800	116.557
1968	285,150	129.000
1969	248,600	112.083
1970	248,600	112.083
1971	234,610	155.500
1972	246,751	152.154

Table 3 Cotton seed productions by years in Adana[a]

Years	Turkey Acreage (ha)	Adana (ha)	%	Turkey product (ton)	Adana (ton)	%	Turkey yield (kg/ha)	Adana yield (kg/ha)
1969	638,000	248,600	3	400,000	328,586	82	641	1,320
1974	838,000	296,760	35	598,000	490,505	82	748	1,653
1980	662,000	200,000	30	500,000	390,900	78	768	1,950
1985	660,000	162,498	25	518,000	378,011	73	832	2,326
1990	641,000	108,614	17	655,000	245,796	38	1,052	2,260
1991	589,000	98,535	17	559,000	234,525	42	946	2,380
1992	638,000	124,000	19	575,000	304,000	53	923	2,457
1993	557,000	76,155	14	602,000	158,758	26	1,092	2,084
1994	581,000	80,043	14	628,000	196,123	31	1,080	2,634
1995	757,000	139,120	18	851,000	366,401	43	1,125	2,630
1996	744,000	121,235	16	784,000	304,945	39	1,054	2,520
1997	722,000	77,712	11	795,000	215,438	27	1,101	2,770
1998	733,000	84,189	11	833,000	231,524	28	1,137	2,750

[a]Adana Provincial Agricultural Directorate Registrations 1997

the city (Fig. 11). A severe *competition between the foreigners* resulted in higher prices and better fibre quality besides better plantation practices. Europeans and their local representatives offered the producers financial assistance and free cotton seeds in contract for product guarantee at a pre-fixed price (Çelmeoğlu April 1998; ADASO 2008; Ener 1986).

The first *ginneries* and *spinning-weaving mills* were established before the end of the nineteenth century. Adana, with its well educated and wealthy foreign citizens and overall prosperity, transformed itself gradually to a European city with paved and

Fig. 11 Soon before the last quarter of the nineteenth century, Adana became a rich and modern city as it was influenced by the Europeans living there

stone-lined clean streets, and developed social and municipal services. The cotton plantation was reduced considerably as the demand from Europe dropped down drastically soon after the American Civil War came to an end. It is worthy to note that the American cotton quality was far superior to Adana varieties not only because of its finer and longer staple but also the better ginning technology.

In 1974, 1,200,000 spindles were running in Adana out of 3,600,000 producing yarn in all over Turkey. Adana also had a remarkably higher yield-per-spindle ratio compared to the Turkish average. The ginning industry was also very important in the region being in operation both with modern roller and saw systems. Most of the ginneries had extensions such as *cottonseed oil extraction, refining unit, soap production* from the substock, which is a final waste at the end of oil bleaching process.

2.12 The Decrease of the Cotton Plantations in Adana

Until the late 1970s, the Turkish Industry and export were depending mainly on the textiles and therefore the labour unions conducting collective agreements with the entrepreneurs were far stronger than the others. Hence, the *Federation of the Syndicates* was controlled by these unions, which meant the power and prestige that could bring a bright political future to the union leaders.

Consequent frictions between the local syndicates lead to a series of *turmoils in Adana*, causing frequent strikes and work slowdowns so that the work-peace was almost non-existent in the field of textiles, in Adana.

With its *once true wealthiness*, Adana was always deprived of all kinds of state incentives, whereas in the other regions of the country the investors were granted with considerable facilities and supports, which reached sometimes as high as 40% of the investment cost besides the tax exemption for some time. At the end, consecutive *lock-outs* and *closing downs* were experienced. While Adana was experiencing such problems, the technology of both the spinning and the weaving advanced considerably by increasing the velocity and the quality at the same time.

The Adana cotton, although responds good to the dyeing process, acquires a moderate staple length and fibre thickness, which is suitable to spin yarn *Ne Count 20* (equivalent of *Metric Count 34*) and coarser, whereas the fashionable goods are woven with finer yarns. On the other hand, short and coarse fibres are not suitable to spin with high velocity. The spindle rotation per minute was fluctuant between 6,000 and 12,000 rpm whereas the late spinning frames were running at over 1,800 rpm for the *ring spinning* (conventional technology) and over 36,000 rpm for the open-end (rotor technology) spinning frames, necessitating finer and longer cotton staple for a satisfactory result, which means a high breaking strength, regularity of the yarn with less slubs and minimum thin places, all depicting the yarn quality which at the end reflects at the finished fabric.

Under these conditions, and in parallel with the rapidly increasing demands from the international markets of the fashionable finished products, the spinners had to shift to the long and fine staple cotton varieties, resulting decrease in the demand for the Adana cotton. At the same time, the cost of *workmanship* had been so high that the local producers were no longer able to compete with imported cotton prices.

Finally, the production went down dramatically. Never the less, as of 2008, a few farmers tried the *machine harvesting* which revealed a satisfactory result thanks to the advanced pneumatical process, new harvesters were bought and encouraged a number of farmers to cultivate again cotton taking care to choose the right varieties regarding the length, strength and the fineness of the fibre without sacrificing the yield per acre. Apparently, during the next 5–6 years, the region will produce more cottton with somewhat better quality compared to the previously sawn varieties.

2.13 Phoenicians Calling at Adana Ports

As there was no remarkable military power in *Mesopotamia* from 1200 to 900 BC, *Phoenicia*, although a small state, was able to prosper and became a well reputed *sea-trader* throughout the Mediterranean Basin. The historians agree that both economic opportunities and the population pressures forced them out into the seas.

One of the best merchandise they could offer was the *cedar tree*, which has been and still is a valuable tree. For some time, Phoenicians were cutting the cedars from their own forests; but later, noticing the destruction of the forests, they decided to gain the product from other sources. These sources were *Cyprus* and *Cilicia* (Fig. 12). The more cedars they could procure, the more active they became.

Fig. 12 Cedar is now under the state protection with purpose to enlarge its plantation area. Besides, millions of seeds are sawn every year

Phoenicians were superior to all other peoples of their time, especially in sailorship by far with dense contacts and colonies established at almost every corner of the Eastern Mediterranean.

In Phoenician, "*willow*" was called "*Adn*", where Adana, the capital of Cilicia, was characterised as an area of very rich water sources where willow trees sprawled over large areas along the rivers and streams. Thus, they described the area as the "*Land of the willow trees*" calling it "*Uri Adn*" or "*Adaniia.*"

Phoenicians, as *smart* overseas traders, included Cilicia into their main routes not only for the cedar trees but also for its various crops such as flaxseed, hemp-ropes, sail fabrics, dates, wheat, barley, rice, honey and dried fruits (Lloyd 1989; Ener 1986, Çelmeoğlu 2010-unpublished book). With the aid of these products, Adana became an *export-minded* city which still is one. The Adana citizens were highly influential all over Cilicia and this was described in the inscriptions crafted in their alphabet which were found at several locations over Cilicia.

2.14 Prognose and Medication

Dioscorides, a native of Adana, was the most reputed Medicinal plant expert in the World of his time. He was a Greek physician and he benefited greatly from the comfort of accessing thousands of plant varieties as well as venomous animals besides inherited knowledge about the medical interventions. Between the ages of 50–70, he wrote his fundamental books known in Latin as "*De Materia Medica.*" This five volume study dealt with the preparation, properties and trial of medicines beyond the hints for diagnosis. This work has been the most recognized farm-acological study in Europe and the Middle East for the next 16 centuries. The area's natural diversification of the regional flora and fauna were probably the reason behind the legendary medical heroes living in the area. Most known of them are *Lokman Hekim*, and *Asclepios*.

It is believed that *Lokman Hekim*, a local physician, whose name is also taking place in the Koran, was able to heal diseases by using the local vegetation and venoms. It is said that he was talented by God to communicate with the plants and therefore he had no difficulty in defining the right part of the right plant to produce medicine. Once, on his way back home from the hills of Misis (*Mopsushestia*) a kind of grass, which he had never met before, called out to him to explain that it was the instrument for the immortality of the human being. Lokman was excited and sat down by this tiny plant to note its decription as to how he could prepare the *elixir*. It was a very complicated explanation but Lokman has been extremely careful to note all the words correctly and missing no one. After he was finished with this plant, he ran down towards the city with delight and reached the Stone bridge over the River Pyramos, the contemporary Ceyhan. At some moment while he was crossing the bridge, the Angel Gabriel came near to him and gave a good wing strike to throw away all the prescriptions into the river (Fig. 13). That had been the end of the hopes for the legendary immortality.

Fig. 13 Angel Gabriel striking with the wing gave an end to the hope for the immortality

Another legendary story involves the Mythological Hero, *Asclepios*. He was an Antique Greek God of Medicine and healing. Being *the son of Apollo*, he was raised by centaur Chiron and become an outstanding physician. Asclepios, in order to practice his profession in the best way he could, decided to live in the Adana Region, between Misis and Yumurtalik (*Aigeai*), where he could find many varieties of plants and venomous animals. Finally he constructed one of the rare *Asclepions* of the world (The Asclepios' Hospital; the other two are located in Bergama-Pergamon – of Turkey and Athens) in Yumurtalık.

Due to his abilities, the God of the underworld, *Hades*, became angry because the number of newcomers to the area was decreasing day by day, scaring him that no more spirits would come there. He complained to *Zeus*, who killed Asclepios with a thunderbolt and placed him amongst the stars as the "*serpent holder*". Not only these but many other legendaries, too, are depicting the fertility of the Adana Region (Çelmeoğlu Jan. 1997, Mar. 1998, Jul. 2000; Can 1946).

2.15 The Reign of the Romans

The Romans sustained the control of the region from 67 BC to 395 AD and were known with their very strong agricultural and industrial workforce. During this period, the Adana region was called to be the source for grains and the *bread basket* of the Roman Empire Also, linen, cotton and wool were produced and processed with very high skills. Most probably, the Roman farm management system, called the "Latifundium" had a great contribution to the successful agriculture practices in the region. It was basically based on the employment of slaves; thus, the overall cost of production was very low. The Romans had adapted this system from the *Carthagians*. Notwithstanding, the agriculture itself was based on several principles which are still in force even today via efficient irrigation, proper product design versus land's characteristics, fertilizer application and use of the ploughs (Çelmeoğlu Mar. 1997, Ener 1986).

Their success of sloppy-colcrete land plantations was also remarkable, where it was only possible to cultivate anything but almonds, figs, grapes, carob beans and olives were remarkable with very high yield and good quality.

A great and continual importance was attached to the olive production because of the increasing demand for olive oil, that was consumed for several purposes. Beyond being a food item, it has been a high quality lubricant, a good fuel for the lamps, and one of the main ingredients of the medicines in history.

When the slopes were steep enough to prevent easy cultivation, the Romans constructed durable and functional *retaining walls and terraces* which are still serving efficiently at the northern and southern fringes of the Taurus mountains and on the coastal strip of the Turkish Mediterranean Region.

Based on the information given above, it is clear that efficient cultivation techniques have been part of the agricultural life in Adana since the early ages. And, thus, this must be the reason that during history, the naval-nations always were spilling blood to settle at Cilician shores. The history is full of pages mentioning a great number of invaders pouring into to the region not only for its unique strategic position from the West to the East but also for its fertile lands.

2.16 The Inherited Practices of the Adana Region

Certain inherited traditional practices are still used in the Adana Province, namely shifting to onions and broad beans (legume inducing nitrogen accumulation in the soil) when the crop yields decline in order to reinforce the next crop. The traditional widespread cotton cultivation in the region would also be a particular example sown at nutrient poor zones (inappropriate lands for cultivation around the lagoons of the Adana wetlands), following the wheat. The wheat, as an inter-crop, was/is sought to reduce the magnitude of over fertilisation of the soil caused by the droppings of the nesting and migrating birds and increase the following cotton yield by impeding the vegetative growth. Concurrently, the wheat yield is enhanced following over-growth via the bird droppings by practicing grazing at the appropriate periods that would increase the tillering per plant.

One of the major disasters occurring at the fields was the *coach grass* invasion. In order to overcome this problem, consecutive deep plowings during the most hot days of the summer was necessary. It was believed that the more the roots of the coach grass are exposed to the sun, the more the relieve was possible. Notwithstanding, if the problem is severer than it is remediable with deep plowing, specialized teams of workers had to be employed. These teams were digging the ground with spades down to the levels at which no more roots of the coach grass could be encountered and eliminating carefully the entire roots from the ground. This, too, would help for a 4–5-year period, where the same problem would repeatedly launch itself in the following years.

One of the unbeatable or hardly beatable diseases was the "Rust", menacing the flax crop, which was persistently demanded by the farmers, but had to vanish following its mass cultivation by the mids of the nineteenth century.

2.17 The Traditional Tomato Crop of Adana

Tomato was a wild vegetable and the fruits were consumed only for medicinal purpose; it was never consumed as food until the second half of the nineteenth century. There is no any written information about the reason or reasons but we can suggest that most probably this was due to the presence of a quite similar plant, which is actually an egg plant, producing a fruit (Fig. 14) very similar to tomatoes.[2]

[2]In my childhood this was grown as an ornamental plant in houses in Adana. I have personally grown several generations until the year 1962, in which we pulled down our old house to construct a modern one, compromising the beautiful small garden. Unfortunately, I have not seen this unique crop in the last 50 years. While drafting this Chapter, I have learned that it has recently been introduced to the US. In the leaflet of a seed producer, this vegetable is referred as "*S.integri-folium – Turkish Orange, Turkish-Italian Orange*". Heirloom originally from Turkey and more recently from Italy. Orange-red fruits look like tomatoes. Tall 4 inch plants are spineless and are very attractive when laden with fruit. (For further information: www.southernxposure.com – Code 45601 and www.rareseeds.com – Turkish Orange Eggplant).

Fig. 14 Ornamental Tomatoes (called recently Turkish Orange) is actually a kind of egg plant known with its poisonous parts

2.18 The Keli-Calcrete/River Terraces

"*Keli*" is a local word used in the Adana Plain. Shortly, it is the name of a narrow, long and low embankment encountered at certain places in the Adana Plain. It is presumed that these embankments have been created naturally by the flood-sedimentation. Most probably, at certain geologic periods materials of the same or similar composition were transported by flood-flows later deposited as the Kelis. The Keli (especially the lime-carbonate-enriched version) is contemporarily designated and described by Kapur et al. (1990) as the Late Pleistocene to Early Holocene mudflow terraces occurring during the pluvial-interpluvial (equivalent periods of the glacial-interglacials of the northern Alps) climatic fluctuations.

Çatalkeli (Bifurcation Keli), *Büyük Keli* (Grand Keli), *Birinci Keli* (First Keli), *İkinci Keli* (Second Keli), *Çakal Kelisi* (Jackal Keli) were probably the most reputed ones. Each one of these kelis were famous with their agricultural products. For instance, the cucumber produced at çatalkeli was number one in terms of taste and smell as well as the long shelf-life. Even after some days after the harvest, the Çatalkeli cucumber was still tender as if it was just picked.

On the other hand, the best kidney beans, chickpeas and okra were produced at the Second Keli. The development of the "Kelis" could be almost the same process as the development of the *natural levees*, i.e., the material carried by the flood such as sand, silt, and clay is deposited on the flood plain over large particles such as stones or boulders. This process, repeated over and over forms low ridges or narrow but long mounds forming the levees. There are two significant differences between the "keli" and a levee. The first one is that the "Keli" is formed over the high plains whereas the levees are met mostly at the river banks. The second and the most important difference is the obvious superiority of the soil along and around the "keli" as it is particularly excellent for one or several plants, reflecting high yield, delicious taste and particular aroma.

References

Abu Nasser (1960) Tarih-el Ziraat-ül Kadim, Beirut (in Arabic).

ADASO (Adana Chamber of Industry) (2008) The Adana History of Industry. Yeniden Group Press, Adana (in Turkish).

Barber EW (1994) Woman's Work: The First 20,000 Years of Woman, Clothing and Society in Early Times. WW Norton and Company, New York, London

Can Ş (1946) Classical Greek Mythology, Maarif Press, İstanbul (in Turkish).

Çelmeoğlu N (2010) The Adana of my childhood (Unpublished article, in Turkish).

Çelmeoğlu N. Güney MEDYA Weekly Magazine Issues 75 (Sep. 1996), 80 (Oct. 1996), 84 (Nov. 1996), 91 (Jan. 1997), 94 (Feb. 1997), 100 (Mar. 1997), 116 (Jul. 1997), 125 (Sep. 1997), 132 (Oct. 1997), 137 (Nov. 1997), 150 (Mar. 1998), 157 (Apr. 1998), 171 (Jul. 1998), 212–213–214–215–216–217–218 (May–Jun. 1999), 243 (Dec. 1999), 254 (Mar. 2000), 267 (Jun. 2000), 270 (Jul. 2000) (in Turkish).

Ener K (1986) An historical overview of the Adana Basin, 7th Edition. Kayı Pres, İstanbul (in Turkish).

Göney S (1976) The Adana Lowlands. İstanbul University Press, Istanbul (in Turkish).

Kapur S, Çavuşgil VS, Şenol M, Gürel N, FitzPatrick EA (1990) Geomorphology and Pedogenic Evolution of Quaternary Calcretes in the Northern Adana Basin of Southern Turkey. Zeitschrift für Geomorphologie, No. 34: 49–59.

Langlois V (1861) Voyage dans la Cilicie et dans les Montagnes du Taurus, Exécuté pendant les années 1852 and 1853, Paris.

Lloyd S (1989) Ancient Turkey. A Traveller's History of Anatolia. Univ. of California Press, Berkeley, CA.

Ünal A (1999) The Hittites-Etis and Anatolian Civilisations, Etibank Publication. Martı Press, İstanbul (in Turkish).

Yurt Encyclopedia (1981) Vol.1, İstanbul (in Turkish)

Impact of Shifting Agriculture on the Sustainability of Anthroscapes in Sarawak, Malaysia

E. Padmanabhan and H. Eswaran

Abstract Shifting cultivation has a negative connotation particularly in western societies. It is viewed as a system that progressively deforests without attempts to regenerate the forests. Although the system is practiced in many developing countries its contributions to the forest ecosystem and the low-input agricultural system is not recognized. A major advantage is the biodiversity that prevails in this system in comparison to that which exists in the mono-clonal plantations. Mismanaged systems can lead to a land becoming a biodiversity-desert. Shifting cultivation, as practiced by the forest communities is a viable, sustainable enterprise but becomes compromised by modern interventions. Despite the negative connotations, the impact of shifting cultivation can be contained in a sustainable land management system. Precise data on spatial distribution of land areas affected by shifting cultivation is needed. The degradation potential of various fragile systems where shifting agriculture has been detected, can be ascertained with greater precision with such precise data.

Keywords Shifting cultivation · Sarawak · Impact analyses · Mismanaged systems

1 Introduction

Shifting agriculture has been linked to deforestation and subsequently leading to land degradation in the tropics (Myers 1993). Lawrence et al. (1998) summarized that shifting cultivation will have a continuing impact on rainforests. Therefore, to evaluate this impact, the rate of forest clearing associated with shifting cultivation has to be determined. In most cases, this is a challenging task as reliable estimates are often unavailable.

E. Padmanabhan (✉)
Department of Science and Mathematics, Curtin University of Technology, Sarawak, Malaysia
e-mail: eswaran_padmanabhan@petronas.com.my

H. Eswaran
United States Department of Agriculture, Natural Resources Conservation Services, Washington, DC, USA

S. Kapur et al. (eds.), *Sustainable Land Management*,
DOI 10.1007/978-3-642-14782-1_12,

A discussion on the impact of shifting agriculture needs to begin with a reflection on the compelling reasons for shifting cultivation. These include, capturing the inherent fertility of the soils (in a zero input situation), counteracting the negative effects of pests and diseases, and responding to the traditions of the society. In these societies, forest products are as important as cultivated agricultural products. Forests provide medicine, food and fuel in addition to materials for the physical infrastructure. Why do farmers abandon a piece of land after having spent so much effort to clear it and make it ready for cultivation? A part of the answer may be the depletion of the inherent fertility. Further, this piece of cleared land becomes an island or focal point for pests and diseases and the farmer is unable to prevent or control this. Onset of pests and diseases is usually a more important reason for abandoning the land. His option is to abandon and to move to another piece of land. With time, this process becomes a tradition; a cultivated piece of land looses its value for barter, wedding gifts or payment for goods and services. Within a community, there is usually some kind of societal control on the clearing of new land. Part of the reason is that land clearing is usually not an individual activity; members of the community contribute to the process. There is an appreciation for the value of the forest and communities invariably try to maintain the steady state. This aspect has not been studied well.

It may be useful to also look at shifting cultivation over time. Prior to the introduction of the saw, machete and other such Western tools, the time-cycle of shifting cultivation was much longer, usually a generation or so. With the introduction of these tools, land clearing was facilitated and cycles were reduced. Simultaneously, the population was growing with a concomitant increase in demand for land. Also, many communities lived at the land-sea interface. With the ability to exploit more land, some members moved inland. In Borneo, communities were close-knit, often living in one house-the long-house. Several generations lived together and knowledge and information were transmitted orally or through imitation. All these changed with the arrival of foreigners as colonizers. Communities on the coast were most impacted while those inland were buffered to some extent. Providing food and infrastructure materials to the colonizers changed these communities drastically. Individual entrepreneurship took the center stage. Later, the introduction of currency totally changed the behavior of these communities. Finally, when western education was introduced, the transformation was almost complete with only the older generation clinging on to traditions, customs and mores.

Modern systems discouraged shifting cultivation and encouraged sedentary forms of agriculture. Most transactions were with currency while bartering became a "within-family" activity. Production and productivity of the land became a question of survival. Though communities maintained their physical entity as a cluster of dwellings, interaction within were more cautious and measured. Competition between individuals and families emerged and governed social behavior. A great contrast developed between the towns and cities and the villages. This marked the beginning of the end of shifting cultivation. Migration of the younger generation to towns meant that no labor was available for shifting cultivation. New land use laws prevented the haphazard clearing of forests.

It is difficult to obtain accurate statistics on the number of households that are involved in shifting agriculture worldwide. In Sarawak, the most important crop grown under this method of agriculture was hill paddy. Shifting agriculture is highly susceptible to population pressure and resilience of the land. Eswaran et al. (2002) concluded that about 76% of the land surface in the ASEAN region comprised moderate, high and very high impact land systems. Apart from other things, they recommended the implementation of an awareness program to educate rural communities on sustainable agricultural practices, and that land use and management has to be treated as a technical and scientific endeavor based on business principles as well as social and legal ethics.

2 Shifting Cultivation

It is an established fact that large tracts of land have been used for shifting cultivation in Sarawak in the past. This type of agricultural system is essentially based on the natural fertility of the soil and the resilience of the soil to revert back to the original state of fertility. The zero-input agricultural system advocates a fallow period to follow the cropping system. However, when the system is practiced over an extended period of time, land degradation is inevitable.

Shifting cultivation here essentially involves felling primary forest with a subsequent burning event. Paddy is then planted. Occasionally, mixed-cropping types are practiced involving crops such as maize. The same land could be used up to 2 seasons before it is left to fallow. The rejuvenated forest is felled and burnt and after which, crops are grown again. However, unlike the original forest stand, rejuvenated forests are dominated by shrubs and weeds such as *Imperata cylindrical*, *Ploiarium alternioliumand Melastoma malabathricum*. This classical example of forest degeneration is the predecessor for the subsequent land degradation.

Therefore, a land that is subject to shifting cultivation displays forests at various stages of growth. A point worthy of note here is that, soil fertility is very low in the soils of Sarawak and nutrients are usually restricted to the topsoil region. Nutrient mining, a common consequent of shifting cultivation (zero-input system), eventually leads to degradation of the non-renewable resource. Apparently, some early work shows no correlation between the fallow period and net accumulation of nutrients. Degradation is a complex issue and encompasses a wide array of possible scenarios. In the current context, degradation refers to loss of nutrients such as N and P, loss of organic carbon and with that the aggregate stability of the top soil resulting in accelerated topsoil erosion. Therefore, it is evident that degradation is inevitable in such systems.

Shifting cultivation, per se, has been credited as the primary cause for a rapid decline of natural vegetation in this region. FAO (1974) estimates that about 1.3 million hectares of land were under various stages of shifting cultivation. FAO (1984) estimated that the area under influence of shifting cultivation increased at the rate of 65,000 ha per annum between the years of 1975 and 1980. State of Sarawak (1976) estimated that in the mid seventies, total hectarage under shifting agriculture was about 38,000 ha. However, current deforestation by logging and the opening up of huge tracts of lands for plantation has relegated the impact of shifting cultivation on the health of the natural rainforests to much lower categories.

3 Impact Analyses

3.1 Land Capability

Despite the apparent lack of population pressure, the carrying capacity and potential of the land mass may have been severely underestimated. This statement is supported by the fact that the majority of the soils in the state of Sarawak could be considered as incapable of sustaining commercial crops at a level less than that possible by high input agricultural systems. In many areas, the land is marginal by nature. It is therefore, not surprising that the economic viability of agricultural systems in such a vast land mass depends very much on several critical factors such as nature of the terrain, local topography and the lateral extent of the particular soil series.

It has been estimated that about 80% of the total land mass in Sarawak is unsuitable for agriculture; of this, approximately 13% is occupied by Histosols and the remainder, by very hilly to steeply dissected land in the interior comprising skeletal and infertile soils. This dictates that the land mass is more suited for forestry rather than agriculture. This also suggests that for low input agriculture to be viable, it must include a strong forestry component.

3.2 Stewardship Versus Ownership

Land ownership is considered as a viable solution to resolve problems created by shifting cultivation. In the absence of cadastral maps and zoning information, communal knowledge and societal trust become important determinants. However, the challenge in sustainable land management strategies in the twenty-first century appears to be linked to inculcating the stewardship concept in the local community; a common issue globally. Evidently, when the benefits of a complex stewardship program are not translated into the financial gains readily perceivable by those involved in shifting cultivation, any framework for sustainable land resource management has a chance of not being successful. This premise has always led to a negative view of shifting cultivation.

3.3 *Degradation of Non-renewable Resources*

In a fragile system, the precarious balance between soil health and degradation can be easily upset. The same applies when the method of agricultural practice is itself not compatible with the ecosystem. Considering the population pressures and the lack of manpower within tribal families as younger generations move out to city areas looking for brighter futures, it is feared that shifting cultivation practices may be evolving into unsustainable practices. One main reason could be the shortening of fallow periods and diversification of crop types within a unit land area. Similarly, food resources could also be severely affected by land degradation. In many cases, the degradation is irreversible, the resultant impact being permanent.

Due to lack of studies, degradation particularly rates of degradation in such systems are still speculative. There is also very little information on the regenerative capacity of such systems. Any assessment of such systems should be larger than the land quality aspect and should include the socioeconomic aspects. Traditions are strong in such societies and so changes cannot be drastic or enforced. The communities must buy into such systems for progress to be made. However, one thing remains quite evident on the impact of shifting cultivation on soil conditions: the detrimental effect shifting cultivation has on soil conditions become magnified when more than one crop type is cultivated in the same type of land.

4 Viable Options

4.1 *Multidisciplinary Approach*

It has long been recognized that the impact of shifting cultivation in any land could be contained to a certain extent; however, the options come with a heavy price. Loss of revenue would be difficult to compensate. Long term solutions may be able to curb some of these negative externalities. A multidisciplinary and collaborative approach has been advocated to tackle this problem. This is particularly true as different departments within a given government may have their own unique Mission statements and objectives.

4.2 *Spatial Data*

Due to the very nature of shifting cultivation in tropical primary forests, it has been well recognized that quantifying the affected area would be a challenge. Reliable estimates of rates of conversion are not available. The challenge of acquiring reliable estimates is compounded by the fact that this activity occurs in very remote areas. This puts a lot of strain on human and capital resources in trying to get the relevant data.

An additional factor to consider would be the availability of reliable soil maps or any kind of maps including cadastral maps. As developmental activities are minimal, governments do not invest in such assessments. Some generalized soil maps may be produced and the reliability of these maps depends on the skill of the surveyor, terrain accessibility, manpower and funds to support a reliable survey. Such assessments are necessary to coordinate development assistance to these communities. Changes to the hydrological regime can be evaluated if detail soil, terrain and hydrology maps are available. In order to evaluate changes in soil chemical fertility, detail sampling, preferably on a grid basis (to avoid complications arising out of spatial variability within a soil category) carried out on a regular and systematic basis would be needed. However, such a program has not been established to date due to budgetary constraints.

One possible option to avoid any negative impacts of anthropogenic factors in pristine or even marginal areas would be to initiate planned settlement of the population. The governing factors here would be a suitable framework that integrates an appropriate land capability classification system with site-specific conditions. Reliable estimates on the number of families involved in shifting cultivation and other forms of anthropogenic activities would be a strong complement for any initiatives for a planned settlement of the population. However, there is also a socio-economic aspect to consider when an option such as re-settlement is to be considered. In some cases, such a move may not be suitable for the families concerned for a variety of reasons. Firstly, the change of venue would undoubtedly have an impact on the income-generation capacity as it may also involve major adjustments to lifestyles. The financial cost of any resettlement program has been a major deterrent. Large government outlays are essential. This was accomplished by the Federal Land Development Authority (FELDA), initially in Peninsular Malaysia, when they developed a program of supervised credit. The tasks of land clearing, planting, and maintaining the fields for a period of five years were undertaken by FELDA. When the plantation was well established, settlers were brought in. For the subsequent 5 years, the settlers worked under the guidance of the staff of FELDA. During the period, the farmers repaid the costs of this effort.

It is recognized that a complementary step in resource inventory would be to collect information, including indigenous knowledge, for specific sites and uses (Lal and Ragland 1993). The same authors have also pointed out the significance of recognizing future trends and changes in farming/cropping systems when defining development priorities. This is important to monitor trends and implement corrective actions when necessary.

A subsidiary consideration is the kind of farming system that is relevant to the system bearing in mind all the drawbacks and pitfalls of shifting cultivation. The ability to maintain sustainability is the major concern. Alley farming has been credited with the ability to maintain high levels of soil productivity and crop production (Atta-Krah and Kang 1993) and thereby avoiding the application of high amounts of fertilizers or even resorting to long fallow periods. Alley farming may be a suitable alternative to shifting cultivation. The bottom line remains the income generating capacity of the system.

4.3 Relocation

Relocation or resettlement of rural people would certainly be a viable option to conserve marginal or fragile areas. Some challenges faced when attempting to relocate any ethnic group would be resentment to the change, implementation of new vocations, relearning trades and sometimes even catering for basic amenities. Relocation also involves a cost which has to be borne by the government. The fate of the original tract of land must also be considered. This land must be rehabilitated and steps taken to ensure that it is not encroached by new settlers. Despite the fact that reliable estimates are difficult to obtain, it is evident that if shifting cultivators were given the option of a comfortable income derived from sustainable forestry practices, forest conservation measures would definitely improve. Sustainable practices such as terracing, no-tillage and water management could be integrated into the shifting cultivation practices. These again involve costs beyond the ability of the farmers and consequently to be borne by the government. Ultimately, the farmers must appreciate the value of these innovations and changes and so education is an integral component of the process.

5 Conclusions

Early recognition of the benefits of managed shifting cultivation practices needs no elaboration. Forest management could also be enhanced through implementation of stewardship programs involving shifting cultivators. Reliable estimates of areas affected by shifting cultivation need to be collected.

Historically, shifting cultivation has a negative connotation. It is viewed as a system that progressively deforests without attempts to regenerate the forests. Although the system is practiced in many developing countries its contributions to the forest ecosystem and the low-input agricultural system is not recognized. A major advantage is the biodiversity that prevails in this system. Mismanaged systems can lead to the land becoming a biodiversity-desert. Shifting cultivation, as practiced by the forest communities is a viable, sustainable enterprise but becomes compromised by modern interventions. The degradation potential of shifting agriculture in fragile ecosystems can only be ascertained with precise spatial data.

References

Atta-Krah AN, Kang BT (1993) Alley Farming as a Potential Agricultural Production System for the Humid and sub-Humid Tropics. In: J. Ragland and R. Lal (Eds). Technologies for Sustainable Agriculture in the Tropics. ASA Sp. Pub. No. 56. Am. Soc. Agron., Inc., Crop Sci. Soc. Am., In., Soil Sci. Soc. Am., Inc., Madison, WI, USA. 67–76.

Eswaran H, Reich PF, Padmanabhan E (2002) Land resources in the Asean region: Impact of Human Activities. In: J. Shamshuddin, J. Hamdan and A.W. Samsuri (Eds). Sustainable Land Management. The M'sian Soc of Soil Sci, Selangor, Malaysia. 125–142.

FAO (1974) The Mixed Dipterocarp Forests of Sarawk and Their Potential for Development – FO: DP/MAL/72/009 (Technical Report 7) – Rome.

FAO (1984) Tropical Forest Resources Assessment Project (in the Framework of the Global Environment Monitoring System – GEMS). Forest Resources of Tropical Asia. UN 32/ 6.1301-78-04. Technical Report 3.

Lal R, Ragland J (1993) Towards Sustaining Agricultural Production in the Tropics: Research and Development Priorities. In: J. Ragland and R. Lal (Eds). Technologies for Sustainable Agriculture in the Tropics. ASA Sp. Pub. No. 56. Am. Soc. Agron., Inc., Crop Sci. Soc. Am., In., Soil Sci. Soc. Am., Inc., Madison, WI, USA. 309–313.

Lawrence D, Peart DR, Leighton M (1998) The Impact of Shifting Cultivation on a Rainforest Landscape in West Kalimantan: Spatial and Temporal Dynamics. Landscape Ecology 13. Kluwer Academic Publishers, Netherlands. 135–148.

Myers N (1993) Tropical forests: the main deforestation fronts. Environmental Conservation 20: 9–16.

State of Sarawak (1976) Annual Report of the Forest Department. Kuching.

Roman Mining Landscapes in the Murcia Region, SE Spain: Risk Assessment of Mine Ponds

A. Faz Cano, A. Zanuzzi, P. Martinez-Pagan, J. Acosta, D. Carmona, S. Martinez-Martinez, and M. Muñoz

Abstract The Mining District of Cartagena-La Unión (SE Spain), mined since the Roman period, exhibits high levels of metal contamination and this poses a great risk to the surrounding environment. This chapter presents the results of the application of geophysical, geochemical and hydrological techniques in two mine ponds located in this area. Its aim is to determine their degree of contamination by heavy metal, to evaluate risks of mobility and dispersion of heavy metals, to study the erosive capacity of the runoff, and to discuss geotechnical stability of the ponds. Metal contents found in the ponds show that although the Descargador pond is more vulnerable to oxidation and leaching processes than Encontrada, they both could be considered polluted due to their high metal concentrations. When materials from the Descargador pond are exposed to adverse weather conditions they can be eroded, while in the Encontrada pond most of the surface is covered by a natural soil layer and vegetation which protects it against erosion. In the former, it is recommended to apply some conservation measures, including the addition of alkaline materials, natural soil and organic matter. In the latter, it is recommended to improve the present conditions in order to control acidification and to minimize risks of landslides and transport of highly contaminated materials.

Keywords Mining ponds · Heavy metals · Geochemistry · Geophysics · Hydrology

---

A.F. Cano (✉), A. Zanuzzi, P. Martinez-Pagan, J. Acosta, D. Carmona, S. Martinez-Martinez, and M. Muñoz
Sustainable Use, Management and Reclamation of Soil and Water Research Group, Agrarian Science and Technology Department, Technical University of Cartagena, Paseo Alfonso XIII, 52, 30203 Cartagena, Murcia, Spain
e-mail: angel.fazcano@upct.es

S. Kapur et al. (eds.), *Sustainable Land Management*,
DOI 10.1007/978-3-642-14782-1_13, © Springer-Verlag Berlin Heidelberg 2011

1 Introduction

There are 85 mining ponds in the Murcia Region generated during the intensive mining activities in the last century, exploited since the Roman period, to extract iron, lead, and zinc. The Cartagena-La Union Mining District is considered to be the most affected area. Mining activities were abandoned several decades ago when the mineral washing processes ceased off, leaving behind legacies of industrial mine tailings. Major iron ores extracted in the zone were iron oxides/hydroxides, sulphides, sulphates, carbonates and silicates; lead and zinc ores were galena, sphalerite, carbonates, sulphates, and lead- and zinc-bearing oxides (Oen et al. 1975).

Due to their peculiar composition and location, environmental risks can negatively affect soils, waters, plants, animals, and human populations, as well as infrastructure. The Ministry of Economy, through the "Dirección General de Política Energética y Minas" compiled an inventory of these abandoned mining ponds in 2003. Its purpose was to establish a classification of the existing ponds in the region according to their potential risk for the environment, human population, and infrastructure. This inventory, which offered an evaluation of the potentially affected receptors, allows an action plan to be established.

To have a detailed knowledge of the current state of mining ponds with the potential environmental risks, studies were conducted, following the recommendations of the above mentioned inventory. Studies were carried out with the application of geophysical, geochemical, and hydrological techniques, which allowed the establishment of a quantitative assessment of the contamination risks. This work shows the results from two representative mining ponds in different states of conservation called the Encontrada and Descargador, located in the Cartagena-La Union Mining District.

2 Materials and Methods

2.1 Study Area

The Mining District of Cartagena-La Union is located in the Murcia Region (SE Spain) (Fig. 1). The climate is a typical Mediterranean with average monthly temperature ranging from 9.3°C in January to 24.4°C in July. The mean annual precipitation is 300 mm, while the potential evapotranspiration reaches 900 mm year^{-1}. For this study two representative mining ponds were selected according to their different current state of conservation:

- In the Encontrada pond, in the northeast part of the Cartagena-La Union Mining District, re-vegetation actions were carried out on the surface. These processes

Fig. 1 Location of the Murcia province

include covering the pond with natural soil and planting some trees and shrubs to protect the materials from erosion.

– The Descargador pond is located, in the north part of the Cartagena-La Union Mining District, where no conservation measures had been taken and the residues applied were exposed to the impact of the atmospheric agents.

2.2 *Geochemical Characterization*

Surface (0–15 cm) and subsurface samples were taken to determine the geochemical properties of mining ponds used in this study. Surface soil samples were collected according to a regular sampling grid covering an area of 100 m^2. The subsurface sampling was performed in a representative area of the pond, and consisted of continuous mechanical drillings.

In the Encontrada pond, the mechanical drilling (core drill) was conducted to a depth of 15 m. Samples were collected at every 1–1.5 m interval. However, in the Descargador pond it was not possible to conduct the drilling activities due to the difficulty encountered in providing access to the heavy machinery. As a consequence, samples were taken every 5 m, along the most representative slope of the pond in order to assess the different characteristics of the layers. Arcview 3.1 was the mapping software used in determining the sampling sites.

Physico-chemical analyses conducted were: pH (Peech 1965), electrical conductivity (Bower and Wilcox 1965), total elements (Pb, Zn, and Cd), DTPA-extractable (bio-available) and water-extractable metals (soluble). Total heavy metals were analyzed after nitric-perchloric acid digestion, using the Risser and Baker's method (1990) (210°C during 1:30 h and addition of 0.1 N HCl). DTPA-extractable metals were determined using the Lindsay and Norvell (1978) and Norvell (1984) methods. Water-extractable metals were obtained by the Ernst (1996) technique, with distilled water addition (proportion 1:5) and agitation for 6 h, under room temperature. Measurements were carried out using an atomic absorption spectrophotometer (UNICAM SOLAAR 969/989, model UK).

2.3 2D Electrical Resistivity Imaging

2D electrical resistivity imaging (ERI) profiles were carried out in the selected two mining ponds to determine the interface between the substratum and tailings and to characterize the substratum according to electrical resistivity values. A Syscal R1 Plus 72 resistivity meter (IRIS Instruments 2001) with multiconductor cable designed for up to 72 electrodes was used. Each steel electrode was 30-cm long and it was connected to the resistivity meter with clips. The coordinates of each electrode placement were recorded using a GPS unit in order to georeference them to the soil samples collected for the geochemical analysis. In the Descargador pond, two 2D ERI profiles were carried out with a lengthwise orientation (Fig. 2), and 54 electrodes per profile with 5-m spacing were installed. In the Encontrada pond, four 2D ERI profiles were studied and their orientations were chosen to guarantee sufficient coverage of the overall pond (Fig. 3). They consisted of one profile of 72, two profiles of 36, and one profile of 54 electrodes. Data acquisition followed the Schlumberger–Wenner (Sharma 1997) configuration. This was due to its sensitivity to vertical and horizontal variations and it is a good choice between the

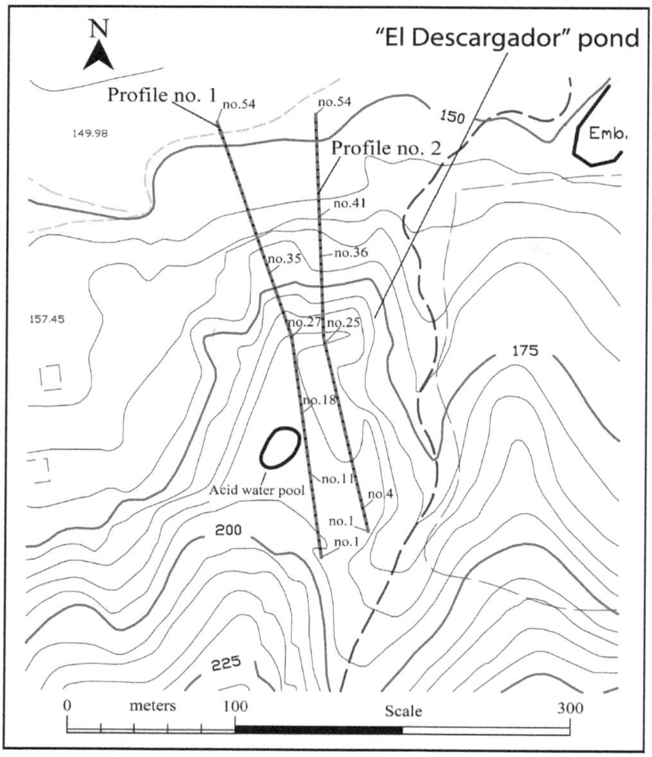

Fig. 2 Locations of the electrodes used for 2D resistivity measurements in the Descargador pond

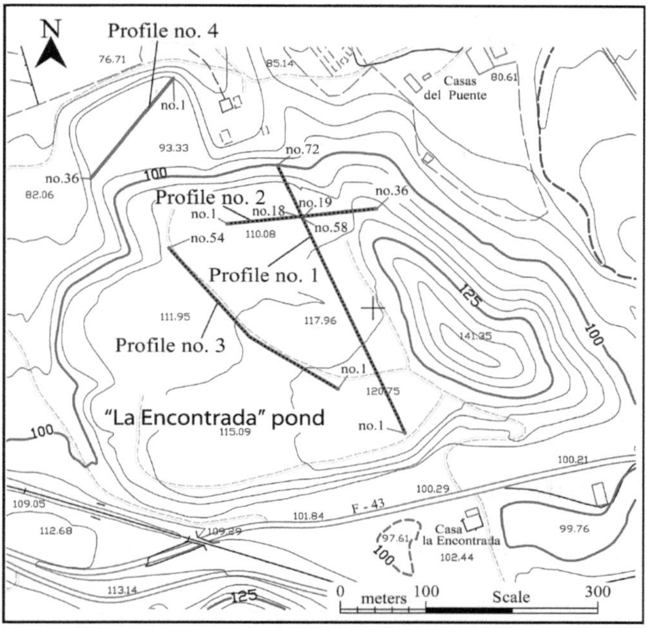

Fig. 3 Locations of the electrodes used for 2D resistivity measurements in the Encontrada pond

Wenner and dipole–dipole array configurations (Pazdirek and Blaha 1996). The electrode spacing and the data acquisition method used in this study provided the resolution and it covered the width (up to 300 m) and depth (up to 30 m) of the mining ponds (Faz et al. 2006).

Electrical measurements collected from each profile were analyzed in a two-stage process, (1) Prosys software (IRIS Instruments, France) was used for initial data processing to remove anomalous values and to make the necessary topographical corrections to the resistivity measurements. (2) The corrected data were used in an inversion process with Res2dinv (Loke 2000) for 2D ERI analyses.

Surfer 8 software (Golden Software, Colorado, USA) was used to calculate the volume of materials stored in each pond. The volume was estimated by using depth to bedrock calculated from the ERI profiles. Surfer calculated the total volume in the pond using three mathematical manipulations: trapezoidal, Simpson, and Simpson 3/8.

2.4 Hydrological Analysis

Restitution and automatic delimitation of the hydrographical drainage were carried out using digital patterns of the land; models of slopes and orientations by means of the GIS. These previous results were checked and confirmed in the field before the final delimitation of the basin. Once delimited the basin, diverse analyses were carried out to determine: (1) the grade of the stability of the materials, when significant hydrological events occurred, (2) the erosive capacity of the surface waters, (3) the capacity of the waters to transport materials, and (4) the risk analysis of the transmittance of the metallic pollutants to the hydrosphere, and the hydrodynamics of the surface waters.

3 Results

3.1 Geochemical Characterization

3.1.1 Surface Metals Distribution

The pH values of the surface samples from the Descargador pond were acidic (USDA 1996), ranging between 2.5 and 4.9 (Table 1), whereas, the pH values in the Encontrada were basic ranging from 7.7 to 8.2, while in KCl pH values were lower in both ponds showing the presence of potential acidity. Electrical conductivity values were also different, where the Descargador pond showed high salinity levels with a mean value of 13.1 dS m^{-1}, while the Encontrada pond exhibited a mean value of 0.4 dS m^{-1}.

Regarding the total heavy metal concentrations in the Descargador pond, the Cd contents ranged from 12.2 to 48.7 mg kg^{-1}; Pb from 2.71 to 11.71 mg kg^{-1} and Zn from 4.10 to 12.66 mg kg^{-1}. In the Encontrada pond, total concentrations were also high, reaching 30.5 mg kg^{-1} for Cd; 21.27 mg kg^{-1} for Pb and 9.81 mg kg^{-1} for Zn (Table 1). Background levels of heavy metals in soils were then compared to the levels established by the Spanish (Basque Country) and the Dutch laws (IHOBE 2000; Ministry of Housing, Netherlands 1994). Taking into account Zn, Pb, and Cd concentrations in the ponds and the values established by law, all the samples were graded as polluted, because Zn, Pb, and Cd contents were higher than 720, 530 and 12 mg kg^{-1}, respectively; however two samples in the Encontrada pond exhibited Cd values lower than the threshold value specified.

As expected, heavy metal mobility was much higher in the acidic pond. The Pb and Zn showed the highest bioavailability (1.86 and 1.30 mg kg^{-1}, respectively) while Zn and Cd reflected the highest solubility (1.65 and 17.2 mg kg^{-1}, respectively) (Table 1). In the basic pond, Pb and Zn were the most mobile elements, where the soluble Pb and Zn reached 6.5 and 5.0 mg kg^{-1}, respectively; however, the bioavailable fraction was up to 975 mg kg^{-1} for Pb and 164 mg kg^{-1} for Zn.

Table 1 pH in water, and in KCl, electrical conductivity (EC), total, soluble and bio-available metals measured in the Descargador and Encontrada ponds

	pH		EC ($dS\ m^{-1}$)	Total metals ($mg\ kg^{-1}$)			Soluble metals ($mg\ kg^{-1}$)			Bioavailable metals ($mg\ kg^{-1}$)		
	H_2O	KCl		Cd	Pb	Zn	Cd	Pb	Zn	Cd	Pb	Zn
n = 15	*Descargador*											
Maximum	4.98	4.68	31.70	48.73	11,711	12,661	17.2	14.5	1,647	12.01	1,856	1,304
Minimum	2.54	2.35	2.67	12.21	2,713	4,102	1.83	0.0	20.5	0.25	1.53	136
Mean	3.19	2.90	13.10	31.33	6,533	6,907	7.44	4.79	704	6.10	353	744
n = 17	*Encontrada*											
Maximum	8.17	7.48	2.49	30.47	21,272	9,813	0.42	6.50	5.01	1.08	975	164
Minimum	7.66	7.29	0.11	9.06	2,840	2,166	0.0	1.09	0.17	0.22	118	9.24
Mean	7.96	7.37	0.36	17.42	5,578	3,861	0.17	2.41	1.86	0.44	304	65.3

3.1.2 Spatial Distribution of Metals

A map was produced in order to study heavy metal distribution in the surface layer of the ponds (Figs. 4 and 5). Metals were associated to pH and EC. High concentrations of total Pb, Zn, and Cd were found in the southwestern part of the Descargador pond (Fig. 4). This area, exposed to erosion processes, is vulnerable to oxidation and hydrolysis processes; therefore mineral dissolution occurs more rapidly when pH decreases. In fact, the most significant contents of Pb and Zn (11.71 and 12.66 mg kg^{-1}, respectively) correspond to an acidic pH range (3.0–3.5).

Figure 4 shows high contents of total and soluble Cd (25–30 and 0.3–0.5 mg kg^{-1}, respectively) in the north and south part of the Encontrada pond. Moreover, Cd exhibited a heterogeneous distribution on the surface of the pond. Total Pb and Zn concentrations followed the same pattern as Cd, with the lowest contents in the northwest part. There was a little difference between soluble Pb and Zn because the highest concentrations were located in the southwest and southeast parts, respectively.

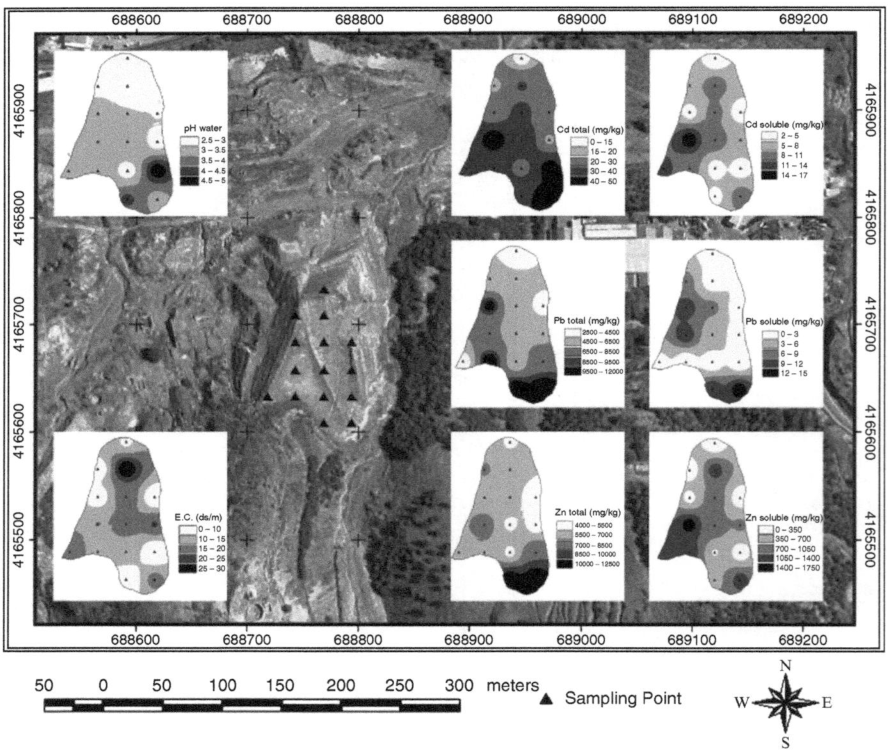

Fig. 4 Spatial distribution of pH, EC, total and soluble Cd, Pb and Zn contents in the Encontrada pond

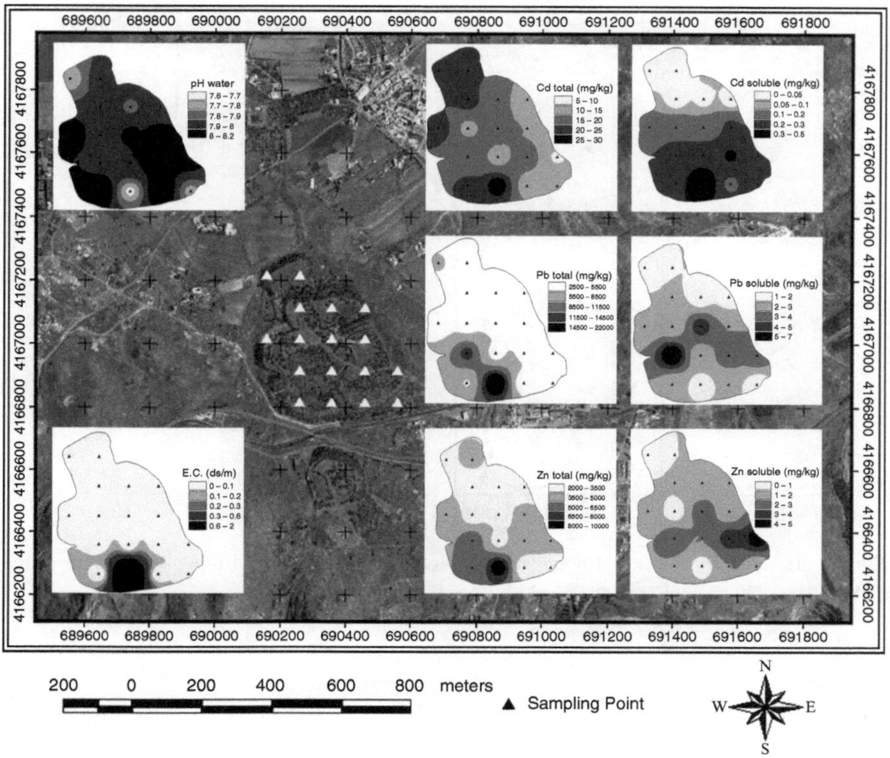

Fig. 5 Spatial distribution of pH, EC, total and soluble Cd, Pb and Zn contents in the Descargador pond

Surface distribution of total and soluble metals in both ponds, such as Cd, Pb, and Zn, allowed us to recognize the area with the highest contents of heavy metals. With this, it was possible to identify which area possesses the highest potential risks associated to heavy metal leaching, run off, and accelerated oxidative/oxidation processes, which increase the secondary mineral accumulation (Lottermoser and Ashley 2006).

3.1.3 Metals Distribution in the Ponds

Figure 5 shows pH values found in the drilling conducted in the Descargador pond. Results confirmed that pond materials are acidic and values range between 3.9 and 1.9, with KCl the values were between 1.9 and 3.7, with the exception of the sample taken at 35 m that showed a higher pH (6.3). Electrical conductivity was very heterogeneous, ranging from 27.5 to 2.2 dS m^{-1}, confirming that all the layers were saline and slightly saline. Figure 6 depicts the opposite behavior between pH and

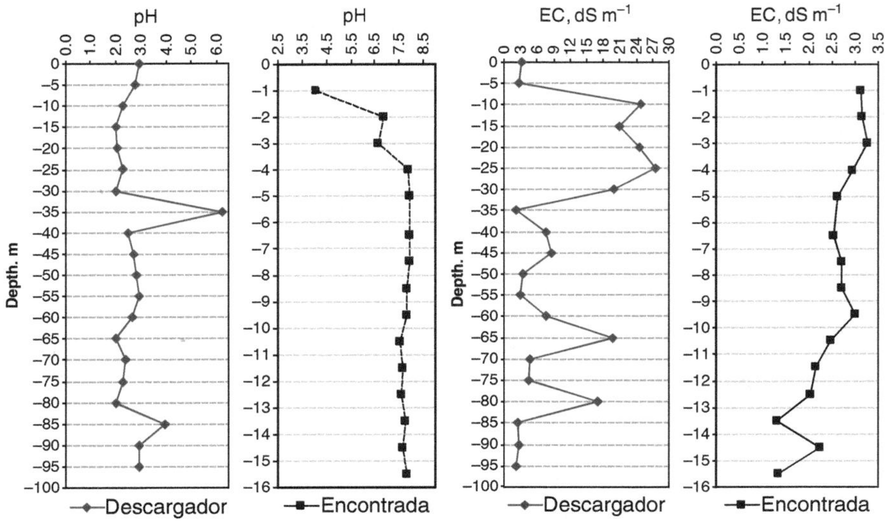

Fig. 6 pH (*left*) and electrical conductivity (*right*) contents of the drilling samples form the Descargador and Encontrada ponds

EC due to the increase of protons in soil solution as a consequence of the dilution of salts.

On the other hand, samples from the drilling carried out in the Encontrada pond showed differences in pH: the layer from 0 to 1 m exhibited an extremely acidic pH that could be related to the acid sulphide weathering and the production of acidity; the layers between 1 and 3 m revealed a neutral pH; and the layers from 3 to 15 m exhibited slightly alkaline pH values. In relation to electrical conductivity, layers from 0 to 3 m showed the highest values and salt concentrations that were decreased with depth (Fig. 6).

According to the total heavy metals, there was not a specific pattern that could explain diverse contents of elements in the samples (Figs. 7–10). Element concentrations vary among different mining soil layers. In relation to the Dutch and Spanish legislation, samples could be considered polluted due to the high contents of Pb and Zn. Based on their Cd contents, layers from 20, 30, 35, 55, 60, 65, 80 to 90 m could also be considered polluted.

3.2 Geo-Electrical Characterization by the Electrical Resistivity Imaging Method

In the Descargador pond the two inverted resistivity pseudo-sections obtained from the ERI profiles accurately showed the region taken by the tailings (Fig. 11).

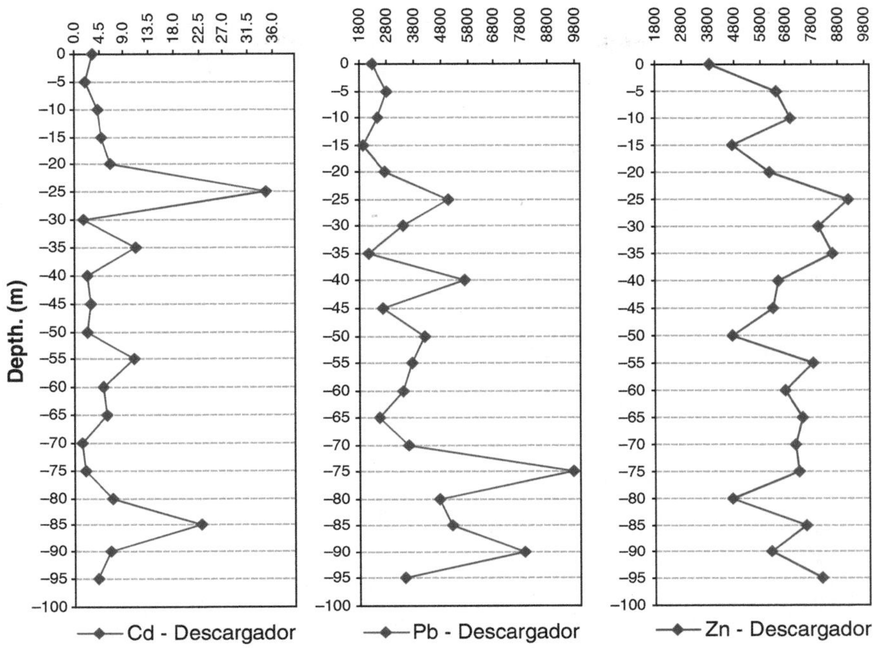

Fig. 7 Total metals concentrations (mg kg^{-1}) measured in the drilling samples taken in the Descargador pond

This region is characterized by electrical resistivity values <8 Ωm. In Fig. 11, this region is delimited by a dotted line. The ERI profiles showed the bedrock with electrical resistivity higher than 30 Ωm. This layer underlay the tailings and reflected that the electrical resistivity values generally increase with depth. In the middle of the electrical pseudo-section, there was a vertical zone where the electrical resistivity values were between 8 and 30 Ωm. This might be due to a fault inside the mining pond which could contribute to the metals leaching. Moreover, on the surface, the geochemical values from the soil samples revealed the highest electrical conductivity values in this region, being in agreement with the geo-electrical data. Tailings were normally hardened on the surface; this feature was shown on the electrical pseudo-sections with electrical resistivity values higher than 8 Ωm.

Under electrode 18 (Fig. 10), a decrease in electrical resistivity (values below 8 Ωm (7–15 m in depth)) was found and this was due to the fact that there were high contents of metals and salts in the soil solution.

In the Encontrada pond, it was not possible to see the tailing outcropping due to the re-vegetation processes. This was reflected in the most sections of the Encontrada pond studied, where there were electrical resistivity values between 30 and 100 Ωm, near the surface, e.g., on profile number 1, from electrode no. 1 to electrode no. 25 (Fig. 12). The pond was overlaid with this filling material which had a thickness that ranged from 2 to 8 m.

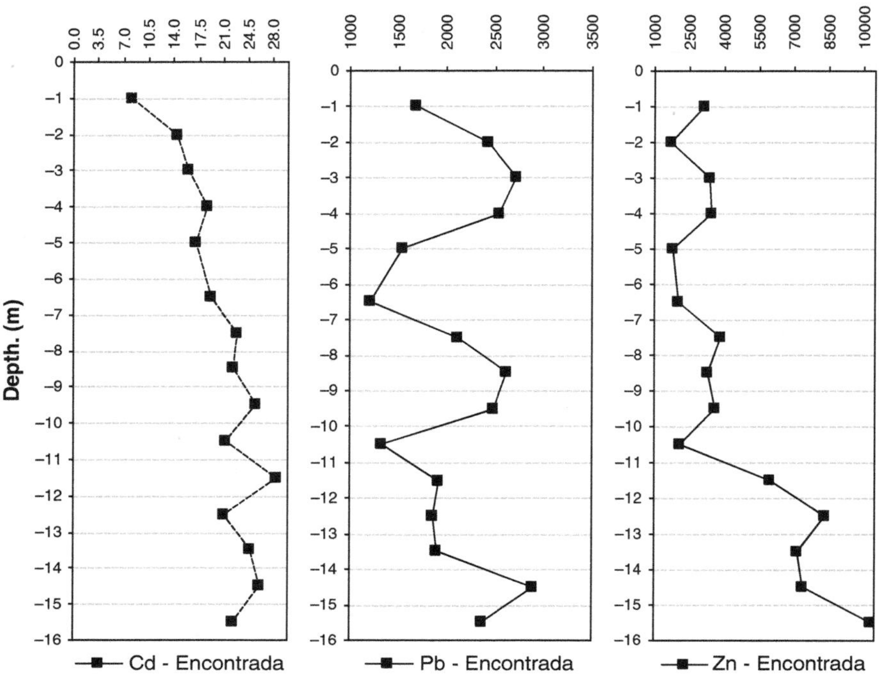

Fig. 8 Total metals concentrations (mg kg^{-1}) measured in the drilling samples taken in the Encontrada pond

The inverted resistivity section of profile number 1 showed both the contact layer tailing-bedrock and some faults inside the bedrock (Fig. 12), which were characterized by electrical resistivity values that ranged from 8 to 60 Ωm. The knowledge obtained for the faults was very important because they were preferential pathways to transport heavy metals or other particulates to the underground, and in some cases they could become the source of the failure of the pond.

From electrode no. 59 to 64 there was a little subsidence zone, highly cracked, and characterized by electrical resistivity values below 8 Ωm (Fig. 12). This figure depicts the crossing point of profile no. 2 with profile no. 1; whose resistivity section was shown in Fig. 13. This pseudo-section showed the calcareous material outcropping characterized by high electrical resistivity values (>65 Ωm), and how it penetrated into the underground to become the bedrock. Under electrode no. 18 of profile no. 2 (Fig. 13), thicker tailings were found (15 m thick) and the subsidence effects could be observed.

Due to the great contrast of the electrical resistivity values among the different materials inside both ponds, it was possible to identify the bedrock-tailings interface in a reliable way with the electrical resistivity imaging method (Fig. 14). These results were corroborated by means of a drill hole carried out in the Encontrada

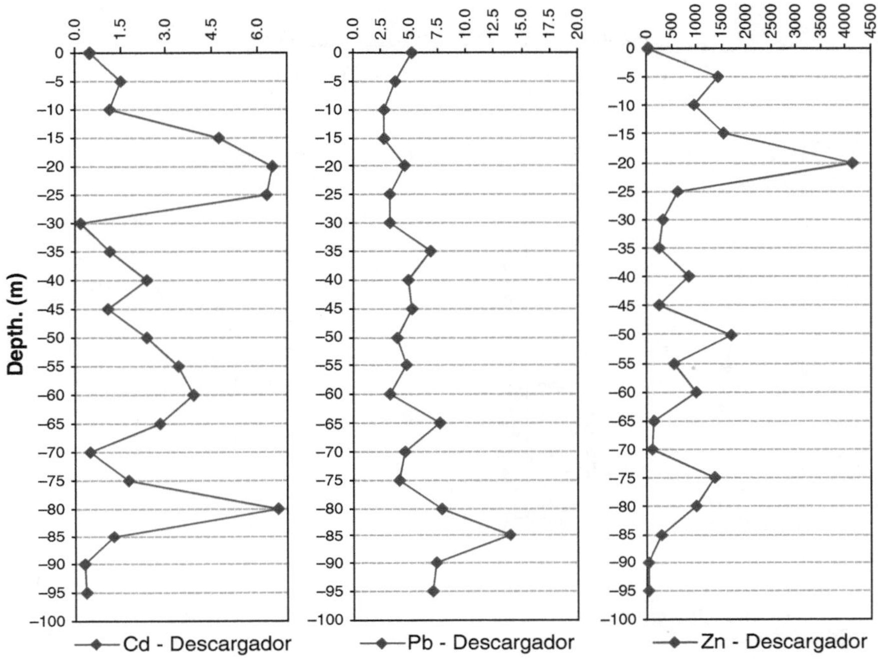

Fig. 9 Soluble metals concentrations (mg kg^{-1}) measured in the drilling samples taken from the Descargador pond

pond and visual inspections on the slopes in the Descargador pond. By knowing the interface position, it was possible to establish the bedrock surface below the study area and to calculate the volume of the poured materials using Surfer 8 software (Golden, Colorado, USA). The volume of materials obtained in the Descargador and in the Encontrada ponds were 184.146 and 1,132,000 m^3, respectively.

3.3 Hydrological Analysis

The surface of the Descargador pond did not have any vegetation and therefore it was exposed to weathering conditions. Evidences of strong human activities were observed in the upper part of the basin, showing a low vegetation cover, which along with the high rainfall events, contributed to the development of a severe soil erosion event. Thus, it was recommended to re-vegetate the area to minimize these effects and accordingly in the surroundings of the pond, a small reforestation program with *Pinus halepensis* was developed. This was accomplished to contribute to reduced erosion, improving water infiltration and stabilizing the hillside. Naturally, these

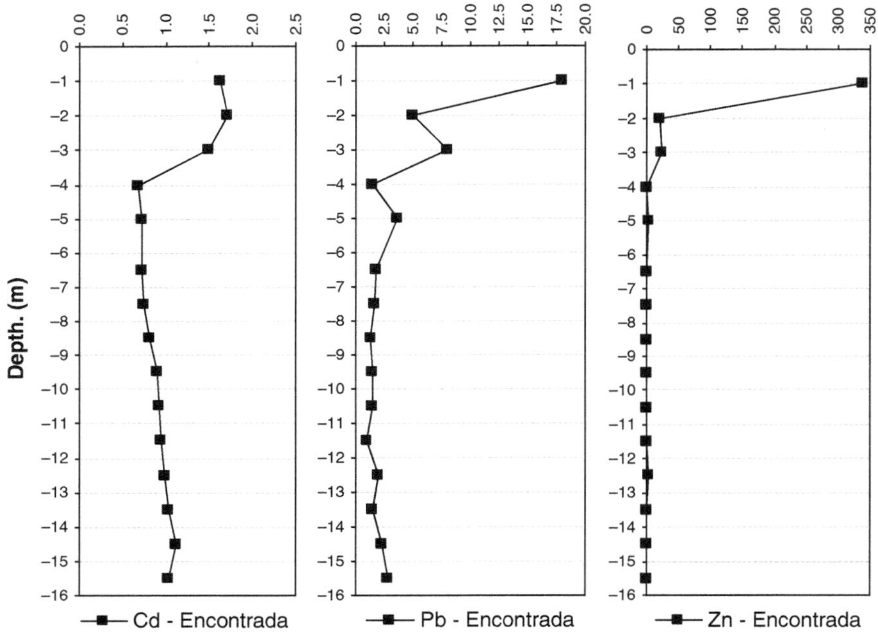

Fig. 10 Soluble metals concentrations (mg kg^{-1}) measured in the drilling samples from the Encontrada pond

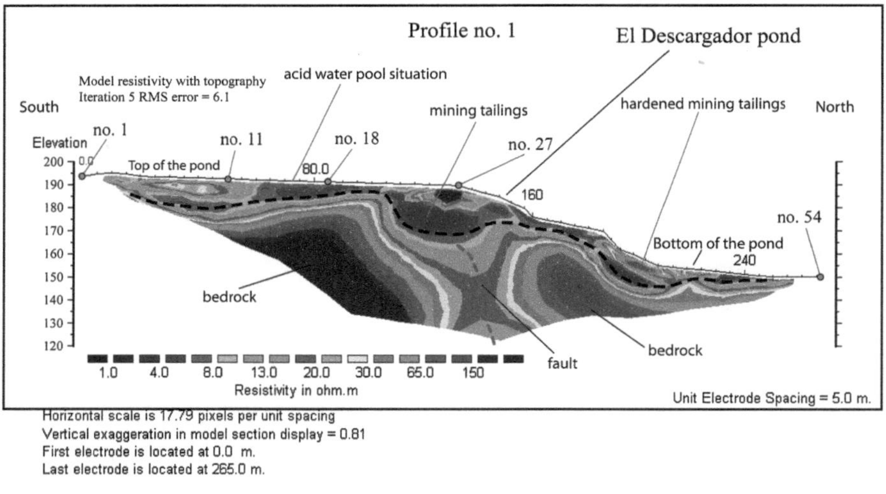

Fig. 11 Resistivity section of profile number 1 in the Descargador pond

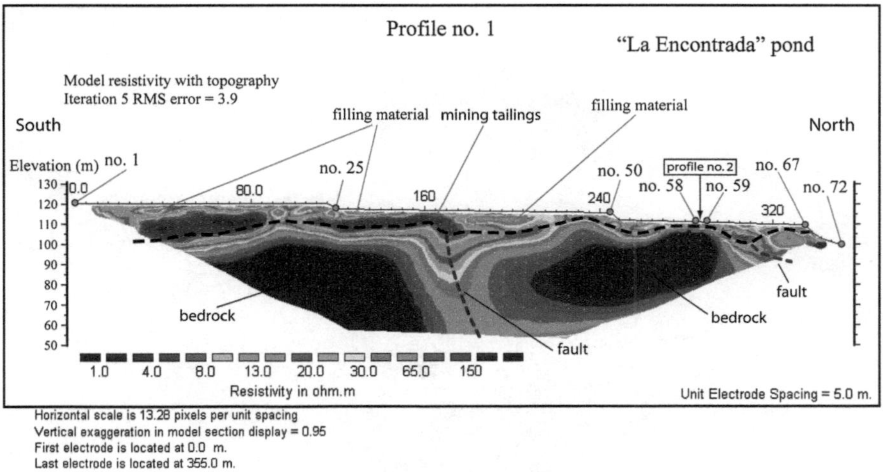

Fig. 12 Resistivity section of profile number 1 in the Encontrada pond

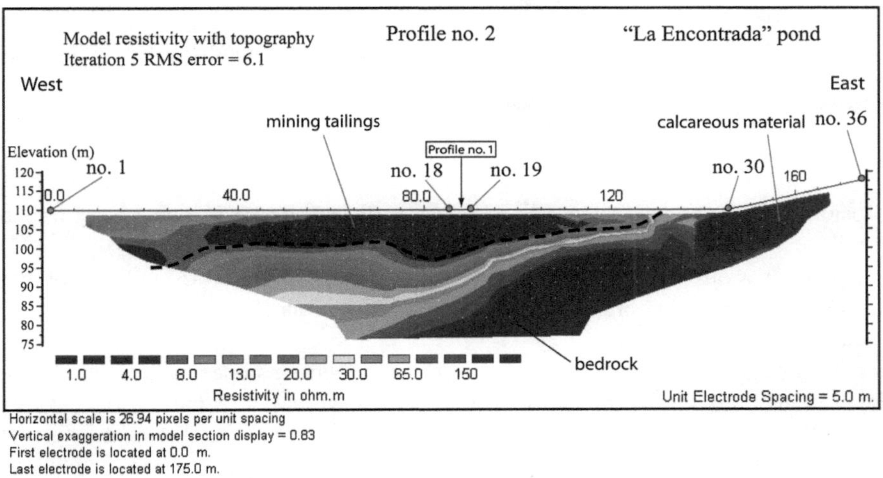

Fig. 13 Resistivity section of profile number 2 in the Encontrada pond

measures have improved the hydrological characteristics of these part of the study site.

Erosion channels produced by water runoff on the slopes of the pond could create a risk of falling of the soil layers. At the top of the pond, there were some signs of water accumulation, which along with other factors, such as erosion and some seismic events, could lead to the collapse of part of the pond. This collapse could also cause the deposition of large amounts of downstream sediments. Such events, influence the water runoff in the area causing the development of significant

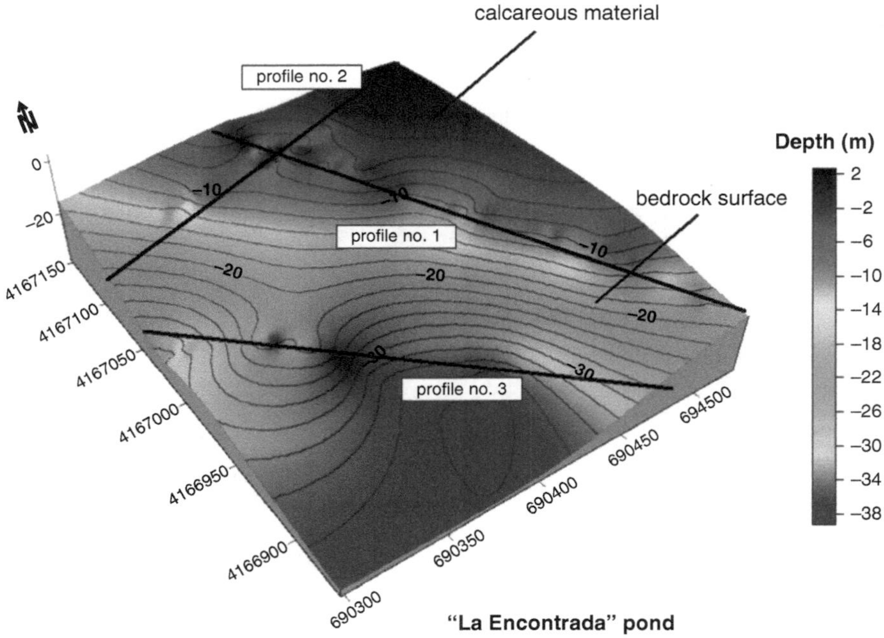

Fig. 14 Spatial view of the tailings-bedrock interface in the Encontrada pond

erosion, which eventually could produce a further collapse in other parts of the pond.

There was also a high amount of solid wastes at the surroundings of the Descargador pond that prevented the natural flow of runoff, with which prevention measures were undertaken.

In the Encontrada pond, most of its surface, including slopes, was covered by a layer of natural soil, as the basis for the restoration activities carried out. In this natural soil, some trees and shrubs were planted, protecting the pond against erosion. Despite this vegetation cover, erosion induced by heavy rain events was consistent in the central part of the pond. Similarly, most of the slopes were covered with vegetation, which increased stability and mitigated erosion. However, the water runoff created erosion channels at locations with some slopes having visible signs of wind and water erosion, probably due to the absence of the natural soil and plant cover. As a consequence, an enhanced risk of contamination of the adjacent soils was encountered in the study area.

At the top of the pond, there were structures for water accumulation with problems of mal-functionality and deterioration. Ultimately, it was recommended to construct new structures and to restore the existing ones in order to prevent the effects and risks of landslides and transportation of highly contaminated materials.

4 Conclusions

Metal contents found in the surface and the deeper parts of the soil samples, allowed us to conclude that the Descargador pond is more vulnerable to oxidation and leaching processes than the Encontrada pond. Due to the absence of vegetation materials, the Descargador pond is more exposed to water and oxygen on the surface than the Encontrada pond. This also accelerates erosion processes and environmental risks, and enhances the release of acidity, sulfates and soluble elements from the surface of the pond. Nevertheless, due to the high metal (Pb, Zn and Cd) concentrations, both mine ponds could be considered polluted.

The results suggested that the mine ponds had a high acid-forming potential. And, that, acidity prevents the colonization of plants, due to the toxicity of heavy metals and salinity. Therefore, applying alkaline materials, such as limestone, or covering the surface with soil and/or placing a barrier layer, would be necessary for controlling the acidification of the tailing ponds and their re-vegetation, as well as for improving physical soil conditions which facilitate plant growth.

The use of the electrical resistivity imaging method, the ERI, confirmed its applicability to identify the nature of materials inside the pond with geological features, such as faults, lenses and the interfaces among different layers. The ERI, is thus considered, to be a useful technique to highlight possible pathways for heavy metal mobility or leaching and to evaluate the geotechnical stability of a pond. This method in combination with the drill holes allow the volume of the tailings below the placement of the profiles to be evaluated, without the need of a great number of drill holes. The ERI, along with geochemical analyses, can also assist in remediation actions to identify or to classify the regions according to their pH values, by knowing that a low pH reduces the electrical resistivity values of the ERI method or vice versa.

The absence of the conservation methods in the Descargador pond caused the development of weathering phenomena on the surface, consistent to the hydrogeology of the area causing erosion by storm events. In fact, a collapse of part of the pond took place some years ago, generating large amounts of materials that were deposited downstream. In the case of the Encontrada pond, most of its surface was covered by a natural soil layer and vegetation which protected it against erosion. Despite this, there are some places in the pond where erosion can be seen and this was probably due to strong rainfall events. It is recommended to carry out some conservation practices in the Descargador pond and to improve the present conditions in the Encontrada pond in order to minimize risks of landslides and transport of highly contaminated materials.

References

Bower CA, Wilcox LV (1965) Soluble salts. In Black CA (ed.). Methods of Soils Analysis. American Society of Agronomy, Madison, WI, 2, 933–940.

Ernst WHO (1996) Bioavailability of heavy metals and decontamination of soils by plants. Appl. Geochem. 11: 163–167.

Faz A, Martínez-Pagán P, Maruri U, Aracil E, Martínez S, Acosta JA (2006) Environmental Risk Assessment of Two Representative Mining Ponds from Cartagena-La Union (SE, Spain) by Using Geochemical and Geophysical Techniques. Sustainable Mineral Supply for Europe from Waste to Resource, Wroclaw (Poland), 115–120.

IHOBE (2000) Valores Indicativos de Evaluación de la Calidad del Suelo. En: Investigación de la Contaminación del Suelo. Manual Práctico IHOBE, 107pp.

IRIS Instruments (2001) Syscal Switch V. 11.4. User's Manual, Orleans, France, 66pp.

Lindsay W, Norvell W (1978) Development of a DTPA soil test for Zn, Fe, Mn, and Cu. Soil Sci. Soc. Am. J. 42: 421–428.

Loke MH (2000) Electrical Imaging Surveys for Environmental and Engineering Studies. A Practical Guide to 2-D and 3-D Surveys, 61pp.

Lottermoser B, Ashley P (2006) Mobility and retention of trace elements in hardpan-cemented cassiterite tailings, north Queensland, Australia. Environ. Geol. 50: 835–846.

Ministry of Housing, Netherlands (1994) Spatial Planning and the Environment. Intervention and Target Values-Soil Quality Standards. Report HSE 94.021, The Netherlands.

Norvell WA (1984) Comparison of chelating agents as extractans for metals in diverse soil material. Soil Sci. Soc. Am. J. 48: 1285–1292.

Oen IS, Fernandez JC, Manteca JI (1975) The lead–zinc and associated ores of La Union, Sierra de Cartagena, Spain. Econ. Geol. 70: 1259–1278.

Pazdirek O, Blaha V (1996) Examples of resistivity imaging using ME-100 resistivity field acquisition system. In EAGE 58th Conference and Technical Exhibition Extended Abstracts, Amsterdam.

Peech M (1965) Hydrogen-ion activity. In Black CA (ed.). Methods of Soil Analysis. American Society of Agronomy, Madison, WI, 2, 914–916.

Risser JA, Baker DE (1990) Testing soils for toxic metals. In Westerman RL (ed.). Soil Testing and Plant Analysis. Soil Science Society of America Special Publication 3, Madison, WI, 3rd ed, 275–298.

Sharma PV (1997) Environmental and Engineering Geophysics. Cambridge University Press, Cambridge, 473pp.

USDA (1996) Soil Survey Laboratory. Methods Manual. Soil Survey Investigations Report. N° 42. Version 3.0. United State Department of Agriculture, 693pp.

Anthroscape of the Mediterranean Coastal Area in the Context of Hydrogeology: Projected Impacts of Climate Change

Katsuyuki Fujinawa

Abstract A common concern currently prevailing around the world is the threat of global climate change, which may cause enormous impacts on ecosystems. Global climate change and its accompanying sea-level rise affect the landscape of the Mediterranean coastal area by changing the local hydrologic cycle. Specifically, future adaptation or mitigation of the impacts of climate change on agricultural practices in the arid coastal region depends on the availability of water resources including groundwater. These human activities would certainly bring major changes to the anthroscape.

SIFEC (Saltwater Intrusion by Finite Elements and Characteristics), a numerical code linking groundwater with lagoon water in terms of water volume and salt mass, was applied to a groundwater system in the Lower Seyhan River Basin along the Mediterranean Sea in southern Turkey. Projected results show that while sea-level rise will not cause substantial saltwater intrusion directly from the seabed during the next 70 years, the combination of sea-level rise with increasing evaporation and decreasing precipitation can cause significant impact on salinity in the lagoon water. The increased salinity of the lagoon may in turn cause saltwater to penetrate into the deeper layers beneath the lagoon and eventually cause a drastic increase in the salinity of groundwater. All of these impacts lead directly to land degradation and desertification through changes in soil–water conditions and the accumulation of salt on the land surface.

Keywords Hydrogeology · Mediterranean · Climate change · Anthroscape

K. Fujinawa
Shinshu University, Wakasato 4-17-1, Nagano 380-8553, Japan
e-mail: fujinawa@shinshu-u.ac.jp

S. Kapur et al. (eds.), *Sustainable Land Management*,
DOI 10.1007/978-3-642-14782-1_14,

1 Introduction

According to the Fourth Assessment Report (AR4) of the Intergovernmental Panel on Climate Change (IPCC), the temperature change between 1990–1999 and 2090–2099 is projected to fall in the range from 1.1°C in the B1 scenario to 6.4°C in the A1FI scenario, and sea-level rise, in the range from 18 cm in the B1 scenario to 59 cm in the A1FI scenario (Solomon et al. 2007). Land use in the coastal area would be strongly affected by climate change and the accompanying sea-level rise, because the area is the most densely populated in the world, provides important habitats for ecosystems, and has large amounts of social capitals. All issues assessed in AR4 (Parry et al. 2007), including beach erosion, increased flooding, increased salinity of estuaries, and degradation of coastal ecosystems, are likely outcomes of sea-level rise. Moreover, sea level is expected to continue rising beyond this century and up to the next millennia, due to the significant inertia of the climate system (Cubasch et al. 2001).

Various case studies on seawater intrusion have been published during the past decade. Among them, Feseker (2007) conducted a hypothetical case study on the impact of climate and land-use changes on salinity distribution in a coastal aquifer in northwestern Germany. He assessed the impacts of a sea-level rise of 0.5 m per century on saltwater intrusion and groundwater discharge to the land surface, and the results after a simulated lapse in time of 50–250 years showed a considerable increase in groundwater salinity in a low-permeability cover layer close to the shoreline. Feseker also found that sea-level rise causes a drastic increase in the amount of groundwater discharge to the land surface. Even after 500 years, salt concentration continues to change slowly, indicating that the system has yet to reach a stable equilibrium. Oude Essink and Schaars (2002) also state that climate change intensifies the groundwater salinisation process.

To evaluate saltwater intrusion, Langevin and Dausman (2005) applied a two-dimensional numerical model for predicting the extent, rate, and lag time of saltwater intrusion to a permeable carbonate aquifer in southern Florida, in the context of various sea-level rise scenarios. They reported that at the fastest rate of sea-level rise estimated in the Third Assessment Report (TAR) of the IPCC (8.8 mm/year), the 250 mg/L isochlor moved inland by about 1,800 m in 100 years, and that it required more than 50 years to reach equilibrium. By comparing observed intrusion rates with simulated rates, they further suggested that sea-level rise will probably result in a slower rate of saltwater intrusion than the intrusion rate caused by extensive drainage.

Sand box experiments conducted by Masuoka et al. (2005) revealed that sea-level rise accelerates saltwater intrusion and surface water discharge in coastal areas, and that preventive measures such as installation of physical barriers against saltwater intrusion are probably effective only for delaying the speed of the intrusion. These results raise special concern about future malfunctions that could occur in mega cities lying below sea level as a result of increased surface water discharge and degradation of groundwater quality in coastal aquifers.

All of the studies suggest that scientists and engineers face an acute task to assess the driving force of the future anthroscape and seek proper methods of adaptation and mitigation in reference to the possible outcomes. In response to TAR's projection that climate change would bring a significant decrease in precipitation and runoff and an increase in evaporation in the area around the Mediterranean Sea, the Research Institute for Humanity and Nature (RIHN), Japan, implemented a comprehensive research project on "Impact of Climate Changes on Agricultural Production System in Arid Areas (ICCAP)" in collaboration with the Scientific and Technological Research Council of Turkey (TUBITAK), to assess impacts of climate change on agricultural production in the eastern Mediterranean region of Turkey. The 5-year ICCAP project was completed in March 2007, and the project outcome was released as the final report of the ICCAP (Watanabe and Kanber 2007).

Although groundwater is an important water resource and a substitute to reduced surface water, only a few of the past studies has taken into account changes in groundwater resources, flow systems, and water quality degradation as a result of climate change and its associated anthropogenic impacts. In reference to the ICCAP project, this paper discusses a long-term projection of the subsurface environment, including groundwater quality, groundwater flow system, water logging, lagoon salinity, and salt accumulation on the land surface, specifically in relation to climate change, accompanying sea-level rise, adapted agricultural activities, and associated management of groundwater resources, within a well-established context area of integrated sustainable management of land and water, the Anthroscape-a human re-shaped landscape unit, for the Seyhan Basin (Kapur et al. 2004).

2 A Mathematical Tool for Evaluating the Hydrogeologic Anthroscape

The subsurface environment in a coastal area can be simulated by using a mathematical model that enables to solve governing equations for density-dependent, saturated–unsaturated, subsurface-water flow coupled with salt transport. There are a variety of numerical codes for handling the mathematical model. Konikow and Bredehoeft (1978) implemented a widely-used Method of Characteristics (MOC). While MOC tracks a large number of moving particles, the Modified Method of Characteristics (MMOC) uses only one particle for each node. Neuman (1984) proposed an Eulerian–Lagrangian scheme based on the method of characteristics and finite elements. Fujinawa et al. (2005) developed a new numerical code named SIFEC (Saltwater Intrusion by Finite Elements and Characteristics), which uses the bi-quadratic interpolation for calculating the concentration of moving particles.

In order to simulate the dynamic interaction between a saline lagoon and a groundwater system, Fujinawa et al. (2009) further modified the SIFEC to take the volume of water and mass of salt in the lagoon into consideration. The modified

SIFEC allows the stage of a lagoon to vary in accordance with a functional relationship between the stage and water volume of the lagoon, and also allows the salt concentration of the lagoon to vary in accordance with the salt budget of the lagoon, including the mechanism of chemical precipitation and dissolution of salt. The updated stage and salt concentration of the lagoon are in turn used as transient boundary conditions for the coupled flow and salt transport model.

The revised SIFEC was finally applied to the eastern Mediterranean coastal region of Turkey to assess the impacts of sea-level rise, global climate change, adapted irrigation practices, enhanced abstraction, and changing inflow conditions along the inland boundary on groundwater resource, groundwater salinity, seawater intrusion, water logging, salt accumulation on the land surface, and lagoon water salinity under various scenarios in terms of sea-level rise, increased evaporation and decreased precipitation, agricultural adaptation, groundwater management, and altered hydrogeologic anthroscape conditions.

3 Description of the Study Area

3.1 *Physiography and Climate*

The study area, the Lower Seyhan River Basin (LSRB), is located in the eastern Mediterranean coastal region of Turkey near the city of Adana (Fig. 1). The LSRB is an extremely flat fan ranging from the Seyhan Dam and the Mediterranean Sea, and is formed by the sedimentation of the rivers Seyhan, Ceyhan, and Berdan that flow from the Taurus Mountain in the north toward the Mediterranean Sea in the south. Adana is situated approximately 50 km north of the coastal line, at an altitude of around 20 m.

There are three lagoons in the LSRB. The largest is the Akyatan, which lies parallel to the coastal line. The Akyatan is separated from the adjacent Mediterranean Sea by a beach dune, and is artificially kept open through a narrow inlet at the southeast. Thus, the lagoon has access to the Mediterranean Sea, while fresh water flows into it through canals (Nazik et al. 1999). The depth of the lagoon slightly varies from 80 to 100 cm depending on the season, and the greatest depths appear in winter and spring due to rainfall and freshwater inflow from drainage/irrigation canals. The sand dune to the south of the Akyatan forms a slight ridge in the otherwise flat coastal plain.

The LSRB is characterized by semi-arid weather with dry summers and rainy winters, as shown in Fig. 2, and is a major agricultural center in Turkey with 213,200 ha of cultivated land. Lower Seyhan Irrigation Projects (LSIPs), divided into four stages, were initialized in the 1960s, and the completed projects now cover three fourths of its arable land. One fourth of the LSRB, corresponding to the fourth stage of the LSIP and located between the Mediterranean Sea and the currently

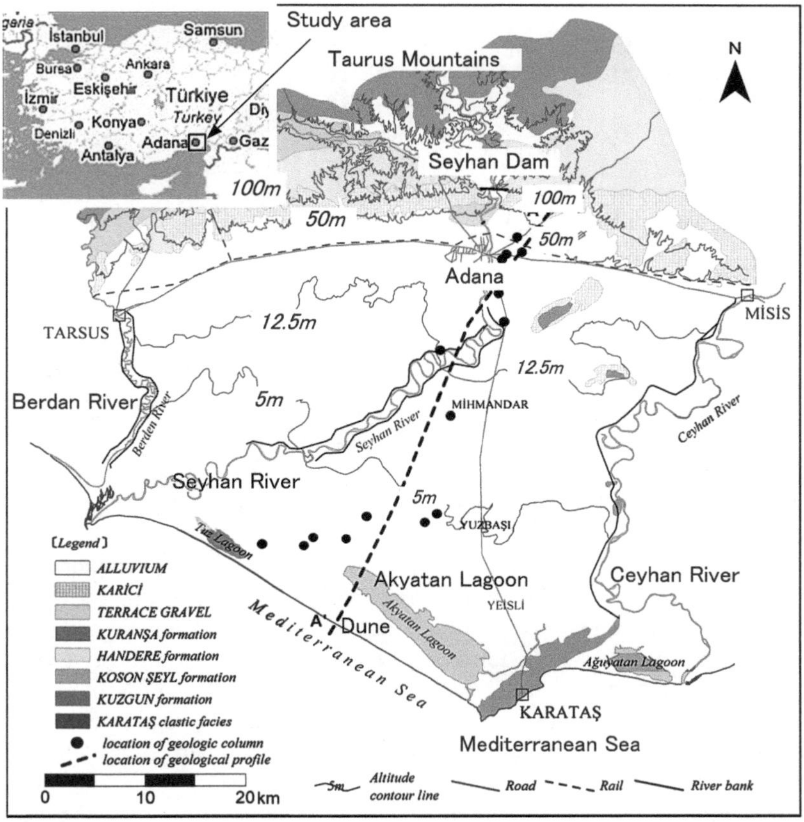

Fig. 1 Map of the Lower Seyhan River Basin showing modeled transect

irrigated northern area, has not been developed due to serious problems of soil salinity and water logging (Çetin and Diker 2003).

Annual precipitation in this area ranges from 600 to 800 mm, and most rainfall events are concentrated during the winter. Therefore, irrigation is practiced during the summer when its water requirement becomes the highest. This situation causes two peaks of almost the same level if fluctuation in the water table. However, the range of the water table fluctuation is relatively small.

3.2 *Hydrogeology and Mathematical Model*

The LSRB is bounded on the south by the Mediterranean Sea and on the north by the Adana Basin, which is one of the largest accumulations of Miocene sediments in southern Anatolia. It is also bounded on the north by the Taurus Mountains (Gürbüz 1999).

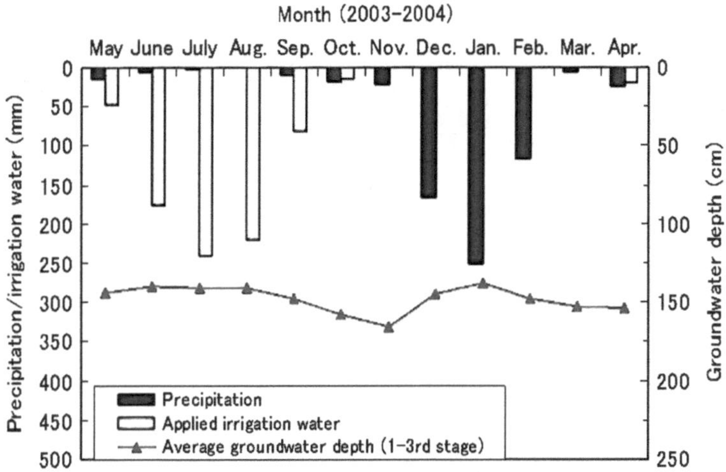

Fig. 2 Monthly precipitation, applied irrigation water and groundwater depth in the LSRB

The quaternary geology of the LSRB is characterized by thick sediments of silty clay on top of an underlying Kurança formation. The Kurança formation, composed of sand, gravel, and conglomerates of the Pleistocene, forms the main aquifer of the LSRB. During the Pleistcene, terrace gravel was deposited on the riverbed to a thickness of 50–100 m, and the Karici formation, composed of calcareous sediment, was formed by the evaporation of groundwater during the dry age. Both the terrace gravel and the Karici formation are restricted to the southern part of the Adana Basin.

A two-dimensional model was constructed to numerically simulate the response of the groundwater system to the aforementioned impacts. Since the north to south flow paths, which are almost perpendicular to the coastal line, are generally expected (Kurttaş and Karahanoğlu 1994), the transect along the A–A′ line in Fig. 1, which passes through Adana, the Seyhan River, the Akyatan Lagoon, and the sand dune, was chosen as a representative cross-section of the model.

A simplified geology of the LSRB along the A–A′ transect is shown in Fig. 3 from logs of wells, whose locations are depicted in Fig. 1. In the northern part of the LSRB near Adana, the Kurança formation is unconfined, and is mainly composed of conglomerates. In most of the other parts of the aquifer, it is composed of sand and gravel, and becomes confined with an overlying clayey deposit of alluvium. The Akyatan Lagoon is floored with silty clay or clayey silt, and the aquifer underneath the lagoon is overlain by a thick low-permeability cover layer. The base of the aquifer is composed of calcareous mudstone or shale of the Neogene inserted by a sandstone layer.

The analytical domain shown in Fig. 3 reaches a depth of 250 m from the mean sea level, and its width, including a distance of 5,000 m offshore into the Mediterranean Sea, is approximately 50 km. Since most of the aquifer is covered by the clayey layer, transboundary inflow from the Adana Basin to the aquifer plays an

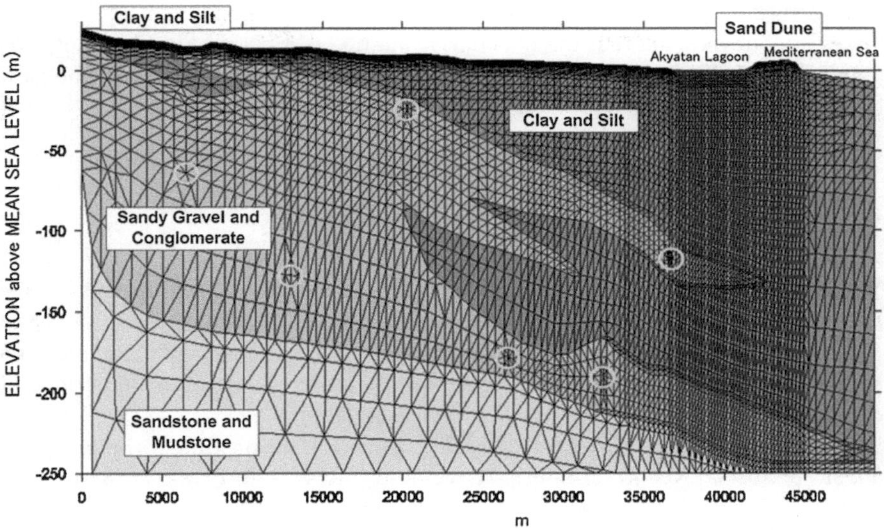

Fig. 3 Geologic profile along transect A–A′ and finite element mesh

important role in the groundwater flow system of the LSRB. The Akyatan is a major exit for discharging groundwater. The modeled domain was discretized by triangular elements. Fine elements were used near the land surface and beneath the lagoon to deal with substantial changes in water saturation in the unsaturated zone and in the concentration and groundwater flow vector in the transition zone. On the other hand, coarser elements were used in other parts of the domain to save the total number of nodes.

Time varying or constant head boundary, $h = H_f(t) - z$, was placed along the left-hand side of the aquifer. Prescribed head $h = (H_s - z)\rho_s/\rho_f$ was assigned along the inclined sea bed. Prescribed head $h = (H_l - z)\rho_l/\rho_f$ was assigned along the lagoon bed. H_f, H_s, and H_l in the above boundary conditions are the elevations of water table at the inland boundary, the sea surface, and the lagoon surface from a reference level. Net recharges were assigned to nodes on the land surface, including the sand dune, at a varying rate depending on projected rainfall and adapted irrigation practices, except the nodes where calculated outward fluxes of groundwater were positive and thus the boundary condition was replaced by $h_f = 0$. No flow was assigned to the rest of the boundaries.

For mass transport, constant concentration boundary $c = 0$ was assigned along the left face of the aquifer, where fresh groundwater flows into the domain. Along the inclined seabed, constant concentration $c = 1.0$ was assigned in cases where groundwater fluxes are inward. Otherwise, the boundary was assumed to be a no-dispersion boundary. Along the lagoon bed, time-variable prescribed concentration $c = (\rho_l - \rho_f)/(\rho_s - \rho_f)$ was assigned in cases where groundwater fluxes are inward. Otherwise the boundary was assumed to be a no-dispersion boundary. The land surface, including the sand dune but excluding the region where groundwater fluxes are outward and thus the boundary condition is replaced by a no-dispersion boundary, was assigned as a prescribed concentration boundary. Along no-flow boundaries, no-dispersion boundaries were assigned.

4 Scenarios

The year 2070 was assigned as the common target year to be projected for the ICCAP. Each scenario for this study was composed of five elements, as shown in Fig. 5; that is, sea-level rise, climate, re-charge from irrigation, abstraction rate, and inland boundary condition. Each element was further divided into three to four components.

According to the IPCC TAR (Cubasch et al. 2001), the maximum rise projected under the Special Report on Emission Scenario (SRES) – A1FI was 0.88 m by the year 2100. Projected sea-level rises were revised in the IPCC AR4 (Solomon et al. 2007), and the maximum rise by 2100 projected under the SRES A2 was 0.51 m. Thus, the sea-level rise element (S) was assigned to three components: 0.0 mm/year (denoted by Sc), 8.8 mm/year (denoted by S3), and 5.1 mm/year (denoted by S4).

According to the IPCC TAR (Cubasch et al. 2001), Coupled atmosphere–ocean General Circulation Models (CGCMs) project that the Mediterranean region including Turkey would be one of the most prominent areas where precipitation will decrease significantly. Since grid intervals of ordinary CGCMs are 100–300 km, its horizontal resolution is not sufficient to estimate regional climate. Therefore, Kimura et al. (2007) developed a downscaling technique to construct a Regional Climate Model (RCM) with a grid interval of 8.3 km by using a Pseudo Global Warming Method that can minimize model bias, and then projected the future precipitation and temperature in the Seyhan River Basin (SRB). Two independent CGCMs, developed by the Meteorological Research Institute (MRI) and the Center for Climate System Research at University of Tokyo and National Institute for Environmental Studies (CCSR/NIES), were used to estimate the change in the regional climate of the SRB in the 2070s under the A2 scenario of the SRES. Evaporation rates were evaluated with the aid of the Penman method, using the temperature projected by the MRI-RCM and the CCSR/NIES-RCM. The current evaporation and the precipitation rate are 1,471.1 and 660.9 mm/year, respectively, while the averaged values of future evaporation and precipitation rate projected for 2070 using the MRI-RCM and the CCSR/NIES-RCM are 1,549.3 and 375.2 mm/year, respectively. The difference between the evaporation and precipitation rate enormously widens from 810.5 to 1,174.1 mm/year during the next 70 years. Since the current maximum depth of the lagoon is less than 1 m, the condensation process of the lagoon is considered to be extremely significant.

Accordingly, three components were assigned to the evaporation of the climate element (E); that is, 1,471.4 mm/year (currently observed, denoted by Ec), 1,549.3 mm/year by 2070 (average of the MRI and CCSR/NIES projections, denoted by Er), and 1,510.4 mm/year by 2070 (central value of the currently observed and the average of the MRI and CCSR projections, denoted by Eh). Likewise, three components were assigned to the precipitation rate in the climate element (P); that is, 660.9 mm/year (currently observed, denoted by Pc), 375.2 mm/year by 2070 (average of the MRI and CCSR projections, denoted by Pr), and 518.1 mm/year by 2070 (central value of the currently observed and the average of the MRI and CCSR projections, denoted by Ph).

The recharge-from-irrigation element (I) has three components; that is, 8.11 mm/year (current rate estimated by the calibration process, denoted by Ic), 12.17 mm/year by 2070 (50% increase of the current rate, denoted by I+), and 4.06 mm/year by 2070 (50% decrease of the current rate, denoted by I−).

The surface hydrology group of the ICCAP assessed the impacts of climate change on water resources for the entire SRB, where the northern SRB is characterized by the continental climate and the southern SRB by the Mediterranean climate. The current average annual inflow to the Seyhan Dam of 5.5 Gm^3, used mainly for irrigation, was estimated to decrease drastically in the 2070s due to a decrease in precipitation by about 160 mm (Fujihara et al. 2007). Thus, it was concluded that global warming and the increased demand for water in the upper basin will lead to water scarcity.

By referring to these projected results and the current irrigation water demand of about 500 mm, four components were assigned to the abstraction-rate element; that is, as usual (current rate estimated by the calibration process, denoted by c), upper LSIP 200 and lower LSIP 100 (denoted by U20 and L10), upper LSIP 150 and lower LSIP 150 (denoted by U15 and L15), and upper LSIP 100 and lower LSIP 200 (denoted by U10 and L20). The numbers 200, 150, and 100 denote the adapted future water requirements for irrigation in the LSIP in millimeter per year, and these data were further converted to abstraction rates of the six representative sinks shown in Fig. 3.

Another potential concern is that climate change may also influence the boundary condition along the northern face of the modeled domain due to the increase in evaporation and the decrease in precipitation in the upper SRB. The present water table height observed in Adana is approximately 19 m above the mean sea level. This level coincides well with the hydraulic head along the northern boundary evaluated from the contour map of the water table (Kurttaş and Karahanoğlu 1994). Therefore, three components were assigned to the inland B.C. element (B) as a time-varying prescribed head boundary; that is, as usual (currently observed at Adana, denoted by Bc), 3 m dropdown (denoted by B3), and 6 m dropdown (denoted by B6) by 2070.

Hereafter, each scenario is given a code using the abbreviations shown in Fig. 4, such as ScEhPhI-U20L10B6, etc. The components of each element that vary temporally in accordance with sea-level rise, climate change, irrigation practices, enhanced abstraction, and changing boundary conditions were all linearly interpolated from the current values to the values of the target year during simulation runs.

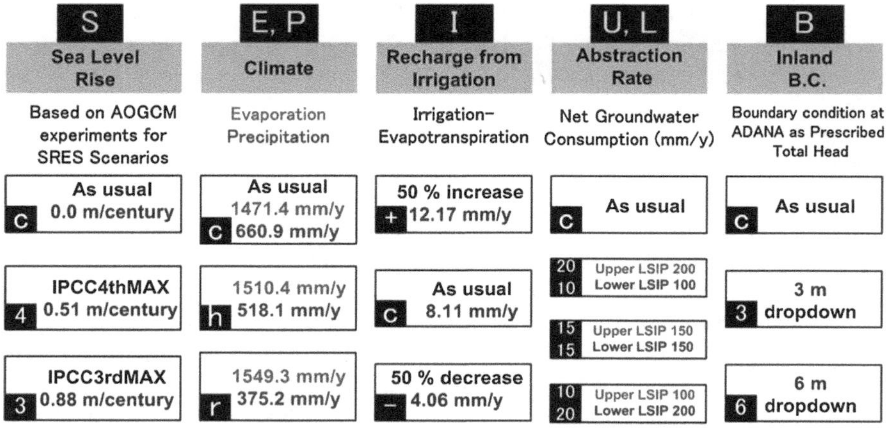

Fig. 4 Scenarios for projecting groundwater environment in 2070 under climate change

5 Calibration of Mathematical Model

Soil samples were collected in the field, and soil–water characteristics were investigated in the laboratory. Hydrogeologic variables to be fixed are hydraulic conductivity, specific storage, effective porosity, and longitudinal and transverse dispersivity. Calibration was performed to fix these parameters for the mathematical model in a manner so that the head values produced by the model match observed values. Field studies by Scheumann (1997), which state the following on groundwater level in the LSRB, were helpful in identifying these parameters:

"In the 1960s, prior to the installation of the irrigation project, rain-fed agriculture was practiced in 95% of the projected land...The groundwater level is high, reaching a maximum level of 0.5–1.0 m below land surface and a minimum of 2–4 m. Large areas become water logged during the winter."

Çetin and Diker (2003) investigated the direction and salinity of groundwater in the LSRB using data collected from 85 observation wells by the branch office of the sixth Regional Directorate of the Sate Hydraulic Works (DSI), which is the authority responsible for the development of land and water resources in Turkey, and found that the groundwater gradient is more or less parallel to the land surface, and that areas with EC of shallow groundwater between 0.0–0.1, 0.1–0.2, and 0.2–0.5 S/m covered 50.0, 45.4, and 4.0%, respectively. In 1999 and 2000, Çetin and Diker (2003) also collected 81 soil samples from depths of 0 to 30 cm in the forth LSIP, and found that soil salinity ranged from 0.48 to 0.30 S/m, with a decreasing trend.

According to the field measurement of surface and subsurface water salinity in the LSRB performed for this study, it turned out that the electrical conductivity of the Akyatan Lagoon was 6.8 S/m, and much higher than the 5.8 S/m of the Mediterranean Sea. The salinities of shallow groundwater were more than 0.02 S/m and less than 0.12 S/m.

The hydrogeologic properties of the geologic formations identified are shown in Table 1. The annual recharge rate from precipitation during the pre-irrigation stage was also identified to be 8.11 mm/year. The annual recharge rate originated from irrigation was assumed to be the same as the natural recharge rate from precipitation, since the contribution of the irrigation to the rise of water table is approximately the same as that of the precipitation.

There are more than 1,000 wells in the LSRB for agricultural, municipal, and industrial uses. These wells are distributed relatively evenly in the LSRB. Therefore, six representative sinks were installed in the conceptual model of the cross section as shown in Fig. 3. Groundwater abstraction rates of the representative sinks were identified, so that the calculated water table coincides with those by observation during the calibration process for the current stage. Table 2 shows the identified abstraction rates of the sinks. The values for U20L10, U15L15, and U10L20 were assigned in accordance with the abstraction rate element of the aforementioned scenarios.

Table 3 shows the vales used as boundary conditions for calibration runs of the pre-irrigation stage and the current stage and for projections under various scenarios.

Table 1 Hydrogeologic parameters

	Sandy gravel and conglomerate	Sand	Clay	Base
Hydraulic conductivity (m/day)	8.6	3.45	0.1	0.04
Specific storage (1/m)	1.3×10^{-5}	4.6×10^{-5}	8.5×10^{-5}	1.0×10^{-6}
Effective porosity	0.22	0.22	0.22	0.22
Longitudinal dispersivity (m)	30	10	7.5	3
Transverse dispersivity (m)	3	1	0.75	0.3

Table 2 Abstraction rate of representative sink

	Abstraction (m^3/day)					
	Sink 1	Sink 2	Sink 3	Sink 4	Sink 5	Sink 6
Pre-irrigation	0.00	0.00	0.00	0.00	0.00	0.00
Current	0.03	0.22	0.92	0.44	0.09	0.03
U20L10	5.31	3.82	1.85	1.67	0.81	0.00
U15L15	3.98	2.87	2.78	2.51	1.22	0.00
U10L20	2.65	1.91	3.70	3.35	1.62	0.00

Table 3 Boundary conditions

	Mountain side (groundwater)	Lagoon (lagoon water)	Sea side (seawater)
Water level Above Sea level (m)	19.0 (Current) 13.0 (in 2070)	0.0 (Current) 0.51 (in 2070) 0.88 (in 2070)	0.0 (Current) 0.51 (in 2070) 0.88 (in 2070)
Density at 20°C (g/cm^3)	1.002	1.035 Variable	1.030

6 Projected Results

6.1 Changes in Lagoon Water Salinity

Salinity of the lagoon water is affected by evaporation, precipitation, run-off/overland flow, and its exchange with seawater and groundwater. According to the field measurement, the electrical conductivity of the lagoon water was higher than that of seawater by 1.0 S/m. This observation leads us to a noteworthy fact that evaporation from the lagoon has been playing an important role in the salinity of the lagoon water. The increase in the salinity of the lagoon water is especially attributed to the inflow of saline groundwater and seawater, as well as to evaporation.

Figure 5 shows projected temporal changes in the salinity of the lagoon water. The numbers attached to ScEcPcIcUcLcBc are $ROF = (\Delta R_o O_v + \Delta R_f L_g)/\Delta R_f L_g$, where $\Delta R_o O_v$ is the volume of water contributed from the canal runoff and the overland flow, and $\Delta R_f L_g$is the rainfall on the lagoon surface. It should be noted that the small change in *ROF* significantly affects salinity as shown by the results for ScErPrIcUcLcBc-2.000 and ScErPrIcUcLcBc-2.140. All the projections under the aforementioned scenarios were performed using the *ROF* of 2.140.

As shown in Fig. 5, reduced evaporation and increased precipitation of the scenario S4EhPhIcUcLcBc compared to S4ErPrIcUcLcBc can slow the process of the salinity increase in the lagoon water. The salinity increase is also delayed by a greater sea-level rise of S3ErPrIcUcLcBc compared to S4ErPrIcUcLcBc, but the salinities for both scenarios eventually reach a similar level.

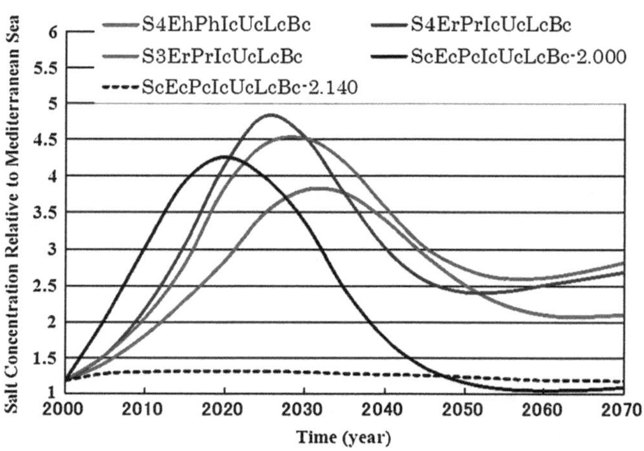

Fig. 5 Projected temporal changes in the salinity of the Akyatan lagoon

6.2 Changes in the Groundwater Table

Figure 6 illustrates the positions of the water table projected under various scenarios, while the top drawing shows the calibrated water table for the pre-irrigation stage, where dots indicate the current levels of the water table at observation wells. As Scheumann (1997) has described, land surface was calculated as the seepage face over the largest area. Due to abstraction from wells, the water table of the current stage declines compared to the pre-irrigation stage. It should be noted that the second simulated water table from the top for the current stage coincides well with the observed water table.

The projected result for S3ErPrIcUcLcBc indicates that a sea-level rise of 0.88 m/century, an increase in evaporation, and a decrease in precipitation cause a slight fall of the water table from the current level. However, the water table beneath the sand dune went up along the lagoon side due to the infiltration of saline lagoon water and the associated shrinkage of fresh water lens. The third and fourth projected results from the top in Fig. 6 are for S3ErPrIcUcLcBc and S4ErPrIcUcLcBc. The S4ErPrIcUcLcBc scenario is different from S3ErPrIcUcLcBc, only in that the sea-level rise for S4 is smaller than that of S3 by 0.37 m/century. Both results indicate that the degree of sea-level rise will not have much impact on the water table.

The result for S4ErPrIcUcLcB6 indicates that the decline in the boundary head of as much as 6 m will cause a remarkable drawdown of the water table in most of the LSRB. According to the result for S4EhPhIcUcLcBc, little increase in evaporation and little decrease in precipitation will not have much impact on the water table. However, the result for S4ErPrI-UcLcBc shows that a 50% reduction of recharge from irrigation may cause a slight drawdown of the water table.

The results for S4ErPrIcU20L10Bc, S4ErPrIcU15L15Bc, and S4ErPrIcU10L20Bc indicate that enhanced groundwater abstraction causes a drastic drawdown of the water table. Furthermore, the shape of the water table and the extent of its drawdown depend largely on the amount of abstraction and its distribution. The increased groundwater consumption of 100 or 200 mm/year in the LSRB means that more than half the irrigation water is supplied from groundwater. If this is the case, water tables are expected to decline to the level of 25 m below sea level by 2070, as seen in the bottom three drawings of Fig. 6.

6.3 Changes in Groundwater Salinity

Figure 7 represents the calculated distribution of groundwater salinity for the current stage. Saltwater intrudes into the aquifer from the Mediterranean seabed, followed by an upward movement of saline groundwater toward the lagoon. Saltwater in the lagoon, having a density higher than the surrounding groundwater, slightly penetrates into the groundwater system from the central part of the lagoon

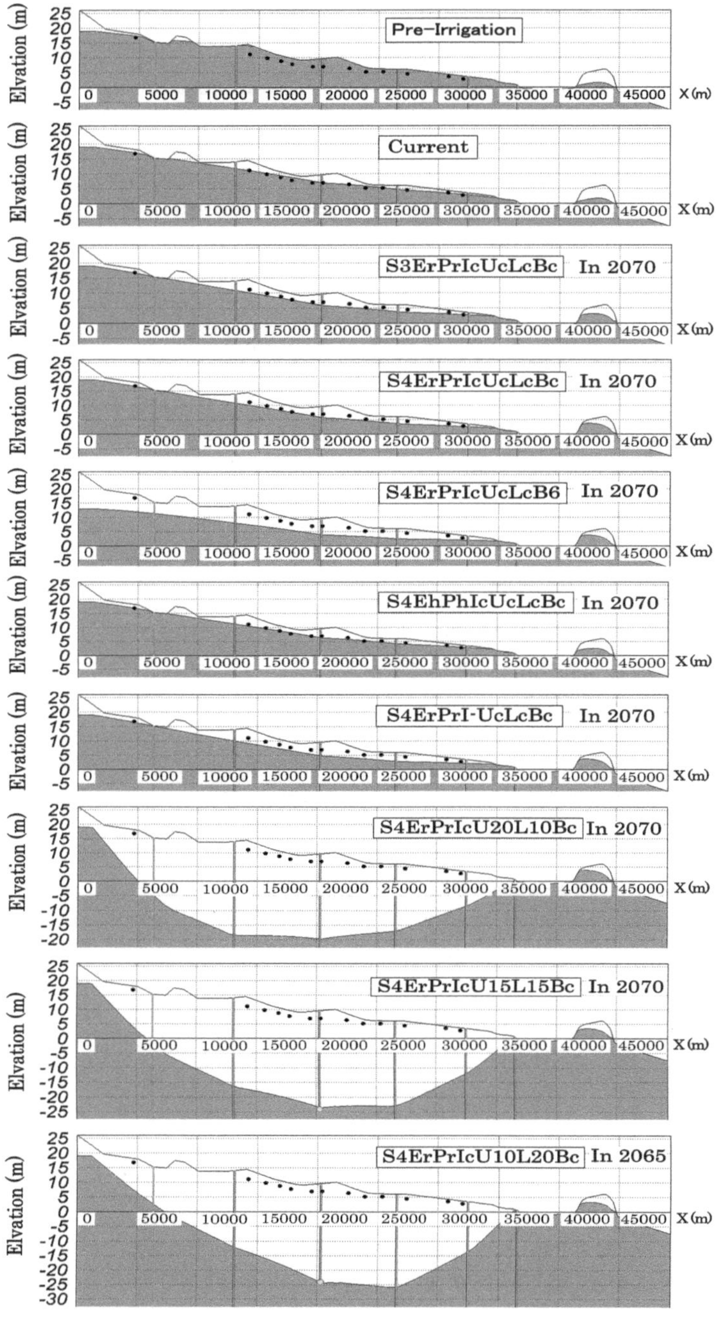

Fig. 6 Locations of the water table during the pre-irrigation and current stages and for various scenarios

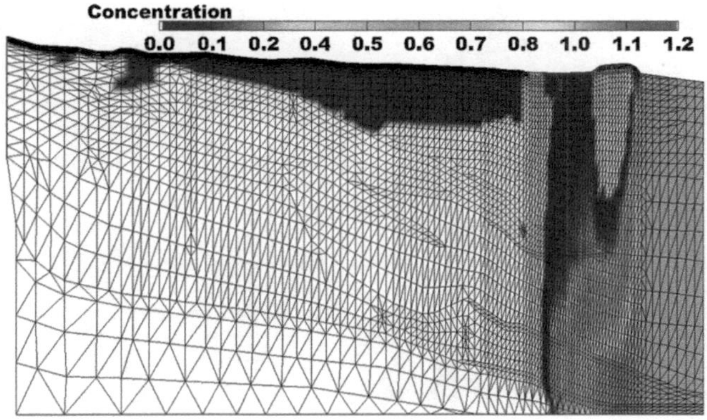

Fig. 7 Distribution of groundwater salinity, relative to the Mediterranean Sea, for the current stage

bed. It should also be noted that a freshwater lens evolves beneath the sand dune, while the raised salinity of shallow groundwater in agricultural areas is attributed to irrigation and groundwater abstraction.

Changes in groundwater salinity are illustrated in Figs. 8 and 9 in terms of the difference in concentration relative to the seawater between the distributions projected under representative scenarios and that of the current stage. The figures show finger-like complicated salinized zones beneath the lagoon.

The impacts of sea-level rise, the dropdown of boundary head, reduced recharge, and reduced precipitation/increased evaporation on groundwater salinity are shown in Fig. 8. A greater sea-level rise, as represented by S3ErPrIcUcLcBc, can cause slightly greater saltwater intrusion as compared to the result of S4ErPrIcUcLcBc in Fig. 9, whereas lesser density of the lagoon water for S3 than for S4 induces less infiltration of saline lagoon water. A lower inland boundary head and reduced recharge from irrigation are expected to cause an expansion of saline zones. However, as shown in the drawings for S4ErPrI-cUcLcB6 and S4ErPrI-UcLcBc in Fig. 8, these influences on the formation of saline zones were not recognizable. On the other hand, reduced evaporation and increased precipitation in S4EhPhIcUcLcBc obviously reduces saline zones due to lower salinity in the lagoon water as compared to S4ErPrIcUcLcBc in Fig. 9.

Impacts of abstraction on groundwater salinity are shown in Fig. 9. Although the distribution and rate of abstraction for S4ErPrIcUcLcBc are the same as the current condition, the salinity zone penetrates deep into the groundwater system beneath the lagoon. This is attributed to the increase in salt concentration of the lagoon water due to increased evaporation and decreased precipitation. On the other hand, the increase in abstraction rate accelerates the infiltration of saline water in the lagoon in accordance with the distribution of abstraction rates from the representative sinks.

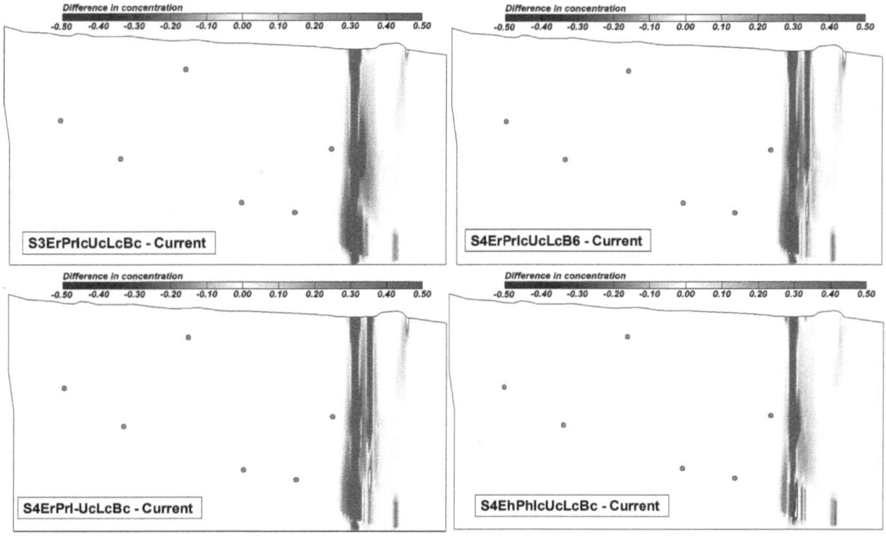

Fig. 8 Difference in groundwater salinity between the current stage and the projection for different sea-level rise, boundary head, recharge, and evaporation/precipitation

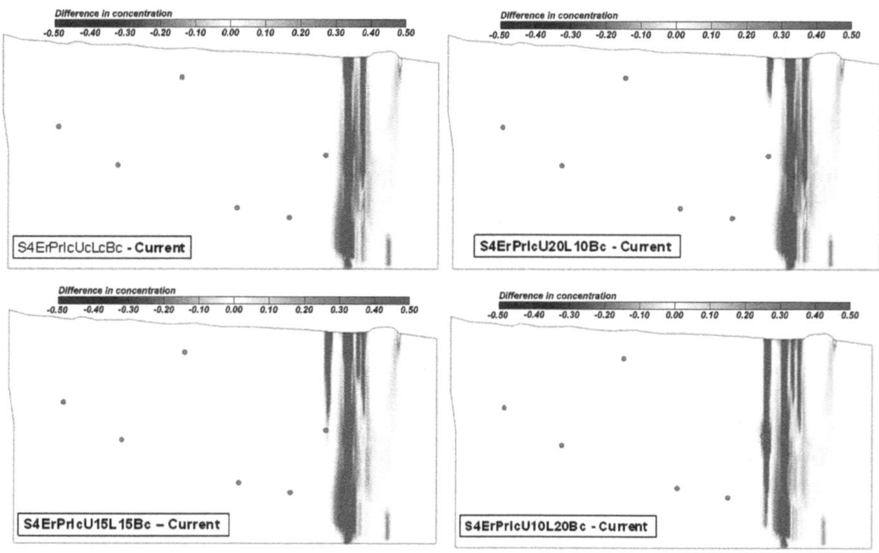

Fig. 9 Difference in groundwater salinity between the current stage and the projection with different abstraction rates

6.4 Changes in Groundwater Velocity Vectors

Figure 10 shows projected vectors of groundwater flow velocity in the entire analytical domain for two scenarios of S4ErPrIcBc. The largest velocity is calculated at the lower part of the aquifer beneath the lagoon in both scenarios. The increased abstraction from the sinks plays an important role in the direction and magnitude of the velocity vectors as seen by S4ErPrIcU10L20Bc. It is also observed that fresh groundwater, supplied along the mountainous boundary, flows mainly in the permeable layer of sandy gravel and conglomerate, toward the Mediterranean Sea.

Figure 11 shows the projected vectors of groundwater flow near the lagoon for the pre-irrigation and the current stages together with two scenarios. During the pre-irrigation stage, groundwater mostly flows upward to the lagoon. Due to the decline of the water table attributed to abstraction for the current stage, the flow direction turns downward beneath the central part of the lagoon. The projection under the scenario S4EhPhIcUcLcBc, having elements of sea level rise and increased-evaporation/decreased-precipitation, resulted in enhanced downward flows. The projected result for S4ErPrIcU15L15B6 exhibits a more complicated flow pattern due to a significant decline of the water table accompanied by enhanced abstraction together with the decline of inland boundary head. The discharge of fresh water into the lagoon from the fresh-water lens beneath the sand dune is also enhanced in both projected scenarios.

6.5 Salt Accumulation on the Land Surface

In the low-lying area near the lagoon where groundwater seeps out, the groundwater transports salt from the subsurface to the surface. The amount of salt can be evaluated by using the calculated discharge and salinity of the groundwater. It is assumed here that the transported salt is not carried away by wind, overland flow, or drainage into open channels, but accumulates there due to evaporation.

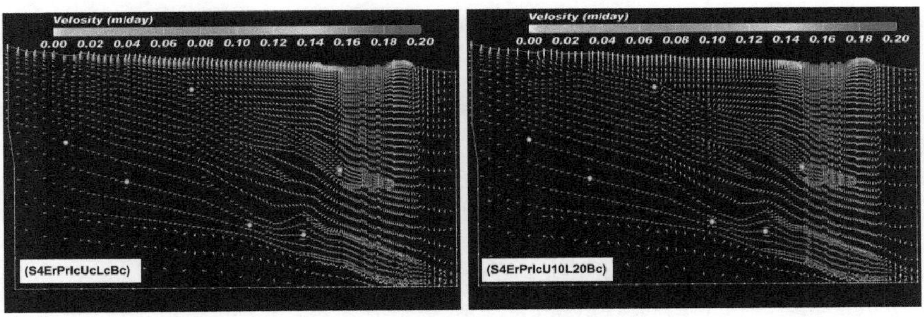

Fig. 10 Projected distribution of global groundwater flow velocity for S4ErPrIcUcLcBc and S4ErPrIcU10L20Bc

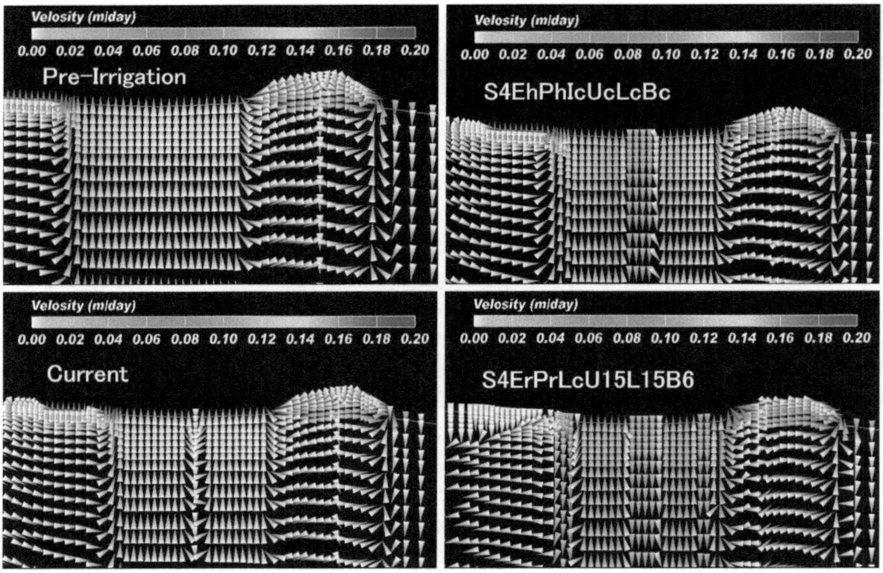

Fig. 11 Distribution of local groundwater flow velocity near the Akyatan Lagoon for S4ErPrI-cUcLcBc and S4ErPrIcU10L20Bc

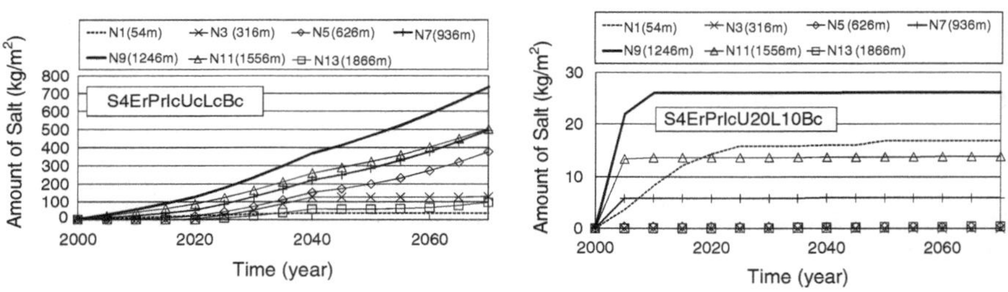

Fig. 12 Projected salt accumulation on the land surface near the Akyatan Lagoon for S4ErPrI-cUcLcBc and S4ErPrIcU10L20Bc

The projected results for scenarios S4ErPrIcUcLcBc and S4ErPrIcU20L10Bc are shown in Fig. 12. The numbers in the bracket are the distance of nodes in meters from the inland shoreline of the lagoon. The maximum amount of accumulated salt for S4ErPrIcUcLcBc in 2070 reaches more than 700 kg/m^2 at node N9 that is located 1,246 m from the shore. Beyond the distance of 1,866 m from the shore, salt does not accumulate, since groundwater does not seep out in that area. On the other hand, the projected result for S4ErPrIcU20L10Bc shows that the amount of accumulated salt is one order less than that for S4ErPrIcUcLcBc, and salt accumulation stops after a certain period of time due to a drawdown of the water table caused by enhanced abstraction.

7 Discussion and Conclusions

This paper analyzes future impacts on groundwater flow, salinity of groundwater and lagoon water, and salt accumulation on the land surface in response to global climate change, sea-level rise, adaptation of irrigation and abstraction practices, and changing boundary head in the Mediterranean coastal area. The long-term projection of the future subsurface water environment was based on rational scenarios obtained from various studies of different disciplines such as meteorology, hydrology, irrigation and drainage engineering, agricultural economy, botany, and groundwater science. Since the exchange of groundwater and surface water in lagoon areas seems to be one of the key processes influencing saltwater intrusion into the groundwater system, the dynamic response of a lagoon to the subsurface environment was incorporated into a conceptual model for the Lower Seyhan River Basin (LSRB) in Turkey.

Part of the past and the current landscapes were computationally reproduced through a calibration process of the mathematical model to represent the pre-irrigation stage and the current stage. Followed by a two-step calibration process, the impacts of climate change on the subsurface water environment in the LSRB were projected under various scenarios consisting of sea-level rise, precipitation and evaporation, recharge from irrigation, groundwater abstraction, and boundary head. The projected results are summarized as follows.

The impact of enhanced abstraction on the water table was significant. If groundwater consumption increases to 100 or 200 mm/year in the LSRB, water tables are expected to drop to levels that are 25 m below sea level by 2070. However, as the LSRB is characterized by a thick clay layer, sea-level rise is not expected to cause substantial saltwater intrusion directly from the bed of the Mediterranean Sea during the next 70 years.

On the contrary, the combination of sea-level rise with increasing evaporation and decreasing precipitation can cause substantial impact on salinity in the lagoon water, and the increased salinity of the lagoon may in turn cause saltwater to penetrate into the deeper layers beneath the lagoon and eventually cause a drastic increase in the salinity of groundwater. The expansion of the saline zone beneath the lagoon may also impede the fresh groundwater flow that would otherwise discharge into the lagoon and cause water logging on the land surface. Moreover, the saline water beneath the lagoon may flow back to the land surface, and water logging and the return flow could cause substantial accumulation of salt on the land surface. Increasing evaporation, decreasing precipitation, and sea-revel rise all contribute to salt accumulation on land.

Based on this study, which has provided information on the dynamic behavior of the lagoon and on the characteristic tendency of salt accumulation on the land surface in response to climate change, sea-level rise, and adapted human interference, Fig. 13 shows specialized features of the future anthroscape of the Mediterranean coastal area. Sea-level rise leads to a rise in the water table and the intrusion of seawater into coastal aquifers. Meteorological future projections of the area by Kimura et al. (2007) have indicated a slight increase in evapotranspiration and a

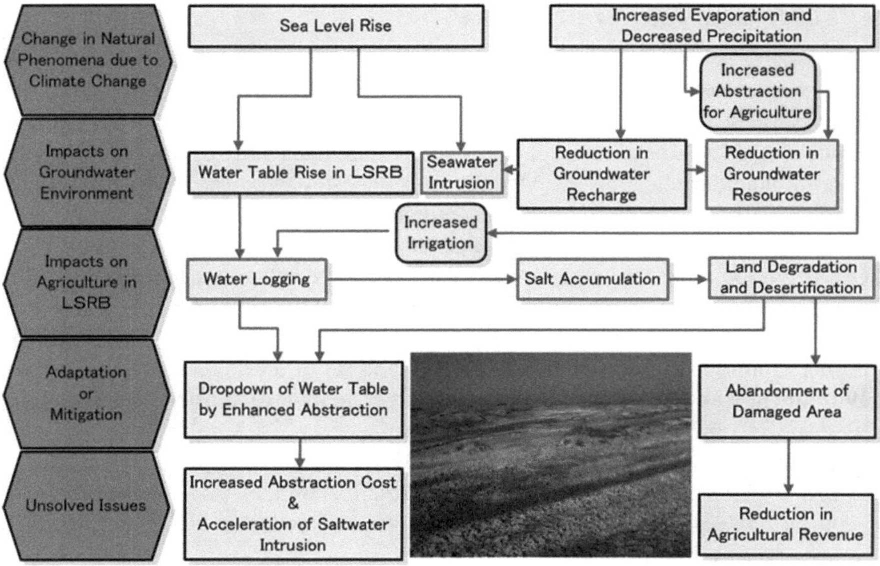

Fig. 13 Projected anthroscape of the Mediterranean coastal area

drastic decrease in precipitation. These hydrologic changes reduce groundwater recharge and induce human activities such as increased abstraction of groundwater for various sectors including agriculture. These changes, in turn, reduce groundwater resources and cause seawater intrusion. In the Mediterranean coastal area, the combined effects of sea-level rise and increased irrigation lead to a rise in the water table, which is an associated cause of water logging, and an accumulation of salt on the land surface. Adaptation to the impacts might be possible by preventing the water table from rising as a result of groundwater abstraction, but doing so may accelerate seawater intrusion into the aquifers and increase the cost of water management. The worst-case scenario for the anthroscape degradation would be the abandonment of damaged areas.

Acknowledgment This research was conducted as part of the ICCAP research project, administered and financially supported by the Research Institute for Humanity and Nature, Japan and the Scientific and Technological Research Council of Turkey. It has also been financially supported in part by the JSPS Grant-in-Aid No.16380164.

References

Çetin M, K Diker (2003) Assessing drainage problem areas by GIS: A case study in the eastern Mediterranean region of Turkey. *Irrigation and Drain.* 52, 343–353.

Cubasch U, X Dai, Y Ding, DJ Griggs, B Hewitson, JT Houghton, I Isaksen, T Karl, M McFarland, VP Meleshkov, JFB Mitchell, M Noguer, BS Nyenzi, M Oppenheimer, JE Penner, S Pollonais,

T Stocker, KE Trenberth (2001) Technical summary. In: Houghton JT, Y Ding, DJ Griggs, M Noguer, PJ van der Linden, X Dai, K Maskell, CA Johnson (eds.) *Climate Change 2001: The Scientific Basis. Contribution of Working Group I to the Third Assessment Report of the Intergovernmental Panel on Climatic Change*. Cambridge University Press, Cambridge.

Feseker T (2007) Numerical studies on saltwater intrusion in a coastal aquifer in northwestern Germany. *Hydrogeology Journal*. 15, 267–179.

Fujihara Y, K Tanaka, T Watanabe, T Kojiri (2007) Assessing the impact of climate change on the water resources of the Seyhan River Basin, Turkey. In: Watanabe T, R Kanber (eds) *The Final Report of ICCAP. ICCAP Publication* 10. Research Institute for Humanity and Nature, Japan, 89–94.

Fujinawa K, K Masuoka, T Nagano, T Watanabe (2005) Numerical simulation modeling for saltwater intrusion in predicting impacts of sea-level rise on areas below sea-level. *Journal of Environmental Systems and Engineering, Japan Society of Civil Engineers*. 790, 35–48 (in Japanese with English abstract).

Fujinawa K, T Iba, Y Fujihira, T Watanabe (2009) Modeling interaction of fluid and salt in an aquifer/lagoon system. *Ground Water*. 47, 35–48. doi:10.1111/j.1745-6584.2008.00482.x.

Gürbüz K (1999) Regional implications of structural and eustatic controls in the evolution of submarine fans: An example from the Miocene Adana Basin, southern Turkey. *Geological Magazine*. 136, no. 3, 311–319.

Kapur S, P Zdruli, E Akça, AH Arnoldussen, B Kapur (2004) Anthroscapes of Turkey: Sites of Historic Sustainable Land Management (SLM). In: van Asselen S, C Boix-Fayos, A Imeson (eds) *'Briefing papers' of the second SCAPE workshop in Cinque Terre (Italy)*, 13–15 April 2004.

Kimura F, A Kitoh, A Sumi, J Asanuma, A Yatagai (2007) Downscaling of the global warming projections to Turkey. In: Watanabe T, R Kanber (eds) *The Final Report of ICCAP. ICCAP Publication 10*. Research Institute for Humanity and Nature, Japan, 21–32.

Konikow LF, JD Bredehoeft (1978) Computer model of two-dimensional solute transport and dispersion in ground water. In: *Techniques of Water-Resources Investigations. Book 7, Chap. C2*. US Geological Survey, Reston, VA.

Kurttaş T, N Karahanoğlu (1994) Modeling of Adana Plain (Turkey) by means of finite difference method. *Proceedings of 15th Anniversary Symposium of the Cukurova University, Adana, Turkey*, 103–122.

Langevin CD, AM Dausman (2005) Numerical simulation of saltwater intrusion in response to sea-level rise. *Proceedings of EWRI '05 World Water & Environmental Resources Congress, Anchorage, Alaska, USA, ASCE*, 1–8.

Masuoka K, K Fujinawa, KM Furukawa, T Nagano, T Watanabe (2005) Experimental studies for identifying the impacts of sea-level rise caused by global warming on groundwater environment in areas below sea-level. *Journal of Groundwater Hydrology, Japanese Association of Groundwater Hydrology*. 47, no.1: 19–28 (in Japanese with English abstract).

Nazik A, G Evans, K Gürbüz (1999) Sedimentology and paleontology with special reference to the ostracoda fauna of Akyatan Lagoon. *Geosound*. 35, 127–147.

Neuman SP (1984) Adaptive Eulerian -Lagrangian finite element method for advection-dispersion. *International Journal for Numerical Methods in Engineering*. 20, 321–337.

Oude Essink GHP, F Schaars (2002) Impact of climate change on the groundwater system of the water board of Rijnland, The Netherlands. *Proceedings of the 17th Salt Water Intrusion Meeting, Delft, The Netherlands*.

Parry ML, OF Canziani, JP Palutikof, et al. (2007) Technical summary. In: Parry ML, OF Canziani, JP Palutikof, PJ van der Linden, CE Hanson (eds) *Climate Change 2007: Impacts, Adaptation and vulnerability. Contribution of Working Group II to the Fourth Assessment Report of the Intergovernmental Panel on Climate Change*. Cambridge University Press, Cambridge, UK, 23–78.

Scheumann W (1997) *Managing Salinization-Institutional Analysis of Public Irrigation Systems*. Springer, Berlin.

Solomon S, D Qin, M Manning, AB Alley, T Berntsen, NL Bindoff, Z Chen, A Chidthaisong, JM Gregory, GC Hegerl, M Heimann, B Hewitson, JB Hoskins, F Joos, J Jouzel, V Kattsov, U Lohmann, T Matsuno, M Molina, N Nicholls, J Overpeck, G Raga, V Ramaswamy, J Ren, M Rusticucci, R Somerville, TF Stocker, P Whetton, RA Wood, D Wratt (2007) Technical summary. In: Solomon S, D Qin, M Manning, Z Chen, M Marquis, KB Averyt, M Tignor, HL Miller (eds) *Climate Change 2007: The Physical Science Base. Contribution of Working Group I to the Fourth Assessment Report of the Intergovernmental Panel on Climate Change*. Cambridge University Press, Cambridge, UK.

Watanabe T, R Kanber (2007) *The Final Report of ICCAP ICCAP Publication 10*. Research Institute for Humanity and Nature, Japan.

Re-evaluating Indigenous Technologies for Sustainable Soil and Water Management in the Sahel: A Case Study from Niger

Takanori Nagano, Haruhiko Horino, and Takashi Kume

1 Introduction

West African Sahel relies much of its food production on rain-fed pearl millet cultivation (*Pennisetum glaucum (L.) R. Br.*). Its productivity is physically constrained by irregular rainfall and the inherent poor nutrient holding capacity of the soil. Extensive and deprival nature of its agriculture was sustained by migrant and semi-nomadic lifestyle of people in the older days (Charlick 1991). However, a great increase of population in the last five decades caused overexploitation of land resources and reduced resilience of the system. Due to erosion and nutrition deficiency, productivity of pearl millet has declined. Today the old paradigm of extensive agriculture is no longer valid for sustainable land management.

Researches to combat desertification in the region became active in the 1970s. Out of many conservation efforts, only a handful number of techniques were actually adopted by the local people. Physical constraints were a lack of capital and a lack of labor. Additionally it was culturally for local people to change the idea of land management from extensive to intensive style.

The author carried out filed trials to re-evaluate indigenous techniques for soil and water conservation in a search for less costly and labor-intensive conservation. Through this process it became clear that nature has a good self-restoration mechanism in this region and enhancing this mechanism by wise use of fauna and flora can ease the burden of conservation.

T. Nagano (✉)
Graduate School of Agricultural Science, Kobe University, 1-1 Rokkodai-cho, Nada-ku, Kobe 657-8501, Japan
e-mail: naganot@ruby.kobe-u.ac.jp

H. Horino
Graduate School of Life and Environmental Sciences, Osakafu University, 1-1 Gakuen-cho, Sakai-shi, Osaka 599-8231, Japan

T. Kume
Research Institute for Humanity and Nature, 457-4 Motoyama-cho, Kamigamo, Kita-ku, Kyoto 603-8047, Japan

S. Kapur et al. (eds.), *Sustainable Land Management*,
DOI 10.1007/978-3-642-14782-1_15, © Springer-Verlag Berlin Heidelberg 2011

In this paper, primarily, the general features of land and millet production in the southwestern Niger and a brief history of desertification and conservation efforts are explained. Working mechanisms of indigenous conservation works evaluated through the field trial are presented second. Thirdly more energy-efficient and sustainable land management in this region by wise use of fauna and flora is discussed.

2 Landscape and Land Management in Southwestern Niger

2.1 *Landscape of Southwestern Niger*

Figure 1 illustrates a typical landscape unit in the southwestern part of Niger. Similar landscapes can be found in northern Burkina Faso and in eastern Mali. The landscape consists of a gently sloping valley and a continuous flat plateau called Continental Terminal. The Continental Terminal is a Miocene deposit which dates back to 25 million years (Wilding and Hossner 1989). The broad and gentle valley starts at the edge of the plateau and continues a few kilometers down to the bottom where a seasonal river or a dallol exists. The slope is generally less than 3% in inclination and the lower part of the slope is covered by a thick layer of an aeolian sand deposit. Villages and fields are located on this sand deposit. With the population growth, villages and fields expanded to the upper slope.

The upper part of the slope and the plateau were covered by savanna vegetation and used as common land for grazing, hunting and fuel wood collection. Nowadays the plateau is very sparse in vegetation, covered by iron stones or a plinthite layer. Plinthite is the horizon of leached sesquioxides 20–100 cm in thickness which formed during the pluvial of the early Quartenary (2 million years BP) (Wilding and Hossner 1989). The plinthite layer is often continuous from the plateau down the slope as shown in Fig. 2. This layer originally developed a few meters below the

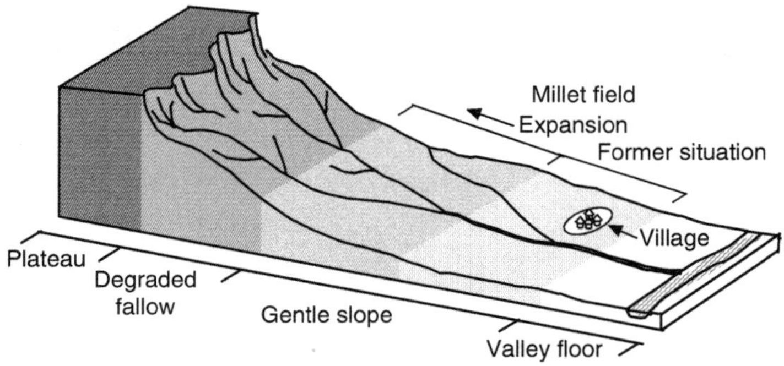

Fig. 1 The landscape unit of the study area in southwestern Niger. Villages are expanding to the upper part of the slope with population growth

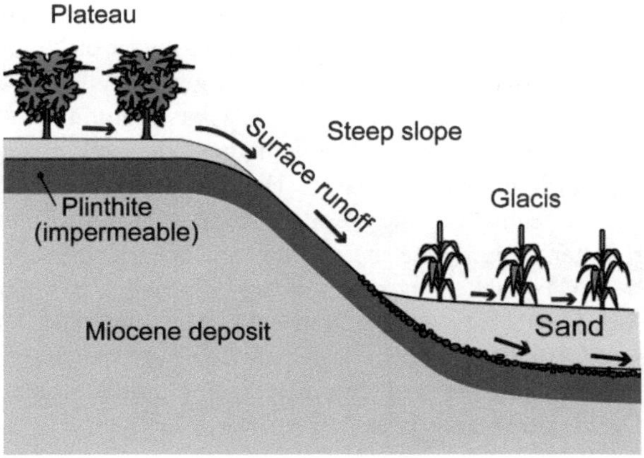

Fig. 2 Plinthite layer in a valley system in Southwestern Niger

Table 1 Physical properties of the sandy soils located near the bottom of the slope and more clayey soils in the upper part of the slope in the south west of Niamey, Niger

Depth (cm)	Texture (%)				pH	Organic matter (%)	CEC (Meq/100 g)	Base saturation (%)
	Coar. sand	Fine sand	Silt	Clay				
Soil in the bottom part of slope (slope <2%)								
0–6	26.1	62.6	8.2	3.1	6.0	0.27	1.59	38.4
6–40	37.1	48.6	9.2	5.1	5.8	0.26	1.33	34.6
40–72	36.2	48.3	8.5	7.0	5.6	0.20	1.41	47.5
72–119	33.3	51.9	7.7	7.1	5.8	0.12	1.18	34.7
Soil in the upper part of slope (slope >3–4%)								
0–10	38.8	55.4	3.6	2.2	5.9	0.33	0.85	88.2
10–52	30.6	56.4	7.2	5.8	5.6	0.27	1.26	52.4
52–117	27.1	50.5	7.3	15.1	5.4	0.16	2.28	65.8
117–152	22.6	59.2	7.0	11.2	6.0	0.09	1.74	82.8

soil surface but in the case of severe erosion especially in the upper part of the slope, it is exposed on the surface and serves as a source area of surface runoff during precipitation.

2.2 *Soil Properties*

Large areas of the Sahelian region comprise red, sandy, strongly leached acid soil resources utilized for pearl millet production. They reflect both a semi-arid tropical environment and a more pluvial paleoclimate. They are generally weakly buffered kaolinitic systems (Ahn 1970), low in organic content (Wilding and Hossner 1989). Arenosols are the widespread soils is in the area. Table 1 shows soil properties of a sandy soil found near the bottom of the slope and more clayey soil found in the upper part of the slope in the south west of Niamey, the capital of Niger. Sandy soils have clay and silt contents generally lower than 10%. The near-surface layer is leached and acidic with low base saturation. Due to the low contents of the clay fraction and organic matter, the soil is poor in nutrient holding capacity and is highly permeable. The formation of the Argillic horizon is one of the specific features of these soils. Clay particles disoriented with depth via infiltrating water accumulate below the surface of the soils. The lower horizons are relatively richer in minerals and nutrients, however insufficient soil moisture limits mobility of the nutrients. The clay and silt contents of the soils increase and they become less permeable at the higher parts of the slope towards the plateau. At high intensity precipitation events, crusts inhibit infiltration and generate surface runoff.

2.3 Pearl Millet Cultivation and Fertility Management

Pearl millet cultivation is carried out in a very extensive style. Farmers start sowing seeds after 20 mm of precipitation at the onset of the rainy season around June. There is no land preparation before the sowing. A farmer walks with a hoe in his hand and digs the soil at every two steps. A follower drops seeds into the hole and closes it with his/her foot. During the growing season, weeding tillage is carried out twice or three times. A hand-harrow is used to cut roots of weeds and to break the crust on the soil surface to enhance infiltration. This weeding is important for crop establishment yet very time-consuming and labor intensive as each farmer has to weed a few hectares of land all by himself. At harvest, the panicles of millets are collected earlier, followed by the crop residue collected for hand-crafts. Left-over residues are eaten by cattle which are left in the field during winter time.

Traditional fertility management in the region is called "Parcage." Farmers provide food and water to herders for bringing cattle to their field and staying overnight for dung deposition. This is a very energy efficient way to harvest nutrients from vast hinterland. Parcage was practiced in the fields near the village by relatively resource-rich farmers. For fields in general, basically no input was provided. Instead when a farmer saw a decline in productivity after a few years of continuous cultivation, the field was left to fallow with a succession of grassland. With higher land pressure nowadays, fallowing is shortened or completely given up.

Presently, phosphorus appears to be the most limiting nutrient for millet production (Bationo et al. 1990; Bationo and Mowunye 1991), followed by nitrogen and organic matter. ICRISAT and INRAN recommend 30 kg N ha^{-1} and 30 kg P_2O_5 ha^{-1} equivalent of chemical fertilizers to be applied for sustainable crop production (Ly et al. 1997).

2.4 Desertification Process

In the precolonial times, people in Niger were mostly semi-nomaic or nomadic. Farmers moved their settlements periodically avoiding excessive exploitation of soil fertility. The colonial control seeking sedentary life, weakened the capacity of the ecological system to renew and sustain itself (Charlick 1991). The main cause of desertification in the region is primarily attributed to the tremendous population growth. Population in Niger was estimated to be around 3 million in 1960, whereas, in 2009, it was estimated to surpass 15 millions (CIA 2009). The growth has been over 3% in the last 50 years. Consequently pressure on land increased and deforestation occurred due to firewood consumption and excessive grazing. The cultivated area expanded to upstream where the condition of the soil was less favorable (less permeable, water deficit-prone) for millet cultivation. The period of fallowing became shorter or was completely abandoned. Competition on resource became apparent among sedentary farmers and nomadic people and their complementary relation was lost.

2.5 Successes and Failures of the New Land Management Efforts

In the Sahel region, Anti-desertification projects were initiated since the 1960s in the Sahel region. Early projects sought soil conservation as their main targets, such as the erosion-control bunds that were built with the use of heavy machinery at a scale of a few thousand hectares. These initial projects did not consider participation of beneficiaries in the planning and had no provision for costs of repair and maintenance. Inevitably this kind of reclamation was not adopted and maintained by local people.

In 1984, a concept of "terroir management" was first introduced at CILSS (Comité inter-états de lutte contre la sécheresse au Sahel). The term "terroir" means the space recognized by the local community as their area of authority or activity. In the terroir management, it is the beneficiaries or villagers themselves who plan and implement the project. The role of the funding body is to enlighten and educate them and to offer technical and economical assistance. This fundamental concept of participatory approach was adapted for all the projects conducted under the umbrella of the "terroir" (Yacouba et al. 1995).

In 1983–1984 severe drought had a devastating impact on food production in the northern Sahel region. The "Food for Work" method was used for large-scale land rehabilitation. This was to motivate participation of beneficiaries in difficulties with provisions of food. This worked as a catalyst for spreading land conservation practices. Although effective, misuse of this strategy could grow dependency of local people for external aid.

Recent projects aim to achieve proactive participation of the beneficiary without external aid. The points below are considered important for the participation to the projects undertaken (Yacouba et al. 1995).

(a) Emphasis on project evaluation by beneficiaries themselves
(b) Priority on farm conservation (outcomes are more apparent than watershed conservation)
(c) Promotion of education for natural resource management
(d) New legislation and reform of law at regional and national level for natural resource management

3 Revealing Mechanisms of Sustainable Conservation Techniques

3.1 Common Conservation Techniques

For resource-poor farmers in this region, the selection of the sustainable conservation techniques is very limited. These techniques are long said to be affordable and simple. As Reij (1988) pointed out the four conditions for a farmer to adopt new conservation techniques lie at the following,

(a) More than 40% of increase in harvest from the first year after application of a new technique.
(b) A new technique can expand the cultivation area. Fertility of degraded farmland and land of low productivity can be rehabilitated.
(c) Harvest becomes stable even in the low precipitation year.
(d) A new technique requires little labor for maintenance.

Some major techniques which have been adopted by farmers are shown in Fig. 3. These are the indigenous and also improved indigenous techniques.

Contour stone bunds and stone line (Fig. 3a, b) use iron stones found on the plateau. Generally stone bunds are used on severely eroded steep slopes for rehabilitating rill and gully erosions. Stone lines are single lines of iron stones half buried in the soil. This technique is used on a moderately eroded surface. Stone bunds and stone lines are both permeable and they do not stop surface runoff completely but slow the speed of the runoff. They can be built with simple leveling techniques to follow the contour. The difficulties lie in the transport of stones. Zaï (Fig. 3c) is a micro-catchment water-harvesting technique. Farmers dig up the soil at intervals of millet planting during the dry season and put manure in each hole at the same time. At each point of the hole, more water and nutrients is available securing the better establishment of the crops (Fatonji et al. 2006). These holes often become buried after the first severe rain-storm of a rainy season. It is labor intensive because the soil is hard to dig when completely dry. Figure 3d shows the mulching. This is just a conduct of leaving crop residues on the field. Residues only need to be protected from complete deprival by men or cattle.

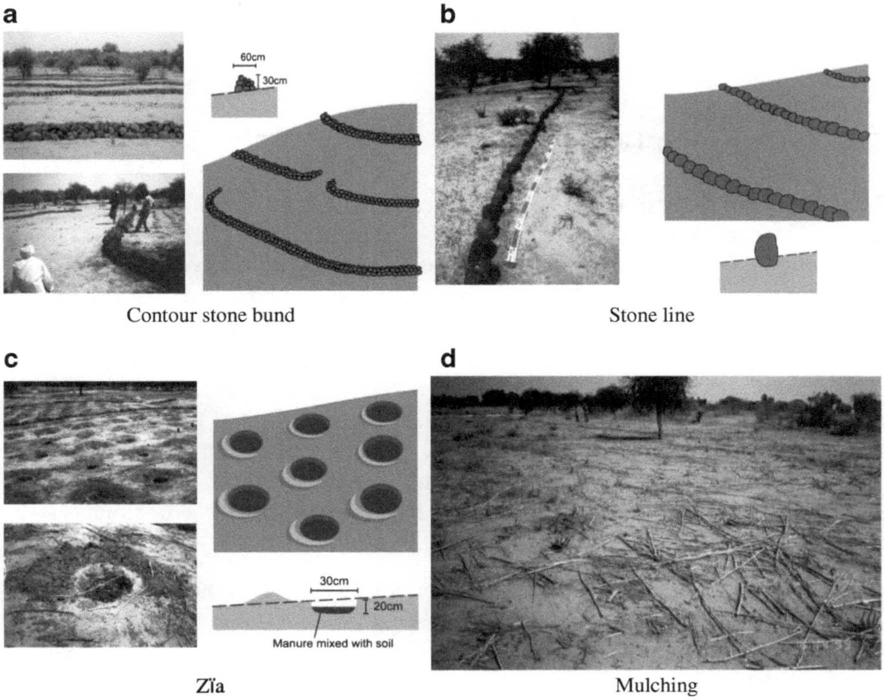

Fig. 3 Common conservation techniques for millet fields in Niger

3.2 Experiments Undertaken on the Working Mechanism of the Conservation Techniques

Although the techniques above are proved effective empirically, their physical effects were not comprehensively described.

These include the basic principles for determining the intervals of the stone lines, degree of water harvesting, and prospects of conservation. Consequently, the principle author of this chapter conducted filed trials aiming to clarify functions of conservation works or techniques on a sloped pearl millet field.

3.2.1 Location of the Study Site

The study was conducted for 3 years, from 1998 to 2000, in the experimental farm of Japan Green Resources Corporation in Magou village in the canton of Torodi, located about 60 km southwest of Niamey, the capital of Niger, West Africa (13° 6′ N, 1° 44′ E). The climate is Sudano-Sahelian with a rainy season generally lasting for 3 months and a half from mid-June to the end of September with the average rainfall of 566.8 mm (1961–1990, Sivakumar et al. 1993).

3.2.2 Design of the Test Field

A test field was created on a moderate slope (2.8%) formerly a millet field as shown in Fig. 4. It was divided into five blocks, 20 m in width and 60 m in slope length. In these blocks, contour soil bunds were built at different intervals so that water harvested at bunds would be of different quantity. Bunds were built using the topsoil in location, wetted and tapped for consolidation. They were 30 cm in height and 60 cm in width when freshly made, and after one rainy season, their height was reduced to about 20 cm. Millet was grown in the field with recommended amounts of manure (1.5 ton ha^{-1}) and fertilizer applied (40 kg P_2O_5 ha^{-1}, 40 kg N ha^{-1}) to avoid the influence of heterogeneity of soil nutrient distribution.

Four runoff plots were created to quantify effect of conservation works on surface runoffs. Each plot was 2 × 20 m in size. Four different treatments were set up where;

MM stood for Millet cultivation with manure and mulch application
M for Millet cultivation with no inputs
B for bare plots
BS for the bare plots with stone lines as tied ridges

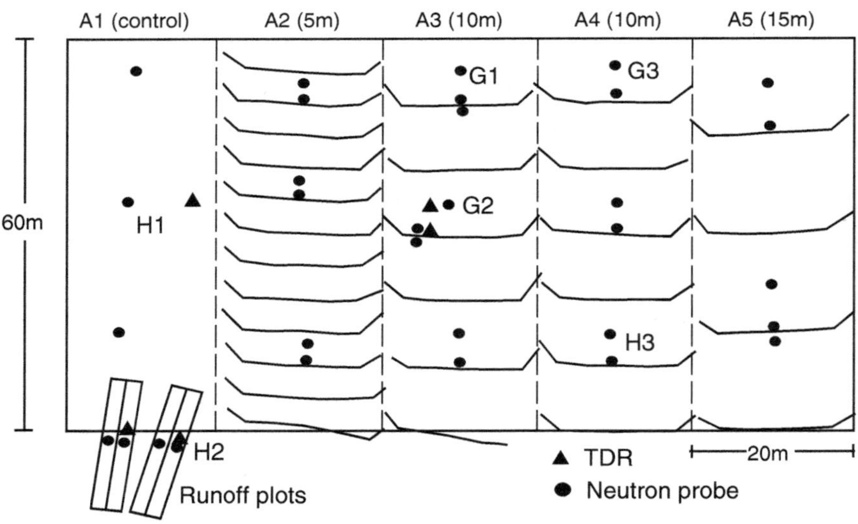

Fig. 4 Cumulative runoff against cumulative effective precipitation (P_r) in 1999 observed in runoff plots with four different treatments (*MM* Millet cultivation with manure and mulch application; *M* Millet cultivation with no inputs; *B* bare plot; *BS* bare plot with stone lines as tied ridges)

3.3 *Results, the Working Mechanisms of Conservation Works*

3.3.1 Effect of Conservation Works on Decreasing Surface Runoffs

In Fig. 4, the cumulative runoffs are plotted against the cumulative P_r, where P_r is defined as the effective precipitation which caused any runoff in either one of the four plots. The figure clearly indicates; (1) a lower runoff rate (runoff/P_r) in the *MM* in the beginning, compared to the *M* or the *B*, until the first weeding tillage, (2) nearly same runoff rates in the *M* and the *B*, until after the first weeding tillage, (3) decrease of runoff rates for all plots after the weeding tillage, (4) a lower runoff rate in the *BS* compared to the *B*.

1. Effect of manure and mulch application
   The lower runoff rate in the *MM* in the beginning of the season was attributed to soil-dwelling termites' activities. When manure was applied before the onset of the season, termites were attracted to the crop residue in the manure and they came up to the surface from their nest, a few meters underground, to ingest cellulose. In their movement, the soil crust, which developed on the surface, was destroyed and vertical pores were created in the soil profile. This greatly increased water infiltration in the beginning of the season. These phenomena had been observed by Mando et al. (1996), Mando (1997) in a series of studies in Northern Burkina Faso, where mulching without the application of manure is also expected to bring the same benefit to the soil (Leonard and Rajot 1997). This enhancement of infiltration in the *MM* diminished after the first weeding tillage due to the disturbance of the top soil.
2. Effect of millet growth
   The same runoff rates of the *M* and the *B* in the beginning of the season confirmed that, until a certain growth stage, the millet field without any input could be assumed as a bare. The difference became significant a few rainstorms after the first weeding tillage onward. This indicated, that the soil water conservation in the beginning of the season was very important when the soil is close to bare.
3. Effect of tillage
   The change in the runoff rate of the *B* reflected the sole effect of the weeding tillage. The break-down of soil crust largely increased infiltration. However, the crust redeveloped after a few rainfall events.
4. Effect of the stone line
   The *BS* showed a lower runoff rate than the *B* and yet the trend of the change was very similar. A good linear relation was found between runoffs from the *BS* and the *B*. A reduction of runoff volume by 35% was achieved by the stone line in the case of this test field. Stone lines do not stop runoff but rather slow the speed of surface runoff and enhance infiltration.
5. Soil erosion
   Table 2 shows the dry mass of water-eroded soil observed in the experimental plots. Soil erosion was generally more severe in the beginning of the season. It

Table 2 Dry mass of water-eroded soil during the growing seasons

		Mass of eroded soil (coarse fraction) ton ha^{-1}			
		MM	M	B	BS
1998	Before first weeding	1.14	1.01	2.44	0.98
	Before second weeding	0.23	0.15	0.80	0.15
	After second weeding	0.40	0.15	0.99	0.20
	Total	1.77	1.31	4.23	1.33
1999	Before first weeding	0.88	1.78	0.69	0.42
	Before second weeding	0.29	0.22	0.37	0.22
	After second weeding	0.19	0.18	1.79	0.21
	Total	1.36	2.18	2.85	0.85

was because of the coarse fraction outcropping on the surface as a result of structural crust formation (van der Watt and Valentin 1992) in the former season. Therefore it could easily be washed off by runoffs in the beginning of the following season when the soil surface was dry. The total amount of erosion was not very high (4 ton ha^{-1}) even in plot *B* because the runoff length was limited to 20 m and the flow speed did not rise as high as to cause a turbulent flow. Despite its small amount, the protection of the coarse sand fraction is very important for the self restoration mechanism of the land surface. This will be discussed in the following sections.

3.3.2 Effect of Water-Harvesting at the Bund

Figure 5 shows a comparison of soil moisture profiles around and in between contour bunds of 10 m intervals in the early growth stage. On the upper side of the bund, the enhanced infiltration due to the water-harvesting effect is clear. On the lower side, soil moisture slightly increased, possibly due to lateral movement of infiltrated water from the upper side of the bund. Growth of millet however, was very poor on the upper side. Soil was too wet for millet growth and nutrients were probably leached with excess water drainage. Changing intervals of the bunds had no significant influence on productivity either. This revealed, that, the main function of the bund was not water-harvesting, but for making optimal use of abundant soil moisture around the bund, where planting of trees or shrubs with deep rooting depth seemed more appropriate.

3.3.3 Spatial Variability of Surface Permeability

Figure 6 shows surface permeability measured at different points in the field (shown in Fig. 7). The measured points are in between the bunds so that influences from

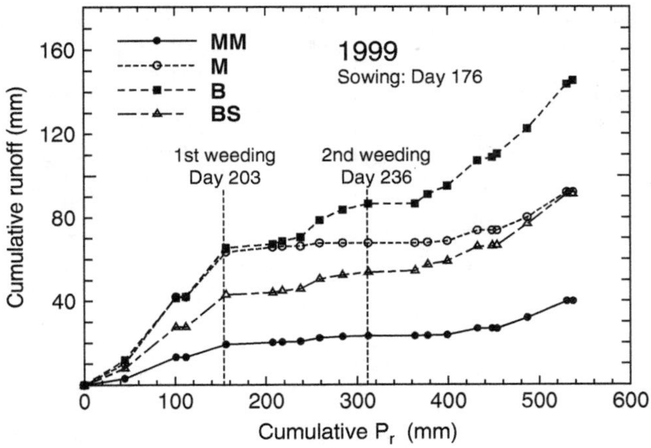

Fig. 5 Comparison of the soil moisture profile around and in between contour-bunds

bunds are minimal. At the time of measurement, crust was fully developed. The final infiltration rate of the crusted surface was 28–40% to that of the underlying soil. Although the clay content of the soil surface was generally very low, its content and infiltration rate seemed to correlate. The less clayey the surface soil, the more permeable it was. Infiltration rates were low at points located in the upper to middle reach of the slope and infiltration rates were higher at points on the lower end of the slope. At points of high infiltration, coarse and loose sand was present on top of the crust. At upper and middle points, the sand layer was absent because of its erosion history. This spatial variability of surface permeability was well reflected on soil water regimes. Figure 8 shows the soil moisture change of six points selected in Fig. 6. In 1998, precipitation was a little below average in quantity and regular in distribution. In the upper 0–60 cm layer, there was no recognizable difference in water storage between the high and low infiltration points. However, the difference was apparent in the 60–120 cm layer. It shows that the difference in the surface infiltration capacity has a crucial influence for the soil water regime in the rainfed agricultural area with high-intensity precipitation.

3.3.4 Growth of Millet in Relation to Water Availability

From the viewpoint of soil water availability, a high infiltration surface seems preferable for millet cultivation. However excessive precipitation can actually bring down growth of millet. Figure 9 shows the inferior growth of millet on high infiltration points in 1999, where, precipitation was abundant at the onset of the season and towards the flowering period. The total seasonal precipitation was

Fig. 6 Final infiltration rate measured by a tension disk infriltrometer (at −294 Pa) and clay content at different points in the field, shown in Fig. 7

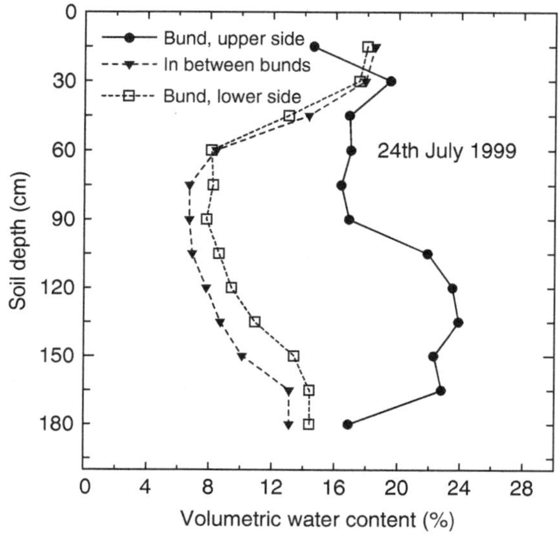

Fig. 7 Design of the test field for assessing the improvements of the soil-water regime by contour bunds

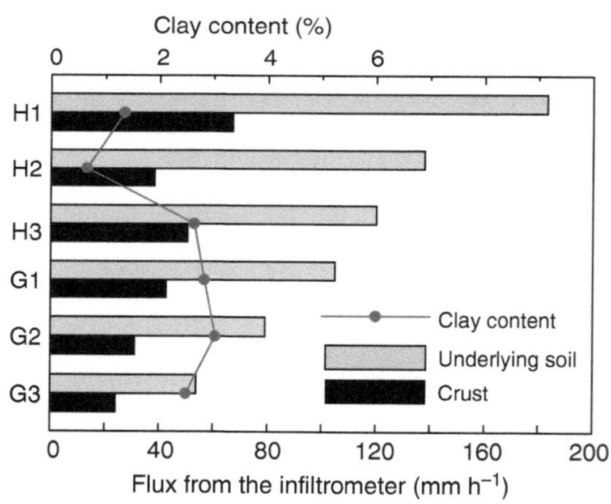

686.3 mm, which exceeded the average by more than 100 mm. Payne et al. (1990) stated that the faster advancements of a wet-front than root-growth can decrease growth of millet due to the leaching of nutrients. This can also explain why millet growth cannot be simply enhanced by means of water-harvesting alone. In the years of abundant precipitation, millet growth in low-infiltration surfaces may turn out better due to less infiltration and better nutrient holding of the soil profile due to higher clay contents.

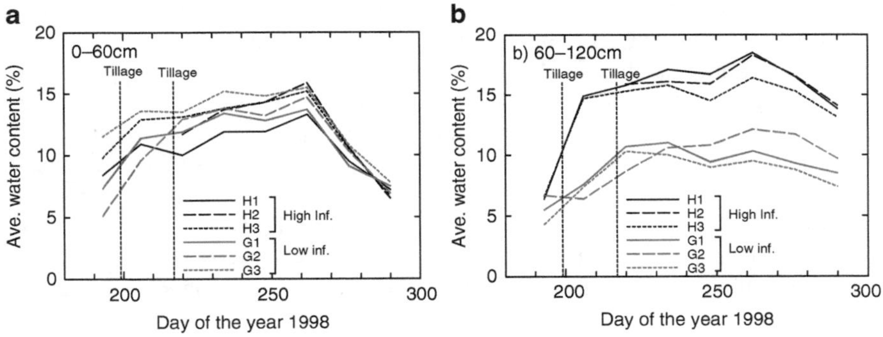

Fig. 8 Change of average soil moisture in (**a**) 0–60 cm and (**b**) 60–120 cm soil profile, of six points in the test field (shown in Fig. 7)

Fig. 9 LAI evolution of pearl millet in different points of the field (shown in Fig. 7) during the growing season of 1999

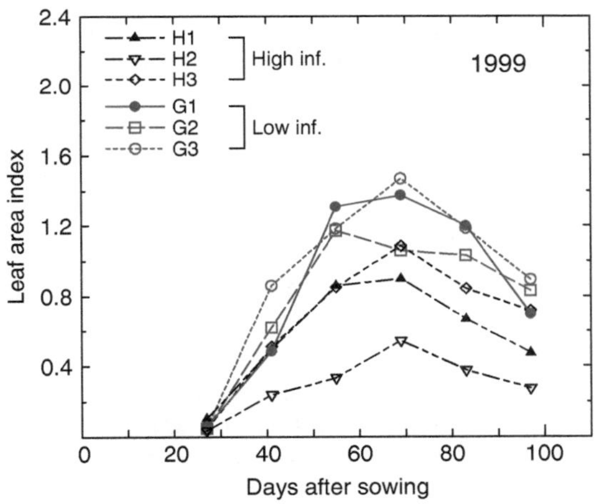

4 Discussion: Wise Natural Resource Management for Sustainable Land Use

4.1 Soil Surface Management

For enhancing infiltration, breaking of the crust and protection of the coarse sand layer on the soil surface are particularly indispensable. The former can be accomplished by a simple conduct of leaving crop residue on soil surface for attracting termites. Mulching is also reported to contribute for better phosphorus uptake by millet (Hafner et al. 1993) and for control of wind erosion (Sterk and Stroosnijder 1997). For the latter, a combination of various techniques is required, depending on the condition of the surface.

For flat to moderately sloping areas with good infiltration, leaving the land to fallow is the most energy efficient strategy for accumulating the coarse sand fraction. The natural grassland can trap moving sand due to its high surface roughness. Sand size fractions in turn facilitate the establishment of grass and this forms a virtuous circle for sand accumulation if continued for a few years.

For a moderate slope with a minor erosion history, it is important to reduce or stop surface runoff from upstream by means of stone bunds or stone lines. Construction of these structures enhances coarse sand accumulation on both upstream and downstream sides of structures. In the rainy season, grass establishes on coarse sand due to high soil moisture availability (Fig. 10) and this enhances more sand accumulation.

For a steep slope with a severe erosion history, the source area of surface runoff in the upstream area must be primarily rehabilitated, due to the hardships encountered in cultivating in a single year. Consequently, planting trees around the bunds is very effective to overcome the constraints in cultivation (the working mechanism for this is explained in the following section). Thus, ultimately, cultivation can be started on the areas where coarse sand layers develop.

4.2 Conserving Trees for Conserving Field

Present day farmers do not prefer to conserve trees in their fields as trees create shades and farmers tend to deem the land susceptible to bird invasion. However trees play important roles in a self-restoration mechanism because they protect the field from wind erosion in the beginning of the rainy season and enhance sand accumulation and contribute to trapping the nutrient-rich clay fraction carried by wind from the Sahara (Harmattan) in the dry season (Stahr et al. 1994). Furthermore, their roots penetrate to deeper layers of soil and absorb leached nutrients. The trees later return nutrients to the soil surface by defoliation and ultimately

Fig. 10 Grass establishment after a bund construction on a severely eroded surface. Grass only establishes where there is a cover of coarse sand fraction on top of the eroded surface

improve soil physical property through the attraction of termites (Fig. 11). Faidherbia albida (Del.) A Chev (syn. Acacia albida) is a well-known tree which defoliates during the rainy season and it is reported to have a positive influence on millet grown under its canopy by supplying additional nitrogen and phosphorus (Kho et al. 2001; Payne et al. 1998). Not only this tree but other species can also contribute to the recycling of leached nutrients to the soil surface if farmers allow their presence in their fields.

4.3 A Strategy to Adapt to Erratic Precipitation

Figure 12 illustrates the infiltration and degree of leaching in relation to topography in the southwestern part of Niger. Infiltration on flatland or towards the bottom of the slope is accumulated deeper in the sand and consequently nutrients are leached and become less available. On the slope, productivity is limited by little infiltration due to the more clayey soil surface, while the soil nutrients are relatively more available in the argillic horizon. Therefore a good strategy to adapt to erratic precipitation is to diversify cultivating fields. In a year with an above-average precipitation, a good harvest would be attained on the upper part of the slope

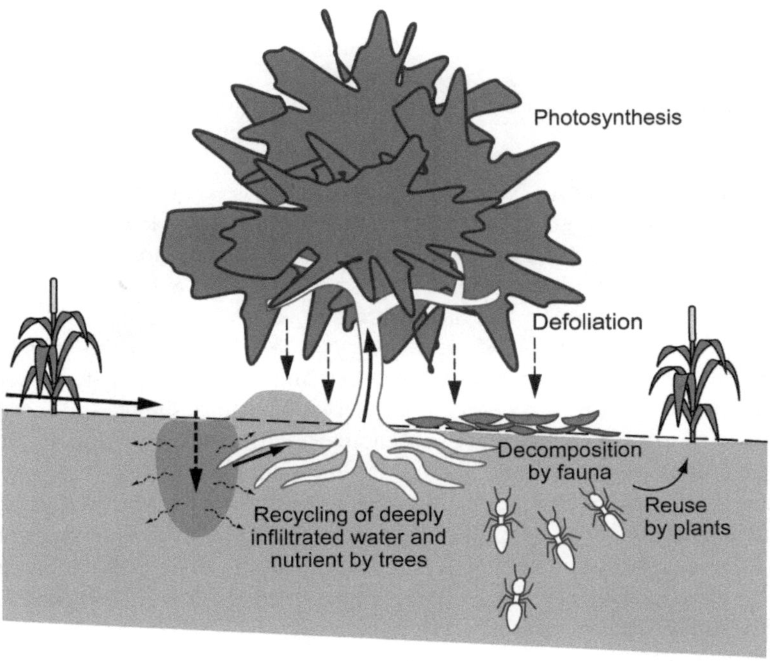

Fig. 11 Ideal image of land management of a sloped millet field utilizing activities of flora and fauna (Nagano et al. 2002)

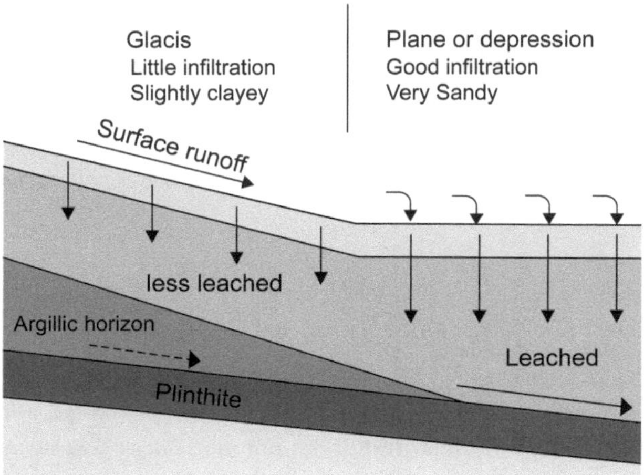

Fig. 12 Relation between topography, infiltration and degree of leaching in southwestern Niger (Nagano et al. 2003)

(given good conservation efforts) while the productivity of the flatland may be low if fertility management is not conducted. On a drought year, the flatland can secure a minimum harvest, while slopes may face severe water deficiencies.

5 Conclusion

In the southwestern part of Niger, a low input and less labor intensive agriculture seems only sustainable given the buffer capacity of the soil is low and the precipitation is erratic complemented by the low calorie diet of the locals. A simple but important lesson learned from this study is that nature has a good self-restoration mechanism for biomass as a fuel and man has to leave some biomass to nature for its return. The accumulation of nutrients would be difficult in the soils of this region, thus people rely on the biomass to determine the degree of fertility of the land and the self-restoration capacity. Conservation efforts should aim for simply increasing or establishing biomass. In another words, once an effort is put into increasing biomass production, this would form a virtuous circle.

The primary role of contour bunds seems to be effective on the conservation of the coarse sand layer to increase the infiltration capacity of the soil surface not only around the bund but ultimately on all the surfaces of the slope they are implemented. Once grass establishes on the hard surface, termites in the dry season further improve the infiltration capacity.

Biomass should be wisely harvested, transferred and supplemented within the command area depending on the intensity of human use. If biomass transfer within the area can modify the nutrient deficit, an efficient transport strategy should be considered. As mentioned in Sect. 2.3, this was already practiced in the "parcage" system. If the deficit is irreversible by biomass transfer alone, the use of the chemical fertilizers or mineral materials should be considered.

There are some legislative problems that limit the flexible transfer of biomass within the watershed. One example is the national law that completely prohibits the cutting of trees. Another imminent problem is the loss of inter-dependency and the increasing conflict between farmers and herders due to land resource shortage. Ultimately, it would be wise to include herders as stakeholders for natural resource management.

References

Ahn PM (1970) West African Soils. West African Agriculture, vol. 1. Oxford University Press, Oxford, 332pp

Bationo A, SH Chien, J Henao, CB Christianson, AU Mokwunye (1990) Agronomic evaluation of two unacidulated and partially acidulated phosphate rocks indigenous to Niger. Soil Science Society of America Journal 54:1772–1777

Bationo A, AU Mowunye (1991) Alleveiation soil fertility constraints to increased crop production in West Africa. Fertilizer Research 29:95–115

CIA (2009) The World Fact Book 2009. https://www.cia.gov/library/publications/the-world-factbook/

Charlick RB (1991) Niger: Personal Rule and Survival in the Sahel. Westview, Dartmouth, 189pp

Fatonji D, C Martius, CL Bielders, PLG Vlek, A Bationo, B Gerard (2006) Effect of planting technique and amendment type on pearl millet yield, nutrient uptake, and water use on degraded land in Niger. Nutrient Cycling in Agroecosystems 76:203–217

Hafner H, E George, A Bationo, H Marschner (1993) Effect of crop residues on root growth and phosphorus acquisition of pearl millet in an acid sandy soil in Niger. Plant and Soil 150:117–127

Kho RM, B Yacouba, M Yaye, B Katkore, A Moussa, A Iktam, A Mayaki (2001) Separating the effects of trees on crops: the case of *Faidherbia albida* and millet in Niger. Agroforestry Systems 52(3):219–238

Leonard J, JL Rajot (1997) Restoration of infiltration properties of crusted soils by mulching. Soil fertility management in West African land use systems: Proceedings of Regional Workshop. University of Hohenheim, ICRISAT Sahelian Center and INRAN: 4–8 March 1997, Margraf, Niamey, Niger:191–196

Ly SA, N van Duivenbooden, CL Bielders, AS Gouro, K Anand Kumar, eds (1997) Technologies diffusables et transférables aux producteurs. Actes des Ateliers cojoins INRAN-ICRISAT sur les technologies diffusqbles et transférables aux producteurs, 21–22 Novembre 1996 et 5–6 Juin 1997. Centre Sahélien de l'ICRISAT, Sadoré, Niger

Mando AL, L Stroosnijder, L Brussaard (1996) Effects of termites on infiltration into crusted soil. Geoderma 74:107–113

Mando A (1997) The impact of termites and mulch on the water balance of crusted Sahelian soil. Soil Technology 11:121–138

Nagano T, H Horino, T Mitsuno (2002) A study on conservation of millet field in the southwestern Niger, West Africa. Proceedings of 12th International Soil Conservation Conference, 26–31 May 2002, Beijing, China

Nagano T, H Horino, T Mitsuno, M Kimura (2003) Growth environment of sloped millet field in the southwestern Niger and effects of introducing contour ridges. Transactions of the Japanese Society of Irrigation, Drainage and Reclamation Engineering 71(2):193–203

Payne WA, CW Wendt, RJ Lascano (1990) Root zone water balances of three low-input millet fields in Niger, West Africa. Agronomy Journal 82:813–819

Payne WA, JH Williams, KAM Moussa, RD Stern (1998) Crop diversification in the Sahel through use of environmental changes near Faidherbia albida (Del) A Chev. Crop Science 38(6): 1585–1591

Reij C (1988) The present state of soil and water conservation in the Sahel. OECD, Paris, 23–26

Sivakumar MVK, A Maidoukia, RD Stern (1993) Agro-climatology of West Africa: Niger. ICRISAT, Patancheru, India, 108pp

Stahr K, L Herrmann, R Jahn (1994) Long distance dust transport in the Sudano-Sahelian zone and the consequence for soil properties. Wind erosion in West Africa: Proceedings of the International Symposium. University of Hohenheim, Germany, 5–7 December 1994, 23–34

Sterk G, L Stroosnijder (1997) Optimizing mulch application for wind erosion protection in the Sahel. Soil fertility management in West African land use systems: Proceedings of Regional Workshop. University of Hohenheim, ICRISAT Sahelian Center and INRAN: 4–8 March 1997. Margraf, Niamey, Niger, 185–189

van der Watt H.v.H, C Valentin (1992) Soil crusting: African view. In: Summers ME, BA Stewart, eds. Soil Crusting. Lewis Publishers 301–329

Wilding LP, LR Hossner (1989) Causes and effects of acidity in Sahelian soils. Soil, crop and water management systems for rainfed agriculture in the Sudano-Saheilan zone. Proceedings of an International Workshop, 11–16 January 1987. ICRISAT Sahelian Center, Niamey, Niger:215–227

Yacouba M, C Reij, R Rochette (1995) Atelier de restitution sur la gestion des terroirs et le developpement local au Sahel, Niamey, 30 Mai-2 Juin 1995. Club de Sahel, SAH/D(95):448

Local Wisdom of Land and Water Management: The Fundamental *Anthroscape* of Japan

Tsugihiro Watanabe

Abstract With the combination of the advanced prediction and the traditional knowledge like *Mitameshi*, we can foster local wisdom for wise land and water management and improvement of the regional environment. Such an approach may be effective in solving the problems caused by global climate changes resulting in living of humans in harmony with nature, which means living in a better anthroscape. Ultimately, the wisdom of Suido is to be sublimed to the context of "*ecosophy*".

Keywords Soil and water management · *Anthroscape* of Japan · Satoyama · Paddy cultivation · Culture · Mitamashi

1 Introduction

In the Edo era of Japan in the early seventeenth century to the mid nineteenth century, they had established a semi-closed sustainable society. In that period, the Japanese Islands were closed except opening one port only for trade with China and Holland. The system was one feudal system with one *Shogun* living in Edo, the former name of Tokyo, and Feudal lords governed the allocated local areas. There was a dispersed sustainable recycling and reusing systems, utilizing local conditions and resources. As the fundamental base, paddy cultivation system had been for producing rice as staple food of the country make the most of land and water in sustainable manner. The adopted and developed land and water management system conform to natural resources or environment of Japan, including climate, topography, soil and hydrological regimes.

T. Watanabe
Research Institute for Humanity and Nature (RIHN), 457-4 Motoyama Kamigamao, Kita-ku, Kyoto 603-8047, Japan
e-mail: nabe@chikyu.ac.jp

S. Kapur et al. (eds.), *Sustainable Land Management*,
DOI 10.1007/978-3-642-14782-1_16,

The system of the Edo in Japan is recently evaluated as one of desirable society of sustainability, which had been actually lasted almost 300 years and of which basic concept and function has continued until now, while its function has been depressed in the process of economic development and globalization of Japan. In this chapter, the implication of fundamental management system of land and water in Japan is reviewed for solving the current global change issues on local resource management including food security and global warming.

2 "*Suido*": Paddy Cultivation Based Land and Water Management System of Japan

2.1 Suido

Everywhere in the world, the human have developed and inherited local system to control natural hydrological regime and use water, even in very dry land like desert environment. In humid area with much water, human have developed the system to utilize water. In the developed system, the art and technology have been built to reclaim and utilize local resources as much as possible, adapting the local conditions. The system is a basic social infrastructure to manage hydrological water cycle including flood and drought as well as water quality and landscape. It is to manage not only water but also land, since essentially they could not be controlled and used independently.

The much stable and long-lasting system of land and water management consists of indigenous knowledge and traditional wisdom established in the long process for adapting and controlling the natural fluctuation and variability of land and water in the region. Historically, human life have been depending on agricultural production, especially in humid region such as the main part of Asia, the core of land and water management has been on use for agriculture, where irrigation and farmland reclamation take major part of it. Here, the whole system for managing land and water in a region, where agricultural management takes major role, is to be referred as "*suido*". "*Suido*" is a Japanese word, which means the total regional system or environment. It could be paraphrased as "climate" in its most wider meaning.

"*Suido*" is a kind of culture as integration of human workings and activities on natural elements, including land and water use, equipments and facilities, design and location of infrastructure, skill and techniques for operation and maintenance, and institutions and organizations as well as human capacity building. "*Sui*" of "*suido*" means water in Japanese, and "*do*" of "*suido*" means soil in Japanese, while "*suido*" does not imply only "water and soil" but the whole management system of local resources, which fundamental elements are surely water and soil.

What is the wisdom in the "*Suido*"? It is specific to some regions and time. It may include what the latest science and technology of modern standards that are functioning, and also traditions and experiences historically inherited. There might be a potential function that has not been demonstrated from some constraints. The wisdom of *suido* is to be a detailed understanding of the natural conditions, like regional climate and topography, geology and hydrological cycle. On the basis of the understandings, it should contain the policy and actual skill and techniques to adapt natural conditions, modify them, and the way to keep these activities. The mechanism is needed in it to prolong the effect of actions. In the case with facilities and equipments, it require persistent feature of the technology for maintenance ensuring the necessary fuel and materials, and development of technology and human resources responsible for it.

The preparation for the future with a view and investment is also indispensable. In addition, process for identifying defects of the structure and mechanism and taking the necessary measures to improve is always asked. Since water use need co-works or cooperation of local people, problems often arise in human relationships in actual water use practices. It is difficult to completely equal the size and distribution of required work and final results, and each of the involved people is also different from perceptions. Then, it often requires mutual support mechanisms to help each other, and sometimes there needs competing and restraining each other in moderate manner to enhance the performance and encouragement.

This wisdom of *suido* can be recognized as a combination of the following seven major knowledge with the aspects and contents described above (JSIDRE, The Japanese Society of Irrigation and Drainage and Rural Engineering 2009).

(a) Identify (Observation: to understand the nature): Assumptions for vision and planning. Understandings on available natural conditions and resources and identifying constraints. Determining the feasibility of reclamation and the possibility of realization of a system.
(b) Exploit (Utilization: to turn environment into resource): Direction and method to maximize local resources sustainably.
(c) Ascertain (Improvement: to test the function): Diagnosis and evaluation of the system on whether it is operating normally expected to comply with the conditions at that time. Implementations of countermeasures to deal with nonconformities and improve the system.
(d) Preserve (Maintenance: to maintain the function): Mechanism and institution to maintain equipments and facilities consciously to keep the system functioning. Continuous improvement of the mechanism and institutions.
(e) Adapt (Adaptation: to respond to changing conditions): "*Mitamesu*" is the introduction to the desired direction according to changes in local conditions and environment and the risk of uncertainty. Also avoiding and reducing the undesirable result. Determining measures against the anticipated risk with feedback of the system. Mechanism and techniques for observation of the system feedbacks and responses.
(f) Forecast (Investment: to plan the future of region): Projection and perspectives on the future, and design of the ideal goals. Necessary preparations towards the goals. Technical education and skill transfer, including saving and insurance for the future.

(g) Collaborate (Coordination: to cooperate the regional society): Mutual assistance, competition and restrains for the system work. This is the basis for regional stability and moral order.

2.2 Formation of Suido *in Japan*

As mentioned above, historically agricultural water use has been located in the hearth of the *suido*, or land and water management system. Irrigation as the water diversion and application system is a major expression of the wisdom of *suido*, where all elements related to the land and water mainly for crop production. Japan, in particular, from the Yayoi period from around 4–5C BC to 4–5C AD, with gradual expanding paddy rice cultivation had developed the basis of the country and culture spinning the wisdom of land and water management that is *suido*. Many of the rice fields in Japan had been reclaimed in the Edo era from 17–19 Cen., establishing the organization of irrigation scheme. During the process, the regime had been set up in which the every available water in rivers in summer time is fundamentally diverted to paddy fields and rural villages. In other words, paddy fields were reclaimed according to the discharge of river during the summer dry spell. The available water resources had diverted to paddy irrigation and rural community that depends on rice production.

The gradually developed irrigation system in Japan, mainly for paddy cultivation, has played many roles not only to supply water to the rice paddy fields but also to support and enrich the rural life. The system has also functioned for flood control and regional drainage. Japan, with this longer history of irrigation development, faced the problems caused by increased demand for water for industries, domestic use, as well as for power generation since the Meiji era from 19C. Almost all available water is used in priority for paddy irrigation, the issues of development of water resources and re-allocation of water resources have been a mainstream of the water management problem over almost hundred years. During this period, while it is hard to say that all conflicts were solved with fine tuning, the outcomes of the long and various experiences have been woven into the wisdom of *suido*.

3 "*Satoyama*": Expression of *Suido* in Japan

3.1 Satoyama

In each region of Japan, Suido was built adapting and modifying local conditions. Its base was rice cultivation and paddy irrigation as described above. The basic structure of paddy irrigation was developed according to regional conditions of topography, geology and hydrology. On the other hand, as a typical and common form of suido, there is satoyama system. “Satoyama” is a Japanese term referred to border domain or zone in mountain foot and arable flat or less inclined land. “Sato” means arable and living area or rural village, and “yama” means just mountain, which is almost equivalent to forest in Japan (Fig. 1).

Fig. 1 Irrigated (**a**) and cropped (**b**) Satoyama

The system was developed within the topographic structure of Japan, the land, water, agriculture, and rural life, formed with paddy fields and surrounding forests. Rice cultivation cannot exist only with paddy "field" and "irrigation", it requires other materials to be used as fertilizer such as grasses taken from surrounding forest. Also, rural life cannot depend on only rice and other field crops but also fruits, nuts, mushrooms and other foods obtained from the forest, as well as small animals for protein. This small, integrated system was the satoyama "system" in Japan which was recognized as a complex land and water use system including paddy fields and "satoyama". The system has been developed from centuries of small scale of agricultural and forestry use, resulting in rich sustainable biodiversity in farmland and forest.

The "satoyama" is originally and literally defined as a forest or mountain managed through local agricultural communities. As mentioned above, during the Edo era before artificial fertilizer was applied, young and fallen leaves were harvested in forests surrounding paddy fields and used in paddy fields. Villagers also used wood or other products of forest for construction, cooking and heating. In those days, the "satoyama" has been defined as a "system" mentioned previously, as the entire landscape of rural agriculture and lives, which embrace a mosaic of mixed forests, paddy fields, upland fields, grasslands, and even streams, ponds, reservoirs and canals for irrigation. In the system farmers have used grasslands to feed their horses and cattle, and irrigation facilities including ponds can provide farmers with fishes as other source of food.

3.2 History of Satoyama

In Japan, it can be traced back to the Jomon period, from 10–11C BC to 3–4C BC, that the hands of the people came to forests, according to the recent excavations. In the historical period, over logging and protection of forest have been repeated in the Satoyama of Japan until today. Overuse of satoyama with deforestation had expanded up to the beginning of 16C, resulting the loss of about one fourth of the forest of Japan was lost, mainly in central part of the country. Deforestation had been advanced into the Edo era, and in the early eighteenth century, and most of the forests of the Honshu, Shikoku, Kyushu, and southern Hokkaido regions were lost. The background of these fierce and continuous deforestations is the increased demand of wood for building materials with rapid growth of the population, and for constructing numerous large castles, temples and shrines.

Thus, until the eighteenth century the forests of Japan had been over used and not developed in sustainable manner. Such widespread destruction of forests caused not only deficit of wood supply but also increase of forest fires, floods, and disasters. The Tokugawa shogunate, the national government of Japan in the Edo era, was concerned about this situation, and in the mid 17C established the forest protection policy, and implemented severe logging regulation and wood transport control.

With these measures, the Japanese forests began to recover, and the sustainable use of Satoyama was realized.

After these recovery processes, in the modern period, satoyama have faced tree crisis in terms of sustainable use. The first crisis took places in the period of just before and after the Meiji Restoration, by illegally logging and uncontrolled deforestation, due to reckless regime change. With stabilization of the society, vegetation of Satoyama had gradually recovered, but, in the world war period, with lack of every materials and fuels, over logging was repeated again as the second crisis resulting in many bare hills over the country. The restoration from over-logging in the war period was carried out in the mid 1950s with national campaign for afforestation.

The third crisis is ongoing problems caused by reclamation of residential land and no-use of Satoyama. Wood or wood charcoals for domestic fuel were completely replaced with fossil fuel in the 1970s, and the spread of chemical fertilizers kicked out natural manure from the forest. Disappearance of animals in rural villages reduced the necessity of forest products. The economic development in rural area, ironically, destroyed economic value of Satoyama, and finally had been reclaimed as residential land rapidly since the 1960s. Also the survived Satoyama are abandoned for the loss of value, changes in vegetation due to the loss of human involvement, resulting in climax phase forest and bamboo intrusion. It is now exposed to pollution by illegally dumped waste.

3.3 Significance of Satoyama System

One of the most important features of Satoyama for rural environment is its wide biodiversity. The Satoyama system with mixed mosaic land use provides wildlife with various habitats with traditional paddy dominant agricultural production system, which facilitates migration of wild animals, birds and insects over various habitats including irrigation ponds, paddy fields, grasslands, forests, and higher mountains. It sometimes becomes inter-village movement. Paddy field and paddy irrigation system, which create seasonal artificial ponding in every year in Satoyama system, plays very significant role in the survival of water dependent species and ecosystem with them, such as flogs, newts and dragonflies, of which early stage of life cycle needs water.

In most of Satoyama, farmers had planted Japanese oaks and chestnut oaks and they were cut down for fuel wood and charcoal every 15–20 years. It maintained the deciduous broad-leaf trees and prevented the succession to dense and dark confer forest. These continuous human activities in the forest maintaining several succession stages related to agricultural practices facilitate the preservation and promotion of biodiversity. In these deciduous forests, most plant and animal species can survive with traditional management practices. Therefore, much more wildlife and rich biodiversity is provided by well managed forests with farmlands rather than dark unmanaged natural forests.

Satoyama is to be recognized as a local "commons", which was a concept referred to as the natural resources that were shared by a particular community. Satoyama was available basically free, but usually limited to the members of the local community for physical access, while others were sometimes placed under the control of local community. Satoyama holded a strong sense of local commons, which was called as "Iriaichi" in Japanese. The most important implication is that the Satoyama has been managed with the people who lived in that place in a mutual cooperation manner. Then, it is a typical common type of "Suido" in Japan, and the Satoyama system is the wisdom of the Suido.

3.4 Conservation of Satoyama

As mentioned in the previous sections, Satoyama in Japan was historically not an ideal space in harmony with nature that could be created and maintained automatically, but very vulnerable zone, which was easily to be over-used and could be maintained by sever regulation and control of government and rural communities. Satoyama in Japan is currently deteriorating with lose of its economic value, while its importance for biodiversity is widely recognized.

In Japan, with wide public recognition of the role of Satoyama, since 1980s the conservation movements have been implemented gradually, hundreds of environmental NPOs are working, including monitoring the tree growth and appropriate logging period, and educating younger generation i.e. primary school students about the ecological and regional importance of the Satoyama. The satoyama system consists of not only Satoyama or forests itself but also paddy fields irrigation facilities, and most of other various elements of rural community. Paddy cultivation system field is shrinking physically and economically. As described below, the area of paddy fields of Japan has decreased rapidly in the last 30–40 years. The conservation of Satoyama should be discussed as the problem of the system, like a crisis of Suido.

4 Depression of *Suido*

In the modernization process, economic development, globalization of the world economy and the deepened and expanded market mechanism has affected the traditional Suido in agriculture and rural area. Most recently, depopulation and with low birth rate have accelerated ill-functioning of the Suido. It is just a crisis of agriculture and culture of Japan.

One of the major changes in Suido was the shrinkage of paddy cultivation. The total area of paddy fields in Japan had a peak as 3.4 million ha in the mid 1960s, and was reduced rapidly and now become 1.7 million ha, which is almost half of the peak. The role and implication of paddy irrigation should be reviewed from its very foundation as the core of Suido and Satoyama systems.

In reviewing the paddy irrigation, it was necessary to introduce two different perspectives to rehabilitate the Suido and reorganize its wisdom. First of all, the exact efficiency of paddy irrigation should be re-evaluated. The paddy irrigation in Japan has been a comprehensive water management of a region functioning as regional environmental management systems. In recent years, however, emphasis on economic efficiency and productivity have turn paddy irrigation system to specialize in supplying water to the fields for rice production. This losing multi-functionality has caused various problems. For example, water in irrigation canals in winter season has disappeared, as the canals were meant for summer paddy irrigation. It made rural villages having less water in winter, and worsened the habitat of wildlife that depended on water. Rapid decrease of firefly population was an impact of this change in water regime. It also resulted in considerable depopulation of dragonflies and frogs, leading to disorder of the food chain with higher organisms such as amphibians and birds.

Review of paddy cultivation and irrigation was needed in the aspects of not only environmental perspective but also the perspective of food production. To increase the very low rate of food self-sufficiency of Japan for enhanced food security, and in terms of safe food supply, there is an urgent need to examine the paddy cultivation and irrigation system in rural areas. It is considered as ways to manage local resources, to renovate local society and culture. It is to be a matter of rebuilding the "anthroscape".

5 Rehabilitation of *Suido*

5.1 Adaptive Management

Another notable wisdom of Suido is "*mitameshi*", is the adaptive management in its wider sense. It is one of the functions of the wisdom of Suido, and is useful to prepare the measures for the unpredictable and uncertain problems like global warming, conservation of biodiversity, and variability of hydrological events. Moreover, observation of changes, identification of necessary actions, ascertain of its effects and impacts, and prediction of the future possible changes are required.

There are many problems on understanding natural events that are difficult to examine quantitatively in the laboratory or simulate and then to respond or react to this kind of issue. One of more effective and feasible solutions to such a dilemma is to take incremental actions in trial-and-error manner, utilizing the best available current information and knowledge of past experiences, and collecting additional information as needed. As an example, an improved relationship between human and natural system including land and water, like the Satoyama system, taking advantage of adaptive management system, the following approaches are recommended to be implemented:

(a) Determine the current land and water use system and its significance in the regional natural regime
(b) Monitor the dynamics between human activities and the regional regime of land, water and ecology
(c) Predict the changes in the environment, including the ecosystems
(d) Employ incremental or gradual development and note feedback responses and
(e) Include all stakeholders in decision making.

In the past, natural events were largely unpredictable and it was only the measures to react them. At present, however, using the advanced observation and modeling technologies, future events could be predicted to some extent. Now, with a combination of advanced prediction and local traditional knowledge, there is the potential for wise management of land and water resources, smart agricultural production, and improvement of regional environments and landscapes.

5.2 "Mitameshi": An Adaptive Management Strategy for Better Human–Nature Relationships

To enhance the sustainability of agriculture and human society, more efficient systems of land and water use should be pursued. Paddy cultivation and irrigation, as an integral component of the Suido and Satoyama systems in Japan, provide the focus for studying and applying sustainable land and water management with improved efficiencies.

The adaptive approach mentioned in the previous section is comparable to the method of "*Mitameshi*" in Japan, an adaptive consensus-building procedure for irrigation development. The "*mi*" of *Mitameshi*i means "to watch" or "to see" in Japanese, and "*tameshi*" means "to try"/"trial" or "example". *Mitameshi* was first introduced in the Edo period from the seventeenth to nineteenth century, in which the people carefully observed nature and the consequences of human actions on water development and management. This method could be reviewed and utilized for addressing current environmental issues, especially for establishing better human–nature relationships that creates innovated anthroscape (Watanabe and Kume 2009).

Mitameshi was a method developed through trial and error to build stakeholder consensus when paddy fields were reclaimed rapidly in the Edo era, to make decisions on issues related to water use for rice production such as irrigation diversion and sequence of water delivery, During this era, communities needed to decide how much paddy field area could be reclaimed and evaluate how they could divert sufficient water to irrigate both new and existing paddy, especially during dry spells and droughts. By this method, they monitored weather and hydrological conditions including fluctuations in source river discharge, for several years. After careful monitoring, they could decide upon the area to be reclaimed and cropped, the form and size of the diversion weir and delivery canals needed, to establish water distribution rules and regulations.

The essence of the *Mitameshi* is the stakeholder consensus based on watching, experiencing, and learning about and implementing incremental change. The essential point is that local knowledge was developed to understand the nature of the region, and wisdom attained to live in harmony with natural fluctuations and variations. Communities recognized, even unconsciously, that water was not only a physical substance but also an object to conserve and utilize wisely for future generations.

Mitameshi is similar to the concept of adaptive management utilized today in the field of natural resource management. This method to address uncertainty in the system and its interactions is essential in ecosystems and human–nature interactions, where resultant changes cannot be simulated and predicted precisely, since repeated experiments are not possible to test (Washitani 2001).

6 Conclusions

Taking into account *Mitameshi* of the wisdom of Suido, or adaptive management, for improved human–nature relationships including the Satoyama system with paddy cultivation and irrigation, the following approaches are recommended:

(a) Diagnose the current system of resource utilization including land and water management and their significance in the regional geological and hydrological regimes
(b) Monitor the dynamics between human activities and the regional natural regime
(c) Predict changes in the regional environment, including the ecosystem
(d) Employ incremental or gradual development and note feedback responses and
(e) Include all stakeholders in decision-making.

Now, with the combination of the advanced prediction and the traditional knowledge like *Mitameshi*, we can foster local wisdom for wise land and water management and improvement of the regional environment. Such an approach may be effective in solving the problems caused by global climate changes resulting in living of humans in harmony with nature, which means living in a better anthroscape. Ultimately, the wisdom of Suido is to be sublimed to the context of "*ecosophy*".

Acknowledgements The concept introduced in this chapter was the products of the Project ICCAP (Impacts of Climate Change on Agricultural Production System in Arid Areas, 2002–2007) of RIHN (The Research Institute for Humanity and Nature, Kyoto, Japan) and the coordinated project of NIHU (National Institutes of Humanities, Japan) on people and water.

It was developed and refined in the process of editing "Reclaiming the *Suido*" (2009) of JSIDRE (Japanese Society of Irrigation, Drainage and Rural Engineering).

References

JSIDRE (2009) Reclaiming the *Suido* – Coordination of Wisdom (Chi no Renkan) (in Japanese)

Washitani I (2001) Reviving Eco-System (Seitaikei wo Yomigaeraseru). NHK-Books No. 916. (in Japanese), ISBN-13: 978-4140019160

Watanabe T, Kume T (2009) A general adaptation strategy for climate change impacts on paddy cultivation: special reference to the Japanese context. *Paddy and Water Environment*. DOI:10.1007/s10333-009-0179-5

Reconstructing the Past by Regenerating Biodiversity: A Treatise on Weed Contribution to Soil Quality at a Post-cultivation Succession

I. Çelik, K.T. Yilmaz, H. Eswaran, A. Mermut, M. Dingil, Z. Kaya, A. Demirbaş, I. Aksit, I. Ortaş, M. Gök, C. Akpınar, T. Nagano, N. Ae, Y.K. Koca, and S. Kapur

Abstract Plant communities associated with the agricultural landscape are gradually getting impoverished and even eradicated and replaced by poor species in many places of the world. A large number of studies are now available about the interactions between root and the soil system. However, influence of plants on aggregate stability remains to be a complex issue. The aim of this work was to determine the changes in the soil quality indicators that have developed at the rhizospheres under the selected weeds in the abandoned land. The results confirmed significant changes on bio-diversity of weed communities in a succession of over a decade. There were only 14 weed species in the abandoned field in the spring of 1998. The survey in the spring of 2008 confirmed the presence of 34 species. A decadal abandonment of the prime soil of the Mediterranean environment proved to increase the amount of the water stable aggregates and the mycorrhizal activity along with the accumulation of the organic and available nitrogen.

I. Çelik (✉), M. Dingil, Z. Kaya, A. Demirbaş, I. Ortaş, M. Gök, C. Akpınar, Y.K. Koca, and S. Kapur
Department of Soil Science, Faculty of Agriculture, University of Çukurova, 01330, Adana, Turkey
e-mail: icelik@cu.edu.tr

K.T. Yilmaz
Department of Landscape Architecture, Faculty of Agriculture, University of Çukurova, 01330, Adana, Turkey

H. Eswaran
United States Department of Agriculture, Washington, DC, USA

A. Mermut
Department of Soil Science, Faculty of Agriculture, University of Harran, Şanlurfa, Turkey

I. Aksit
University of Erciyes, Kayseri, Turkey

T. Nagano and N. Ae
Kobe University, Kobe, Japan

S. Kapur et al. (eds.), *Sustainable Land Management*,
DOI 10.1007/978-3-642-14782-1_17,

Keywords Long term experiment · Soil quality · Weeds · Biodiversity · Çukurova region

1 Introduction

1.1 The Weeds

A great number of plant communities associated with the agricultural landscape were impoverished or eradicated and replaced by species of poor woodland communities in many places of the world (Brakenhielm 2000). Spontaneous vegetation comprising the ruderal and segetal communities are extreme floristic drivers and consist of weeds. The term weed defines plants which grow spontaneously in a habitat that have been greatly modified by human action. Weeds or segetal plants cover a great number of species which occur in cultivated habitats, and are grouped into three categories, namely the xero-segetals, hydro-segetals, and the halo-segetals. A reference source on the subject matter, for the definition, classification and floristic composition, of the Middle Eastern segetal vegetation can be found in the review published by Zohary (1973).

In some Middle Eastern countries with relatively smaller floras, the segetal flora may amount up to 20% of the total number of the indigenous species. Not less than 5% of the indigenous plants occur year after year among crops or in fallow fields. Most of the families of the higher plants have representatives in the segetal flora. Outstanding among segetals are species of Cruciferae, Leguminosae, Umbelliferae, Compositae, and Graminae. About 80% of the segetals are annuals, whereas there are a series of woody plants among crops that could be termed as relic weeds.

Several studies revealed the change of plant communities occurring in post-cultivated successions. For example, in the savanna ecosystem the post-cultivation succession is initiated by the primarily establishing ruderal and annual herbaceous species of small size and of short life-span, which are later replaced by annual herbaceous species larger in size and with a longer life-span. Generally after 10 years, very competitive perennial grasses develop and become dominant (Devineau and Fournier 2007).

El-Sheikh (2005) reported that abandoned fields, called "old-fields," are appropriate sites for investigating the general theory of vegetation succession. According to this study, dense therophytic communities were established soon after abandonment, which rapidly changed in approximately 1–2 years, and the basic shifts in the species composition were almost completed within 25 years of abandonment at a site along the Nile valley. It was also stated, that the outcomes of successional studies can improve our understanding of secondary processes and provide a basis for strategic conservation and land management planning. A context of conservation management of the land may also be developed for future predictive assessments and mitigation of the consequences of ecological and human changes on cropland biota.

1.2 Studies on Soils

The chemical, physical, and biological interactions that occur between roots and the surrounding environment of the soil are stated to be the most complex processes. A large number of studies are now available about the interactions between roots and soil microbes, roots and soil fauna, roots and other roots, as well as signals relating the nature of local soil chemical and physical properties to nearby roots (McCully 1999). Recent soil aggregate stability studies vs. different plant rhizospheres were conducted with results suggesting that the growth and activities of living roots may be a major factor controlling the overall direction and magnitude of changes in aggregate stability under arable or ley crops (Reid and Goss 2006).

Intra-porosity together with pore and aggregate shape studies of water stable aggregates of long-cultivated dryland soils of the Middle East were conducted by Kapur et al. (2007), in order to determine the contribution of the long-term effects of legume and fallow field practices through the micromorphological studies. A pot experiment, was carried out to determine the development of soil structure using the relative importance of roots and AM-fungi, through a mycorrhizae-defective tomato mutant and a wild-type tomato (*Solanum lycopersicum* L.) (Hallet et al. 2009). Results of the experiment revealed that, although the AM fungi accentuated soil stability, changes were not linked to the fungal biomass, but to the presence of plants, regardless of AM fungi, which appear to have the greatest impact on increasing soil stability.

The effect of the phenolic compounds on aggregate stability has been highlighted by Osazawa et al. (1992), which was later complemented by organic nitrogen (N_{org}) studies (phosphate extracted organic nitrogen form compounds with phenolic derivatives) by Matsumoto and Ae (2004) and Matsumoto et al. (2000a, b). This study was conducted to monitor the secondary succession of plant communities (weeds), and the parallel soil quality increase via physical, chemical and micromorphological (sub-microscopic) properties of the dominant water stable aggregates (WSA) obtained from the weed rhizospheres. The aim of the study was to reconstruct the past natural vegetation by regenerating the biodiversity and determination of the parallel increase in soil quality within the root zone.

2 Materials and Methods

2.1 Materials

2.1.1 Location and Soils

Soil sampling was conducted at the spring of 2008 in the long-cultivated (about a century for traditional cereals) one hectare parcel of prime soil, widely distributed Arık soil series-Vertisol-Typic Haploxerert in the Çukurova Region, Southern

Turkey, abandoned since 1998 in the experimental plot of the Soil Science Department, University of Çukurova Farm. The two other presently cultivated adjacent parcels, one in the Field Crops experimental field (cultivated for about 40 years) and the other in the Soil Science experimental field (cultivated for about 20 years) of the same soil sampled for comparison of the rhizosphere characteristics.

The cultivated plots were fertilized and pesticides were applied in the regular farmer's style without using manure and burning the stubble. Compound NPK-fertilizers were applied in the seedbed at rates of 172 kg N ha^{-1} and 55 kg P_2O_5 ha^{-1} for wheat, 250 kg N ha^{-1} and 60 kg P_2O_5 ha^{-1} for maize at both plots. The major crops produced at these plots were wheat with maize planted once every 2–3 years. The stubble of the wheat and maize were chopped and mixed with the soil, in contrast to the widespread practice in the region, seeking to constantly increase the organic matter content by over-ground and especially below-ground biomass with dense root distribution, and creating a uniform cultivated layer for soil sampling.

2.1.2 Weeds

The weed species composition was determined within the abandoned field at the beginning of 1998 and at the end of the observation period in 2008.

2.2 *Methods*

2.2.1 Weeds

The weed colonization was studied at a representative 20 × 20 m plot within a one hectare abandoned field to identify the species composition of plant communities in the post cultivation succession pathway. Occurrence of each species was recorded according to the abundance/coverage values determined using Braun-Blanquet (1964). The root zones of the dominant annual (*Latryus cicera*, *Crepis sancta*, *Medicago scutella*, *Bromus* sp., *Ainsworthia trachycarpa*, *Fumaria densiflora*, *Sinapis arvensis*, *Phacelia tanacetifolia*, spontaneously regenerated at the same point location) and perennial weeds (*Ornithogalum orthophyllum*, *Sorghum halepense*, *Convolvulus arvensis*, *Echinops viscosus*) were selected for soil sampling and quality determination.

2.2.2 Soils

The approach for sampling the root zone soils of the abandoned field, for analyzing the water stable aggregates (WSA) as well as counting mycorrhizae spores and conducting physical and chemical analyses, was primarily based on the selection of the dominant weed within the widespread plant community (some forming hummocks on the surface) and carefully sampling the soil/aggregates clinging on the fine roots and the surrounding soil (Fig. 1).

Microstructure characterizations related to the root/hyphae distributions and orientations were also conducted via sub-microscopy (SEM) seeking to determine their morphologic/physical contributions within the aggregates (especially within the pore-spaces and to a lesser extent in the soil/aggregate matrix) of the root zones of the weeds. Water stable aggregates of the soil samples from the two adjacent long-cultivated fields of the same soil (sampled from the long-cultivated uniform surface horizon) were also determined for comparison with the weed root zones of the abandoned parcel.

The wet sieving method of Kemper and Rosenau (1986) was used to determine the mean weight diameter (MWD) as an index of soil aggregation of the soils collected from the rhizospheres of the selected weeds. A set of sieves of 4, 2, 1, and 0.5 mm diameters were used after the soil samples were passed through 8 mm. Approximately 50 g of the soil was put on the first sieve of the set and gently moistened to avoid a sudden rupture of aggregates. Once the soil had been moistened, the set was sieved in distilled water at 30 oscillations per minute. With 10 min of oscillation, the soil remaining on each sieve was dried, and then sand and aggregates were separated (Gee and Bauder 1986). The mean weight diameter was calculated as follows: $\mathrm{MWD} = \sum_{i=1}^{n} XiWi$ where, MWD is the mean weight diameter of water stable aggregates, Xi is the mean diameter of each size fraction (mm) and Wi is the proportion of the total sample mass in the corresponding size fraction after the mass of stones deducted (upon dispersion and passing through the same sieve). Ultimately the size fraction of 2.1 mm was selected for SEM analysis.

Soil organic matter was determined according to Nelson and Sommers (1996), total nitrogen according to Bremner and Mulvaney (1982), phosphate extractable

Fig. 1 Spring time weed community in the experimental field

organic nitrogen (PEON) by Matsumoto et al. (2000b). The descending order of the C/N ratios was used as the main variable in the tabulation of the results regarding the significance of the quality of the organic matter.

Spores of mycorrhizae were determined according to Gerdemann and Nicolson (1963) by isolation of wet sieving and decanting methods. The sub-microscopic (scanning Electron microscopy-SEM) analysis was conducted, at a Carl Zeiss EVO 40 apparatus coupled with a Bruker AXS Microanalysis-XFlash 4010 unit seeking to reveal the occurrence and formation of the aggregates at the root zones of selected/dominant weeds.

3 Results

3.1 The Weed Succession from 1998 to 2008

The results revealed significant changes on bio-diversity of weed communities in a succession of over a 10-year period, where the plant species were determined at a spring time. However, field surveys will continue at seasonal intervals in order to determine the annual cyclical changes of the plant communities for a long-term.

There were only 14 weed species in the abandoned field when the initial survey was carried out in the spring of 1998. This record was utilized as baseline data for comparative evaluation. The recent survey was performed in the spring of 2008, confirmed the presence of 34 species altogether.

To compare species composition of the abandoned field with the cultivated land, a contemporary survey was carried out in the neighboring wheat field. Eighteen weed species, comprising only one perennial (*Convolvulus arvensis*) which is a constant weed in the study site, were recorded in the wheat crop. This shows the state of the first year succession of the weed flora within the field. Although there is a floristic heterogeneity within the three weed communities recorded, six species fell into a high constancy class. These were *Sinapis arvensis*, *Convolvulus arvensis*, *Latyrus cicera*, *Senecio vernalis*, *Melilotus indica*, and *Fumaria densiflora*. A distinct group of pioneer weeds, which occur only in the first year succession in the wheat field, can be clearly seen in Table 1.

A total of 45 weed species were recorded within the abandoned experimental field allocated for wheat. They were distributed among 15 families, including 2 monocots. The most important families were Compositae (14 species) and Leguminosae (9 species). Umbelliferae (4 species), Graminae (3 species), Cruciferae (3 species), Papaveraceae (2 species), and Liliaceae (2 species), while the remaining families had only one species.

The initial colonization by short-lived non-woody vegetation (except only one chamaephyt and one hemicryptophyt) is replaced by a relatively diverse herbaceous community with longer-lived species. Results show that almost all of the species of the initial community in the abandoned field were constant, whereas additional one chamaephyt, one hemicryptophyt and two geophytes were associated with the recent

Table 1 Occurrence, abundance/coverage values, families, and plant life forms of weeds recorded in the observed plots

Plant taxa	Number of taxa			Plant family	Plant life form
	Abandoned field		Wheat field		
	Spring 1998	Spring 2008			
Total: 45	14	34	18		
Ainsworthia trachycarpa	2	2		Umbelliferae	Th (A)
Medicago orbicularis	r	+		Leguminosae	Th (A)
Vicia angustifolia	r	+		Leguminosae	Th (A)
Bromus sp.	+	2		Graminae	Th (A)
Echinops viscosus	3	+		Compositae	H (P)
Sonchus oleraceus	+		+	Compositae	Th (A)
Anthemis cotula	+			Compositae	Th (A)
Senecio vernalis	+	+	+	Compositae	Th (A)
Sinapis arvensis	1	+	1	Cruciferae	Th (A)
Melilotus indica	2	+	+	Leguminosae	Th (A)
Latyrus cicera	+	+	+	Leguminosae	Th (A)
Convolvulus arvensis	+	1	1	Convolvulaceae	Ch (P)
Fumaria densiflora	r	+	3	Papaveraceae	Th (A)
Crepis sancta	r	2		Compositae	Th (A)
C. aspera		+		Compositae	Th (A)
C. paludosa		+		Compositae	Th (A)
Veronica sp.		+	1	Scrophulariaceae	Th (A)
Anagallis arvensis		+	1	Primulaceae	Th (A)
Alopecurus myosuroides		+		Graminae	Th (A)
Sorghum halepense		+		Graminae	H (P)
Ornithogalum orthophyllum		+		Liliaceae	G (P)
Allium sp.		r		Liliaceae	G (P)
Prosopis farcta		+		Leguminosae	Ch (P)
Trifolium sp.		+		Leguminosae	Th (A)
Cornilla scorpioides		+		Leguminosae	Th (A)
Vicia sp.		2		Leguminosae	Th (A)
Daucus guttatus		+		Umbelliferae	Th (A)
D. carota		+		Umbelliferae	Th (A)
Ammi majus		r		Umbelliferae	Th (A)
Geranium dissectum		+		Geraniaceae	Th (A)
Ranuculus arvensis		r		Ranunculaceae	Th (A)
Helminthoteca echioides		+		Compositae	Th (A)
Cichorium intybus		+		Compositae	Th (A)
Hedypnois cretica		+		Compositae	Th (A)
Myagrum perfoliatum		+		Cruciferae	Th (A)
Aster squamatus		+		Compositae	Th (A)
Poligonum aviculare			r	Poligonaceae	Th (A)
Trifolium sp.			+	Leguminosae	Th (A)
Lamium amplexicaule			1	Labiatae	Th (A)
Papaver rhoeas			r	Papaveraceae	Th (A)
Chenopodium album			+	Chenopodiaceae	Th (A)
Capsella bursa-pastoris			+	Cruciferae	Th (A)
Crysanthemum coronarium			+	Compositae	Th (A)
Calendula arvensis			+	Compositae	Th (A)
Carduus tenuiflorus			r	Compositae	Th (A)

Ch chamaephyt, *H* hemicryptophyt, *G* geophyt, *Th* therophyt, *A* annual, *P* perennial
Relative coverage 3: 75–51%; 2: 50–25%; 1: 25%; Abundance +: abundant; r: rare

community. These are *Prosopis farcta* (characteristic for segetal communities in the Middle East), *Sorghum halepense*, *Ornithogalum orthophyllum*, and *Allium* sp. However, the dominant species in this community were the therophytes (39 species) followed by hemicryptophytes (2 species), chamaephytes (2 species) and geophytes (2 species). The determination of the plant species of the abandoned field revealed a noteworthy change in the composition of the weed communities in the 10-year period. This change represents a net increase in the number of species (from 14 to 34), and the number of perennials within the weed community. In regard to the life form spectra, the numbers of perennials have increased from the first year succession to a 10-year period (2 spp. in 1998 and 6 spp. in 2008) respectively.

3.2 Physical and Chemical Properties of the Water Stable Aggregates (WSA) Within the Weed Rhizospheres

The mean weight diameter (MWD) distribution of the water stable aggregates (WSA) collected from the rhizospheres of the widely distributed plant communities revealed some significant differences (Fig. 2 and Table 1), and did not correlate with the organic matter, total nitrogen contents and the C/N ratios of the abandoned

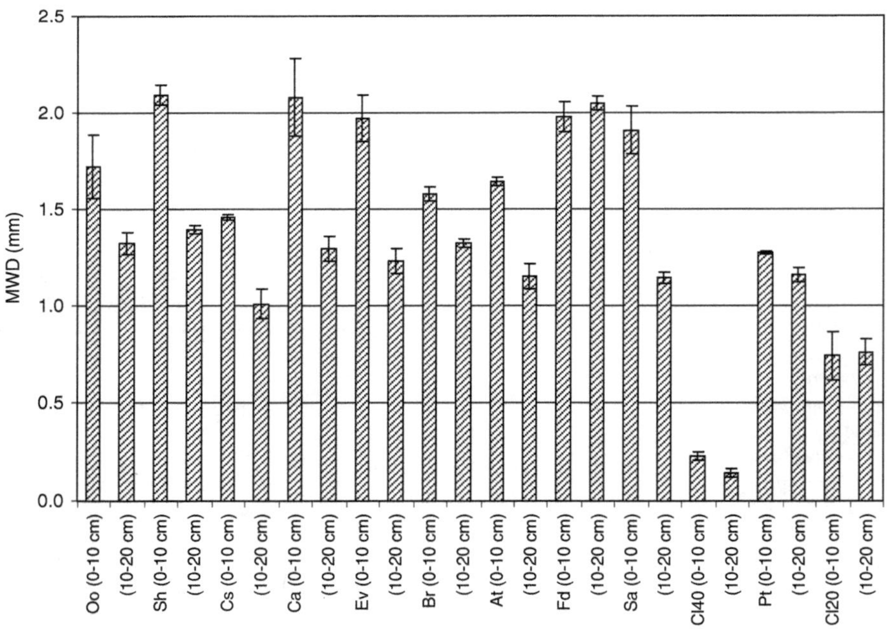

Ornithogalum orthophyllum **(Oo)**, Sorghum halepense **(Sh)**, Crepis sancta **(Cs)**, Convolvulus arvensis **(Ca)**, Echinops viscosus **(Ev)**, Bromus sp.**(Br)**, Ainsworthia trachycarpa **(At)**, Fumaria densiflora **(Fd)**, Sinapis arvensis **(Sa)**, Cultivated land **(Cl40)**, Phacelia tanacetifolia **(Pt)**, Cultivated land **(Cl20)**.

Fig. 2 Water stable aggregates of the weed rhizosphere soils site, but inversely correlated with the increasing amounts of PEON (Fig. 3). This may indicate that the increase in aggregate stability could retard the microbial activity, thus causing the decrease in the amounts of PEON. The abrupt decrease in the WSA of the two cultivated sites has documented the well known fact of soil structure degradation due to the increasing tillage operations. The size of organic molecule was larger (Table 2) in the 40-year cultivated field than the 20-year

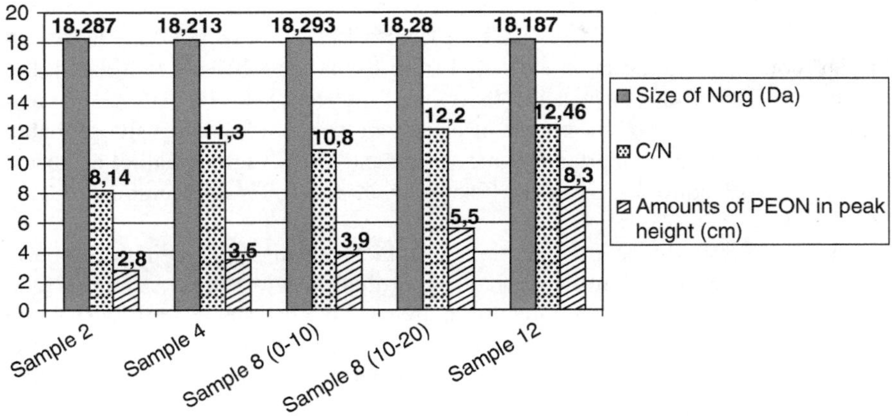

Fig. 3 Trends of some physical and chemical properties of selected water stable aggregates of weed rhizospheres and of a cultivated field. Samples 2, 4, 8 and 12 are aggregates of the *Sorghum halepense*, *Crepis sancta*, *Fumaria densiflora* and the cultivated land for 20 years respectively

Table 2 Some chemical analyses of the water stable aggregates (WSA)

Weed species	MWD (mm)	Number of mycorrhizal spore (10 g soil)	Organic matter (%)	Organic C (%)	N (%)	Size of organic N molecule in Dalton (Da)	C/N
Sorghum halepense	2.2	25	1.97	1.14	0.14	18.287	8.14
Convolvulus arvensis	2.1	38	1.94	1.13	0.10	18.213	11.30
Fumaria densiflora (10–20 cm)	2.1	65	1.85	1.08	0.10	18.293	10.80
Fumaria densiflora (0–10 cm)	2.0	35	2.10	1.22	0.10	18.280	12.20
Crepis sancta	1.5	13	1.90	1.10	0.07	18.287	15.71
Cultivated land							
20 years (Soil Science Dept. Exp. Field)	1.0	18	2.33	1.62	0.11	18.187	12.46
40 years (Field Crops Dept. Exp. Field)	0.2	16	2.1	1.22	0.12	18.230	10.16

cultivated one and this was most likely due to the accumulation of undecomposed large organic molecules in the longer time period.

The mycorrhizal communities (number of spores) fluctuated with the increase in organic matter and N contents and did not correlate to the WSA (Table 2). This most probably was due to the differences in the contributions of the root zone processes, bound to the microbial activities, of the different weed species. However, the C/N ratio standing as the primary indicator for the humus/organic matter quality, was the lowest at the rhizospheres of the *Sorghum halepense* (Sh) and *Fumaria densiflora* (Fd) (10–20 cm), with PEON extracted N_{org} higher in the Sh (compared to the Peon extracted from the 20 and 40-year cultivated, the Fd and the Convolvulus arvensis WSAs), and highest in the Fd (10–20 cm) and total N% Highest in Sh and higher in Fd among the weeds, reflect the probable high amount of organic matter decomposition and likely high uptake of adsorbed N_{org} from soil colloidal surfaces (Okamoto et al. 2003; Matsumoto and Ae 2004).

The WSA level was determined highest in the Sorghum and high in the Fumaria rhizospheres pointing out to the probable contribution of the high content of total organic nitrogen to the development of aggregate stability and N fertility (Numan et al. 1998). The uniform and high molecular weights (around 18.000 Da) of the N_{org} of the studied WSAs of the weed rhizospheres were most likely the protein-like compounds forming from rapidly decomposing organic matter to very uniform organic entities of complex proteins and high amounts of humic acids as compared to the 800–9,000 Da level, as stated to be the limit of fulvic acid content by Schactschabel et al. (1998).

The uniformity of the size of the organic molecules of the WSAs of the experimental field soils with minor variations also shows the absence of the spatial variability of the field composed of a single soil series. The N_{org}-the available N fraction in the soil is stated to be finally adsorbed through Al and Fe on clay mineral surfaces with high surface areas. [The soil of the experimental field was determined to contain high amounts of clay size particles and dominant smectite, Miltner and Zech 1999 (Table 2)]. Thus, as commonly known, the higher humic acid contents (low C/N ratio) together with the higher amounts of $N_{org,}$ determined in the study, are the indicators of the higher soil organic/chemical quality and fertility attained after the abandonment of the land to the natural vegetation.

The highest and the lowest amounts of PEON were determined in the WSAs of the soils of the 20-year cultivated field and the Fd rhizospheres respectively, which seem to indicate the effect of cultivation that increases the amount of PEON by increasing microbial activity (Fig. 4) and in turn decreasing the size o the N_{org} molecule (Fig. 4c). The other rhizosphere WSAs did not yield any relationship indicating the complexity of the root zone processes. The sizes of the N_{org} molecules of the cultivated fields were lower than the N_{org} molecule sizes of the weeds (except the N_{org} molecules of the WSAs of the Convolvulus arvensis root zone), also indicating an increase in microbial activity (Table 1).

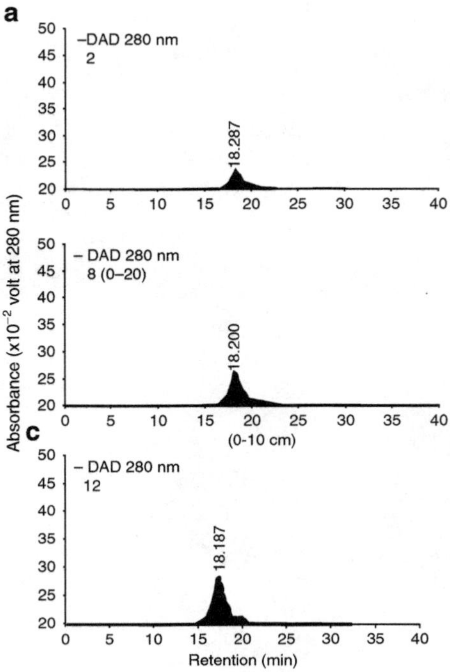

Fig. 4 (**a**) Norg content of *Sorghum halepense* in rhizosphere of WSA. (**b**) N_{org} content of *Convolvukus arvensis* in rhizosphere of WSA; (a: 0–10 cm), (b: 10–20 cm) N_{org} content of *Fumaria densiflora* in rhizosphere of WSA; (**c**) N_{org} content of the WSA of soil of 20-year cultivation

3.3 Micromorphological (Sub-microscopic) Properties of the WSA

The SEM images revealed the development of mycorrhizal hyphae in the decomposed root cavities of the WSA from the *Sorghum halepense* rhizosphere, by partly coating the inner walls of the lignified oval cells (Fig. 5), whereas, lignified and/or highly decomposed fine roots in the WSAs of the *Convolvulus arvensis* rhizosphere were covered and/or impregnated by clay mineral laths (Fig. 6). Highly decomposed *Fumaria densiflora* fine root residues are incorporated into the matrix with high shrink-swell capacity of the dominant smectite clay mineral (Fig. 7). Partly decomposed residues of fine roots (root envelope) and fresh hyphae seem to induce the development of rounded smaller aggregates, within the matrix of the WSA, probably forming from excretions of the mycorrhizae populations in the 20-year cultivated field (Fig. 8).

Despite of vague presence of the mycorrhizal features and the root residues determined in the root zone of the 20-year cultivated field, Fig. 9 reflects, the uniformity of the soil matrix attained by the long cultivation practiced in the 40-year

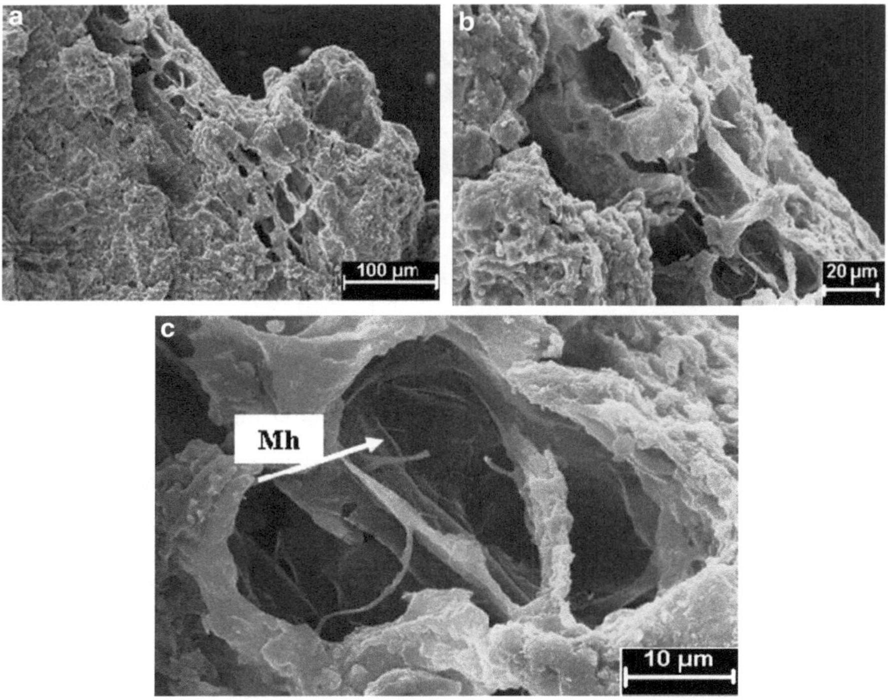

Fig. 5 Mycorrhizal hyphae (Mh) in decomposed root cavities of the WSA from the *Sorghum halepense* (same feature in order of increasing magnification (**a–c**))

field and the absence of the mycorrhizal hyphae and decomposed plant remains. Given both fields, occurring on the same soil type/series, the absence of these features at the 40-year cultivated field may be attributed to the longer period of cultivation.

4 Conclusions

The increase in the weed species within the 10-year period of abandonment of the Ark soil has confirmed the importance of the natural succession of the soil and vegetation communities. As expected for the Mediterranean environments, the dominant families in the systematic spectrum were Compositae and Leguminacae.

The root zones of the selected weed species revealed significant differences in WSA, namely between the weed and the cultivated field aggregates. The highest WSA was determined in the root zones of the *Sorghum halepense* and the *Convolvulus arvensis*, whereas the lowest were determined at the 20 and 40-year cultivated fields, indicating the structural degradation of the soil by agricultural practices.

The SEM images, which reveal, the mycorrhizae hyphae coating pores/void spaces of plant residues and the decomposed *Convolvulus arvensis* root remains with clay aggregate linings/coatings manifest their contribution to the stability of

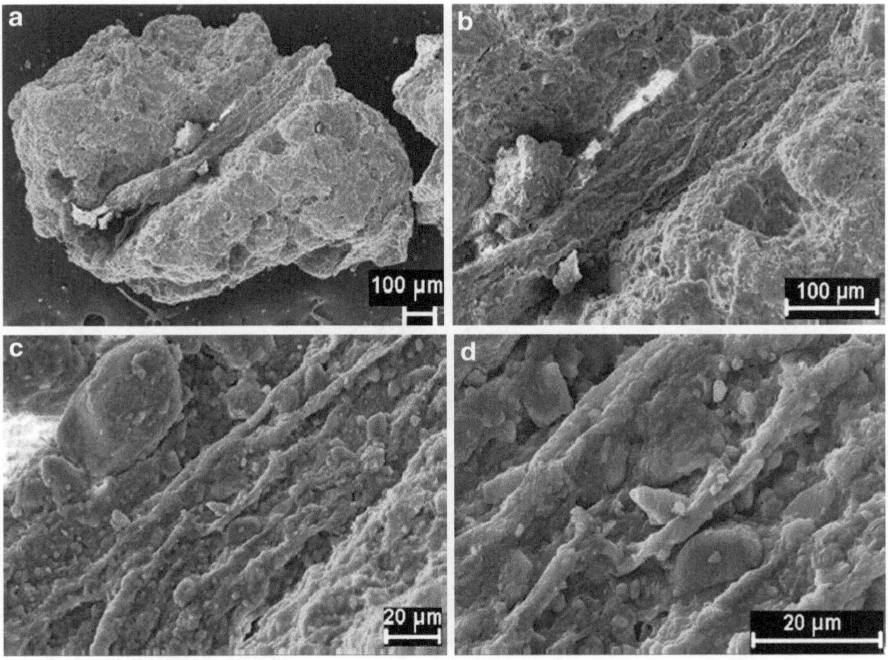

Fig. 6 Lignified highly decomposed fine root, coated by clay aggregates, in the WSA of the *Convolvulus arvensis* rhizosphere (same feature in order of increasing magnification (**a**–**d**))

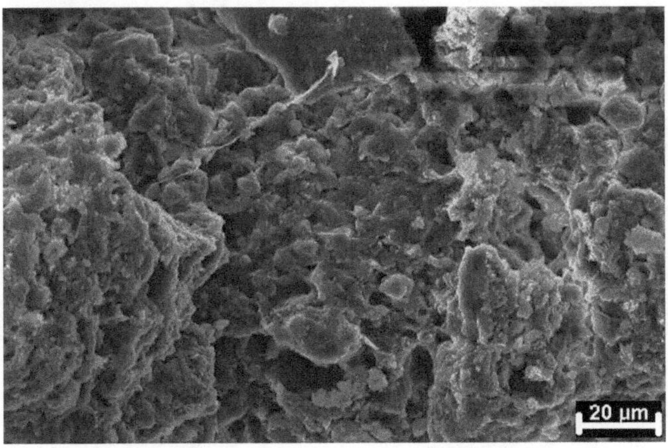

Fig. 7 Shrink-swell features of *Fumaria densiflora rhizosphere* WSA

the microstructural units. The shrink-swell reshaped matrix, of the smectitic clay minerals, inducing the formation of the capillary channels/cracks/vertical pore spaces is the specific contribution of the *Fumaria densiflora* root zone activity.

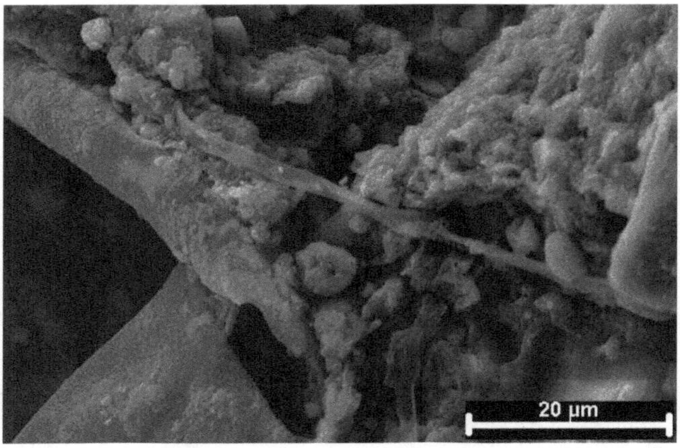

Fig. 8 Partly decomposed residues of fine roots and fresh hyphae in WSA of cultivated land (20 years)

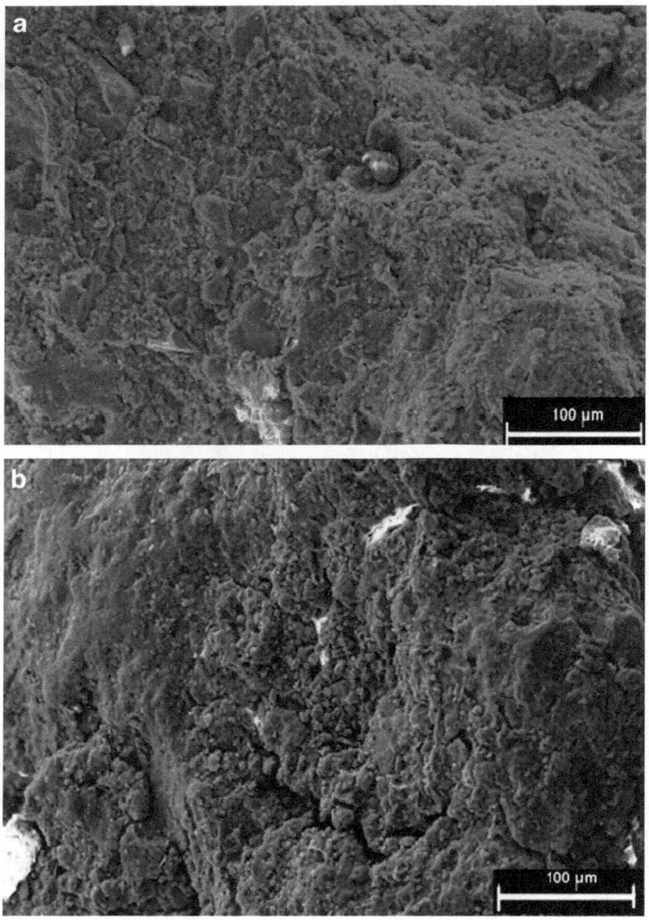

Fig. 9 The uniform soil matrix of the 40-year cultivated field

The fresh/slightly decomposed plant residues of the 20-year cultivated field, void of strongly decomposed root residues, along with the longer tilled 40-year cultivated field root zone aggregates with a uniform matrix, would most likely stand for the long term degradation effect on land, exerted by the agricultural practices. The largest N_{org} molecule was determined in the root zone of the *Fumaria densiflora* (0–10 cm depth) followed by *Sorghum halepense* and the *Convolvulus arvensis*, where the smallest N_{org} was of the 20-year cultivated field, probably indicating the effect of tillage on the high rate decomposition of the organic matter. The C/N ratio revealed a parallel increase with PEON documenting the close relation between the total N and the N_{org} extracted by the phosphate buffer standing for the plant available N.

A decadal abandonment of the prime soil of the Mediterranean environment proved to increase the amount of the water stable aggregates and the mycorrhizal activity along with the accumulation of the organic/available nitrogen.

References

Brakenhielm S (2000) Plant succession on afforested farmland. Proceedings of the IAVS Symposium, 41, 63–66

Braun-Blanquet J (1964) Pflanzensoziologie. Springer, New York, 865pp

Bremner J, Mulvaney C (1982) Nitrogen-total. Chemical and microbiological properties. ASA and SSSA, Madison, pp. 595–624

Devineau JL, Fournier A (2007) Integrating environmental and sociological approaches to assess the ecology and diversity of herbaceous species in a Sudan-type savanna (Bondoukuy, western Burkina Faso). Flora, 202(5), 350–370

El-Sheikh MA (2005) Plant succession on abandoned fields after 25 years of shifting cultivation in Assuit, Egypt. Journal of Arid Environments, 61(3), 461–481

Gee GW, Bauder JW (1986) Particle size analysis. In: Klute A (Ed.), Methods of soil analysis. Part 2, 2nd Edition. Agronomy Monograph 9. ASA and SSSA, Madison, pp. 383–411

Gerdemann JW, Nicolson TH (1963) Spores of mycorrhizal endogamy species extracted from soil by wet sieving and decanting. Transactions of the British Mycological Society, 46, 235–244

Hallet PD, Feeney DS, Bengough AG, Rillig M, Scrimgeour CM, Young IA (2009) Disentangling the impact of AM fungi versus roots on soil structure and water transport. Plant and Soil, 314, 183–196

Kapur S, Ryan J, Akca E, Çelik İ, Pagliai M, Tulun Y (2007) Influence of Mediterranean cereal-based rotations on soil micromorphological characteristics. Geoderma, 142, 318–324

Kemper WD, Rosenau RC (1986) Aggregate stability and size distribution. In: Klute A (Ed.), Methods of soil analysis. Part 1. Physical and mineralogical methods, 2nd Edition. Agronomy Monograph 9, Soil Science Society of America, Madison, pp. 425–442

Matsumoto S, Ae N (2004) Characteristics of extractable soil organic nitrogen determined by using various chemical solutions and its significance for nitrogen uptake by crops. Soil Science and Plant Nutrition, 50(1), 1–9

Matsumoto S, Ae N, Yamagata M (2000a) The status and origin of available nitrogen in soils. Soil Science and Plant Nutrition, 46(1), 139–149

Matsumoto S, Ae N, Yamagata M (2000b) Extraction of mineralizable organic nitrogen from soils by a neutral phosphate buffer solution. Soil Biology and Biochemistry, 32, 1293–1299

McCully ME (1999) Roots in soil: unearthing the complexities of roots and their rhizospheres. Annual Review of Plant Physiology and Plant Molecular Biology, 50, 695–718

Miltner A, Zech W (1999) Microbial degradation and resynthesis of proteins during incubation of beech leaf litter in the presence of mineral phases. Biology and Fertility of Soils, 30, 48–51

Nelson DW, Sommers LE (1996) Total carbon, organic carbon, and organic matter. In: Sparks DL (Ed.), Methods of soil analysis. Chemical methods. American Society of Agronomy, Madison, pp. 961–1010

Numan N, Morgan MA, Herlihy M (1998) Ultraviolet absorbance (280 nm) of compounds release from soil during chloroform fumigation as an estimate of the microbial biomass. Soil Biology and Biochemistry, 30, 1599–1603

Okamoto M, Okada K, Ae N (2003) Growth responses of cereal crops to organic nitrogen in the field. Soil Science and Plant Nutrition, 49, 445–458

Osazawa S, Ae N, Hasegawa S, Kobuta T (1992) The effect of phenolic compounds on soil aggregate stability. 6th International Soil Conservation Conference of the International Soil Conservation Organisation (ISCO). In: Hurni H, Kebede T (Eds.), Erosion, conservation, and small-scale farming, Ethiopia and Kenya, 1989, November 6–18 , Bern: Geographica Bernensia ISCOWASWC, pp. 273–284

Reid JB, Goss MJ (2006) Effect of living roots of different plant species on the aggregate stability of two arable soils. European Journal of Soil Science, 32(4), 521–541

Schactschabel P, Blume, HP, Brümmer G, Hartge KH, Schwertmann U (1998) Lehrbuch der Bodenkunde. Vol. 14, Auflage, Ferdinand Enke Verlag, Stuttgart, 494pp

Zohary M (1973) Geobotanical foundations of the Middle East. Vol. 1–2, Gustav Fischer Verlag, Stuttgart, 738pp

Economic and Social Impact of the Degraded Antroscapes

Nejat Erk

Abstract The study at hand started with the modest goal of discussing economic and social implications of degraded Antroscape, for the upstream, transitory and downstream zones. Secondarily, the study aims to set up the necessary tools which can be used in the assessment of degraded Antroscape and also in developing policy tools for sustainable optimal solutions. Recent literature on degraded antroscape lacks perspectives with alternative approaches which could complement to the efforts of soil engineers, environmentalists as well as economists. While, the work at hand incorporates economic growth perspective with sustainable land use; alternative measurement techniques are also discussed where each fulfills the need for a holistic and integrative approach in quantifying assessment and solutions. Current literature require further alternative approaches of handling a complex issue like environment that its inhabitants sharing and degrading the existing antroscape.

Keywords Cost of degradation · Degraded anthroscapes · Quantification of environmental degradation · Sustainability

1 Introduction

If the sustainable economic development path can be defined as an overall development path where the choices for today's generation are growing without limiting the choices of future generations, global resource issues asks for an overlapping generations decision making.

In most societies broadening income differences between traditional and modern sectors, have to be resolved for the sustainability of resource issues as well as social ones. This brings in an unresolved debate on the economic growth rates of the developed and developing world. For decades economists wondered and tried to

N. Erk (✉)
Department of Economics, University of Çukurova, 01330 Balcalı, Adana, Turkey
e-mail: erk@cu.edu.tr

S. Kapur et al. (eds.), *Sustainable Land Management*,
DOI 10.1007/978-3-642-14782-1_18,

find a rationale why high income countries grow slower than low income countries. Although there are diminishing returns and convergence theories which tries to fill in the gap, yet no consensus has even been achieved. This part of the Anthroscape study attempts to explore whether the following assessment tells us more than a statistical observation. Looking at an historical perspective, resource sustainability seems to be more discussed after the late 1970s. But with the beginning of the twenty-first century global warming, water resource availability and similar issues became far more of a hot topic for the global world.

Turning back to the global economic growth rate disparities, which have been well known since the 1950s, could have been an indicator showing the pressures on resource use and resource sustainability. Moving from this fact, as the economic growth rate increases which in time leads to additional consumption and resource absorption levels, puts a pressure on sustainability of the structure by restricting future economic growth rates.

The main factor that concealed this interaction could be that, we have looked at the economic growth issue more from the demand dimension, while the critical issue stemmed from the supply side. In a historical perspective, where frontier advantage concealed supplies restrictions, most of us believed that economic growth also ensures resource supply. While that fact was, supply of a good is demand driven while the resources used in the supply process is an issue of sustainability. Very common in economic thinking, we had an overall misunderstanding of the short-run vs. the long run. This could be an explanation why more and more economists have intensified their research on strategic issues. If an important element of strategy is taking

$$A = TP = f(L, K)$$

or

$$A = TP = f(L)$$

where K is fixed

$$TP/L = AP = f(L)/L = h(L).$$

Therefore,

$$A = AP.L = h(L).L.$$

Hence,

$$MP = d(TP)/dL = h(L) + h(L) + Lh'(L).$$

Or in more open format,

$$A = TP = f(L) = -\ L3 + XL2 - YL + Z,$$

$$A'(L) = MP = 3L2 + 2XL,$$

$$\mathrm{A}''(L) = 6\mathrm{L} + 2\mathrm{X},$$

$$\mathrm{A}''(L) = 0.$$

Therefore,

$$\mathrm{L}<2 \quad \mathrm{F}''(L) > 0,$$

$$\mathrm{L}<2 \quad \mathrm{F}''(L) > 0,$$

the economic growth rate vs. GDP would reveal the trend below.
Economic growth rate

$$A = \overline{TP} = f(L, K)$$

or

$$A = TP = f(L) \text{ where } K \text{ is fixed}$$

$$\frac{TP}{L} = AP = \frac{f(L)}{L} = h(L).$$

Therefore,

$$A = AP.L = h(L).L.$$

Hence,

$$MP = \frac{d(TP)}{dL} = h(L) + h(L) + Lh'(L).$$

Or in a more open format;

$$A = AP = f(L) = -L3 + XL2 - YL + Z,$$

$$A'(L) + MP = 3L2 + 2XL,$$

$$A''(L) = 6L + 2X,$$

$$A''(L) = 0.$$

Therefore

$$L < 2 \quad F''(L) > 0,$$

$$L < 2 \quad F''(L) > 0.$$

The economic growth rate vs. GDP would reveal the trend below.

2 Brief History of the Traditional Economic Growth Theory

Starting with Adam Smith, Thomas Malthus, David Ricardo and more recently Frank Ramsey, Allyn Young, Joseph Schumpeter, Frank Night, E. Domar contributed to the economic theory in terms of competiveness issues and the magnificent dynamics of market equilibrium. One thing that is common to all scholars is a key ingredient of economics known as diminishing marginal returns. Although classical economists look at market issues solely from a supply view of point, they have always questioned the sources of economic growth and the rules for competitiveness (Smith 1776; Malthus 1798; Ricardo 1817; Ramsey 1928; Young 1928; Schumpeter 1934; Domar 1946).

But the modern origin of the economic growth theory relies on the neoclassical economic thought which takes us to the late 1950s. This is another turning point for more mathematical contributions, the economic theory and to the theory of economic growth. To be able to capture economic foundations for 1950s we have to go back to the 1920s capturing Ramsey and Fisher which look at consumption patterns, business cycles and the optimality conditions for economic growth. Nonetheless, Ramsey's conditions for economic growth have not even been accepted as a contribution since the late 1960s. Following a Keynesian tradition, Harrod and Domar implemented Keynesian dynamics to economic growth. In this respect aggregate production functions has been adopted arguing that the great depression in 1929 could have been avoided with the appropriate government interference to economic life. This modeling invited several scholars contributing on the topic, but nonetheless more discussion on similar arguments is also ruling today's way of thinking. Following the chronological pattern Solow and Swan, imposing constant returns to scale and diminishing marginal returns showed the interaction between production and savings to capture the general equilibrium in an economy. This contribution is also well known as the convergence theory where low per capita GDP countries grow at a faster rate with respect to high per capita countries under a steady state (Ramsey 1928; Fisher 1930; Harrod 1939; Domar 1946; Solow 1956; Swan 1956).

Although we are going to be more focusing on this unresolved convergence debate through the rest of the chapter to have a continuity of thought, we will continue with the chronological developments in economic growth theory. Solow and Swan, simply argue that technology is the main source for economic growth where growth in per capita GDP ceases if technological improvements stop. Tracing the historical events towards economic growth show very common features carried from 1600s from the real classical time. Thomas Malthus and David Ricardo as in the case of Solow and Swan, restrict economic growth to technological improvements (Solow 1956; Swan 1956; Ricardo 1817; Malthus 1798).

Collapse of neoclassical growth thought occurred in the early 1960s, after capturing technological improvements not being an exogenously determined growth factor. Economists of that age argued that profit motive and market dynamics induces technological improvements, thus technology is no longer a random issue causing large fluctuations in per capita GDP. Cass and Koopmans interlinked saving rates and consumer optimization behavior in determining the long run per capita growth rate (Cass 1965; Koopmans 1965).

Given arguments related to technological improvements is not very easy to handle due to the fact that beyond supply factors market structure factors such as market concentration rate has a key role on the speed of technological improvements. Once believed that a very competitive market leads to a desired optimality collapsed with the stagnating economic growth stemming from zero economic profit rates in the long run. Up to modern times almost no economist questioned whether the pace of resource usage could be recapitulated by productivity growth. Even under these circumstances, green revolution of 1960s and 1970s did not help much in terms of sustainability of economic growth. Most of us thought that artificial inputs could be a source of sustainability in terms of nutrition, technological advances and income growth. Arrow discusses the externalities of production and technology assuming that knowledge is non-rival. These arguments are still under debate with Romer where extensions from competitive markets to imperfect competition changes income patterns and income growth (Arrow 1962; Romer 1987, 1990).

The 1960s being prosperity years and the non-measurable reflections of economic growth theory led to the diminishing interest in economic growth theory during the 1970s. A revival of the economic growth theory came back with Romer and Lucas. These economists captured the importance of business cycle theory; supply and demand originated asking for governmental active economic policies to damp the size of the economic fluctuations (Romer 1986; Lucas 1988).

Economic growth and resource availability had been within the main agenda of sustainability issues for the last decade. Looking carefully at the sustainability of economic growth, the last 50 years had been spent looking at whether technological developments or improvements can make up for the relative resource shortage. Recent performance shows that technological lead is not the constraint about the resource shortage or in fact became another constraint. For several decades development theorists believed that as long as the technological innovation catches up with resource depletion economic growth becomes sustainable.

But we neglect the fact that technological improvements face difficulties at different levels. This is especially important at a development level. In fact the issue is not the foundations of technology creation but an issue known as the absorption and diffusion of technology. Thus availability of a technology is not the answer for sustainability but on the contrary as the speed of technological development increases its phase of change diffusion and absorption of that technology becomes far more difficult. This reminds me about my conference visit program to Frankfurt where we had to take a train back to our hotel at one o'clock in the morning. Having no train ticket personnel we ended up using the ticket machine by inserting two destinations and inserting the money. At that moment we saw three elderly ladies in a panicky mood not knowing how to buy a ticket from the ticket machine. This is a very good example for the problems faced at the level of diffusion of technology. This is another way of saying that creation of a technology will not mean much if we cannot diffuse or absorb it. As we all know diffusion of a technology simply means how wide spread is the available technology while absorption of a technology shows how much in depth the society is exhausting the technology.

All above arguments show that basic technology availability could have created no difficulties for all society segments to absorb and diffuse in the past not hindering economic growth burdened by resource exhaustion. But fast technology change era (1980 to present) broke up the link between technology itself and different segments of the society. It is an inevitable fact that diffusion and absorption is a function of educational attainment besides the existing cultural norms in that society. Looking at the rural society we can divide the society as upstream (least technological diffusion and absorption) transition zone (partially aware of diffusion and absorption of technology) and the downstream society having a higher educational attainment thus having the least difficulty in the diffusion and absorption of technology. One more time we witness the fact that human capital factor is in fact the main determinant of economic growth. Looking at the growth issue intuitively, whether the growth model is endogenous or not all three determinants of economic growth, technology, fixed capital increases and human capital improvements all can be translated with a common the nominator, human capital. R&D without a human capital dimension, fixed investment increases to be sustainable also asks for human capital contributions.

May be this is why overall national educational attainments are under criticism in terms of years and quality dimensions. There are few countries where university teaching is made out of lecturing of professors. Active learning (student centered learning) prepares students for lifelong learning and independent research. But again upstream and transitory zone lacks such skills leading to income and knowledge base differences. This could be a very sound explanation of income inequalities in the globalized world, while overall national GDP's are increasing. A similar study conducted also compliments our observations on, economic growth.

At this stage, we believe that the role of diffusion and absorption of diffusion needs further to be underlined. Existing learning paths simply shows us that without diffusion and absorption of technology, innovation at different levels stagnates and in fact as in the case of real business cycles limits economic development or economic growth. Although there are cases where the market system solves the diffusion and absorption problem, there are cases for state-directed diffusion of technology (Pomfret 2002). This literature raises the question whether intended diffusion of technology can be premature. If the technology diffusion had been too rapid, the mechanization continued due to difficulties in diffusion. This observation is applicable for all levels of manufacturing as well as agriculture (Sassenrath et al. 2008). Thus looking at resource and environmental issues just in terms of existing resources will be a fallacy if the end users are not exhausting the technology to its limits (absorption) while at the same time how much spread there is of the technology (diffusion) will determine the balance between existing resources and demand for these existing resources. Maybe this might be a rationale for a new definition of sustainability. If absorption or diffusion of technology is a function of education, increasing high unemployment levels informs us that resource availability education attainment spread and depth of technology will be the determinant of economic growth for the future. The level of difficulty in terms of absorption and diffusion is directly linked to the pace of technological change in that sector.

Osama Michail argues that adaptation of technology in the field of clinical technologies asks for lifelong learning (LLL) to assess the level of productive capabilities. In this sense, cost of premature adaptation of technologies which simply means substantially delayed adaptation is as important as resource constraints (Mikhail et al. 1999). In the past when pace of change did not create any difficulties breakthrough invention and innovations were taking place in a lengthy time period. This allowed end users to absorb and diffuse technology in a considerably long time period nurturing productivity gains. This asks for management of technology which is far more in depth covered under the heading of knowledge management. Apart from the technical dimension, following criteria's strongly influence technology adaptation (s.258). (1) And these are regulatory requirements, (2) technological challenges, (3) degree of benefit perceived by end users, (4) nature of trials to improve benefit and restrict complications, (5) payback period of the technology and ease of use. This is true for all high technology and traditional production techniques while it is also applicable to the service sector like health and research. Following the route for computer usage, we see that ease of use is as important as technological limitations.

3 Back to Convergence in Economic Growth

Previous research conducted by Erk et al. (1998) shows that the complexity of a technology does not guarantee that productivity levels can be nurtured in short time intervals. It is possible that nations with relatively backward technologies can compete with nations with higher patent attainments due to the fact that their education attainment matches with physical capital requirements within a sector or an economy. The common argument why developing nations grow faster than economically advanced countries could be explained by this paradox. Examining the existing education and training techniques simply shows us that high level technology is in no way a guarantee for faster economic growth or sustainable production levels if the education level (primary, secondary, tertiary) is not appropriate for faster economic growth. Erk et al. (1998) shows us that this is not consistent solely for developing nations but also consistent for developed ones.

The above diagrams show us that, the level and trend-wise capital stock growth in the developing world, grows faster than the developed world along with a higher relative human capital.

3.1 The Antroscape and Environmental Externalities

Griffin and Steele, stated that, the external costs exist when "the private calculation of benefits or costs differs from the society's valuation of benefits or costs" (Griffin and Harry 1986). This simple observation shows us that there are wealth transfers among individuals, regions and countries. These are areas where market failure dominates in the field of economics. It is very unfortunate that, externalities have been captured as an chapter where positive and negative externalities are given without considering its impact on relative wealth, relative income, relative economic growth rates and whether this transfer issue is sustainable or not. Turning back to the externalities issue, the most prominent reflection of "externalizes existence" brings in a problem where the price mechanism cannot handle. For example, the western world's export of pollutant chemicals to Africa cannot be rationalized by the export incomes of African nations. The logic is very simple, although an special import activity increases the national incomes as in the case of exports of the African countries, unforeseen underground clear water loss, loss of agricultural output, change in forestry and climatic changes that follow is the real cost of the income received in terms of these pollutants.

Whenever we look at the issue from an individualistic country perspective externalities seem to be far more simple than imagined. In a case of multiple nations, externalities deviate income growth among nations. If economic growth can be defined as increasing the quantity and productivity of factors of production, the process turns out to be a zero sum game increasing a countries income while decreasing the others not directly reflected in national accounts.

The model below, aims to look at income differentials among selected countries, while considering carbon dioxide emission, income inequality measured as the GINI coefficient and changes in forestry areas in explaining income differentials among nations (Table 1).

On the same lines, economic growth accounting and other measures related to economic growth will be very biased without the contribution of economic externalities. Some argue that the total factor productivity in relevance with economic growth showing the technology level will be biased without externalities inclusion. Thus the above figure simply reflects, that ranking countries in terms of income, shows that CO_2 per capita increases as the nations get richer and as the level of income increases inequality in income within a country increases. Thus in this respect we can talk about two different negative externalities former being among rich and poor and the later being among high and low income people within the same country. Although producers are different in both cases they harm the low income group more than others. On the contrary in terms of end users both parties pay the same price which is another source of inequality created by the negative externalities. These are typical biases that we come across as in the case of government subsidies. But this time the net return of an action has been biased from the beginning. That is the reason why, in terms of additional educational opportunities should be transferred to the three alternative income strata to narrow

Table 1 Income inequality and environmental decay

Countries	GINI coefficient	CO_2 per capita	Changes in forestry	GDP per capita
Norway	25.8	19,100	0,200	63,918
Australia	35.2	16,200	−0,200	36,032
Ireland	34.3	10,500	3,400	48,524
Switzerland	33.7	5,400	0,400	49,351
Netherlands	30.9	8,700	0,400	38,248
France	32.7	6,000	0,500	34,936
Finland	26.9	12,600	0,100	36,820
United States	40.8	20,600	0,100	41,890
Spain	34.7	7,600	2,200	25,914
Denmark	24.7	9,800	0,800	47,769
Austria	29.1	8,600	0,200	37,175
United Kingdom	36.0	9,800	0,600	36,509
Belgium	33.0	9,700	−0,100	35,389
New Zealand	36.2	7,700	0,500	26,664
Italy	36.0	7,800	1,300	30,073
Germany	28.3	9,800	0,200	33,890
Israel	39.2	10,400	0,700	17,828
Greece	34.3	8,800	0,900	20,282
Singapore	42.5	12,300	0,000	26,893
Korea (Republic of)	31.6	9,700	−0,100	16,309
Slovenia	28.4	8,100	0,400	17,173
Portugal	38.5	5,600	1,500	17,376
Hungary	26.9	5,600	0,600	10,830
Poland	34.5	8,000	0,200	7,945
Argentina	51.3	3,700	−0,400	4,728
Chile	54.9	3,900	0,400	7,073
Lithuania	36.0	3,800	0,500	7,505
Estonia	35.8	14,000	0,400	9,733
Latvia	37.7	3,000	0,400	6,879
Uruguay	44.9	1,600	4,400	4,848
Croatia	29.0	5,300	0,100	8,666
Costa Rica	49.8	1,500	−0,400	4,627
Mexico	46.1	4,200	−0,500	7,454
Bulgaria	29.2	5,500	0,600	3,443
Trinidad and Tobago	38.9	24,900	−0,300	11,000
Romania	31.0	4,200	0,000	4,556
Panama	56.1	1,800	−0,100	4,786
Malaysia	49.2	7,500	−0,400	5,142
Belarus	29.7	6,600	0,500	3,024
Bosnia and Herzegovina	26.2	4,000	−0,100	2,546
Russian Federation	39.9	10,600	0,000	5,336
Macedonia (TFYR)	39.0	5,100	0,000	2,835
Brazil	57.0	1,800	−0,500	4,271

the income inequalities. To reflect this as a representative of different income levels, income groups of upstream, transitional and downstream locations of a basin have been formed within the context of the anthroscape (the human reshaped basin-wide land unit-The Seyhan Basin, South Turkey) (Kapur et al. 2004). With the existing educational attainment and skill differences, direct income transfer programs have been proven inefficient and more distorting productivity gains. The argument put forward in the beginning section, proves that independent of the technology level, expertise in technology use leads to higher productivity gains due to absorption and diffusion of technology. Upstream, transitional and downstream showing different level of human capital accumulation, asks for distinctive policy recommendations towards economic growth.

Below we find the OLS relationship between GDP level being the dependent variable and CO_2 per capita, the GINI coefficient and changes in forestry areas (Table 2; Fig. 1).

Estimates show us that, the CO_2 increases as the nations get richer, which is consistent with the negative externalities hypothesis we made. As the income per capita increases, the GINI coefficient gets smaller reflecting a better distribution of income. We cannot go to the extreme arguing that high inequality in low GDP countries is solely linked to negative externalities from high income countries (Fig. 2). Because the improved tax systems, efficient public spending and equal opportunity creation also improves income inequality. But one can also argue that,

Table 2 OLS estimates of income inequality, carbon emission and forest decay

Model summary (B)

Model	R	R square	Adjusted R square	Std. error of the estimate
1	0.632 (a)	0.399	0.353	13.26013

(a) Predictors: (Constant), FOREST, GINI, CO_2
(b) Dependent variable: GDP

Anova(b)

Model	Sum of squares	df	Mean square	F	Sig.
1. Regression	4,561.831	3	1,520.610	8.648	0.000(a)
Residual	6,857.410	39	175.831		
Total	11,419.242	42			

(a) Predictors: (Constant), FOREST, GINI, CO_2
(b) Dependent variable: GD

Coefficients (A)

Model	Unstandardized coefficients		Standardized coefficients	t	Sig.	95% Confidence interval for B	
	B	Std. error	Beta	Lower bound	Upper bound	B	Std. error
1 (Constant)	26.283	11.540		2.278	0.028	2.941	49.625
GINI	−0.546	0.259	−0.279	−2.110	0.041	−1.069	−0.023
CO_2	1.505	0.431	0.462	3.493	0.001	0.633	2.376
Forest	3.459	2.217	0.198	1.560	0.127	−1.025	7.943

Dependent variable: GDP

Fig. 1 CO_2 and forestry change among selected countries (UNSD, 2006)

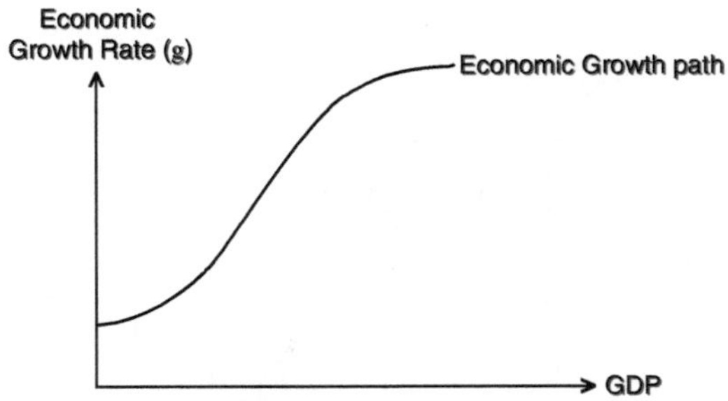

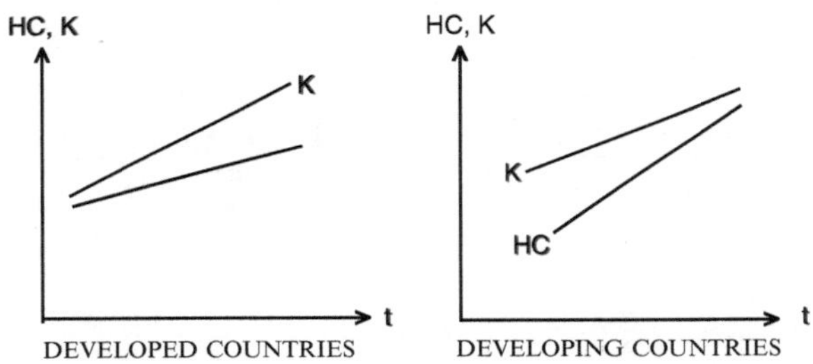

Fig. 2 GDP level and the economic growth rate

negative externalities in terms of wealth creation in low income countries could also be another source of income distribution distortion. Other independent variable forestry areas shows that consciousness related to conservation of green at least have improved (although slightly) for all countries in the analysis. Thinking in terms of the model being used, the high absolute value for the constant term can be interpreted as more explanatory variables needed in explaining income differences among selected countries. We should not neglect the fact that not only environmentally crucial factors but also determinants like investment in human capital and capital accumulation also explains the GDP differences.

Rodriguez (2004) argues that, beyond other factors there are also learning externalities which strongly influence economic growth. Thus, this is a world, where human capital properties of the leader country can be transferred to other nations without any cost (pure public good) in learning or knowledge externalities that are expected to occur. Rodriguez (2004) shows this as a different rationale for GDP conversion among developed and developing nations.

The figure above (Fig. 3) reflects the fact that knowledge and learning externalities are also a key determinant in economic growth. And the trend shows us that doubling time of GDP is on a decline.

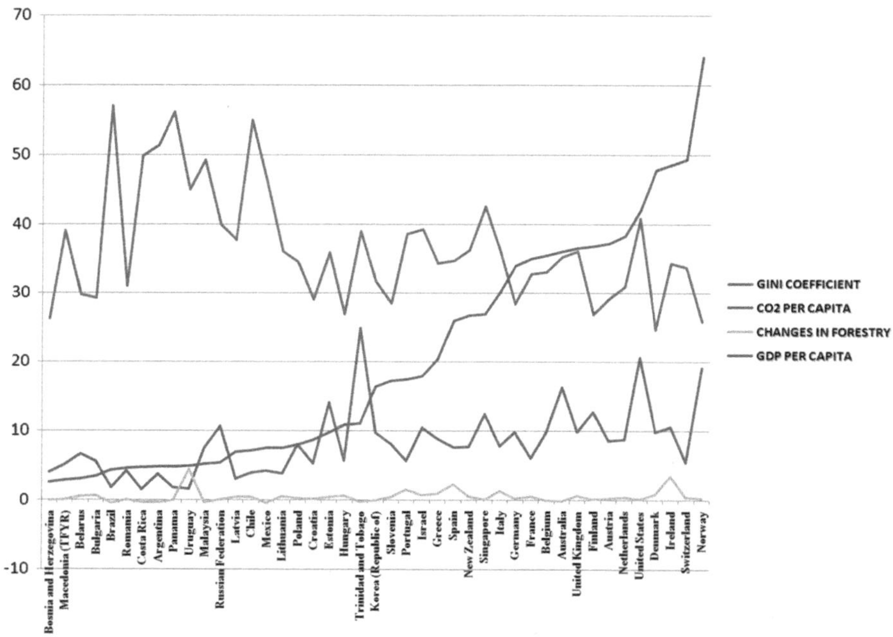

Fig. 3 Alternative convergence and divergence approaches

4 The Degraded Antroscape

Keeping in mind the theoretical foundations of economic growth of nations, now we come to the worsened economic conditions of farmland due to climate change. Keeping the same line of thought, the most important asset of the degraded antroscape is the accumulated knowhow and experience gained by learning and doing in production, distribution or consumption. A structural change in economic sense asks for a very fast adaptation to reduce negative externalities Fig. 2 can be a promising agenda, where inhabitants of the degraded areas can learn from others to improve the adaptation phase and period. We believe that the necessary human capital transfer could be done in alternative ways, each leading to the very different outcomes stated below (Table 3).

1. Developing a knowledge base to initiate a production style abolishing the traditional knowhow
2. Moving down the stream to find equivalent zones for living and production
3. Moving to cities to startup a new way of life
4. Improving the knowledge base within habituated zone by adapting to the new production and living conditions

Following the above statements, we would initially like to start by differentiating degradation in humid and dry lands (Blay et al. 2004).

Table 3 The strategic cycle

Strategic cycle	
Mission	Our objective, what we should do
Vision	Where we would like to see ourselves in the future
Strategic scope	Evaluation of project outcomes
Strategic themes and outcomes	Concentration areas and outcomes
Objectives	Method of evaluation of project outcomes
Strategy map	How could project outcomes be improved
Performance indicators	Are the outcomes as desired?
Strategic steps	Which steps and projects enable desired outcomes?

- Shortages of firewood and other wood
- Shortages of non-timber forest products
- Increased sediment deposits, floods and land slides
- Drying of springs and water bodies, siltation of dams
- Increased evidence of water-borne diseases
- Loss of biodiversity
- Climate change and desertification

All these trends are threats to food production as well as to livelihood conditions. Thus among the alternatives stated above adaptation seems to be a common rationale which endows economic and social rationale. We will far more concentrate on the economic consequences of degradation and adaptation. To our observation one major difficulty comes from the failure of the information on degradation (Turner et al. 2000), which simply means that the amount of degradation is not captured and awareness is very limited.

Another group of studies focus on alternative cost calculations given the assumption that land is degraded and local people can no longer cultivate or grow what they used to (Katyal and Vlek 2000). There are numerous in-depth studies for almost all continents. This approach is very valid if we are tackling cases where contamination of the problem is not the case. During the global era of global warming unfortunately this is not the case. Thus a more holistic and strategic approach should be welcomed.

The word strategy endows the concept of long range planning. The original concept belongs to the Second World War. But when events started changing faster after 1980s, the concept was once more vitalized and became part of every critical event. Strategic planning at the same time looks at the system as a whole looking at the implications of an action on other sub elements of the system as direct and indirect costs. This seems to fit to the case of land degradation because a unique adaptation or a solution for a region will always have direct or indirect implication for other regions. If the crop mix of a region changes as a part of adaptation policy in a region it is inevitable that will have positive or negative impact on nearby regions service and manufacturing industry as well as the agricultural sector as a whole. This asks for a more rigorous detailed analysis of sectoral interactions to maximize the aimed territory. We should not forget that although the overall demand elasticity for the most agricultural products is price inelasticity, important supply increases will inevitably decrease farm the income. This will make the adaptation solution questionable from the very beginning. In a world of globalized markets, supply shifts not only effect the market price but the volatility itself is an unwanted development to make the system sustainable. Unfortunately this is mostly true for all agricultural sectors. This is why the developed world, has shifted away from agriculture towards agribusiness and manufacturing and recently to service industry. Beyond these observations it is time to state the formal methodology of strategy setting as is in the following.

1. Sustainability strategy related to sectoral growth for post adaptation period
2. Strategy towards inter-sectoral linkages in terms of supply of goods and services
3. Strategy towards education and skills needed

The strategic planning process is a consensus that is created on priorities on an issue among the stakeholders. Which simply states that in the solution of a problem, (e.g., climate change) in terms of what should be done, priorities should be established with the consensus created among stakeholders. This is a road map, for long term commitment on going from one status to another. The key question is, not only doing the things right, but on the contrary doing the right things (priority issue). Examination of most strategic plans shows us that, efforts have been towards the endorsement of the existing actions. We have to start with whether we are going to be more prosperous, if we follow the ongoing road map.

Strategy is not a game that, if you supply these resources I will accomplish the following. It is an action where given the available resources, by changing the resource allocation how could the beneficiaries get better in a sustainable way. It can only be accepted as a success story, by measuring the performance indicators. Towards this end, every project should have a clear mission not on issues of what we are doing but an amendment of what we should be doing. If we have not launched such a project, what would the economy, society or region will lose would be the mission of the project. Vision, is where we would like to project to the project area within 5 years given the available resources.

Mission should carry certain themes which will motivate and activate the stakeholders of the project (Fig. 4). Some examples could be

- Doubling the regions per capita income
- Creating an entrepreneurial society from a farmland
- Empowering 90% of woman in new economic activities
- Reeducating and retraining 95% of the stakeholders within the coming 5 years
- Improving citizen satisfaction in the project region to 80% etc

All submissions for different actions should be consistent with the overall mission of the project seeking to develop the strategy map.

The project strategy map endorses actions, participants and outcomes in terms of performance indicators. It relies on the model that strategy realization can never be without performance measurement. Only the performance indicator can be part of Kaisen, i.e., continuous improvement. To be able to see the overall project cycle the following presentation could be used (Fig. 5).

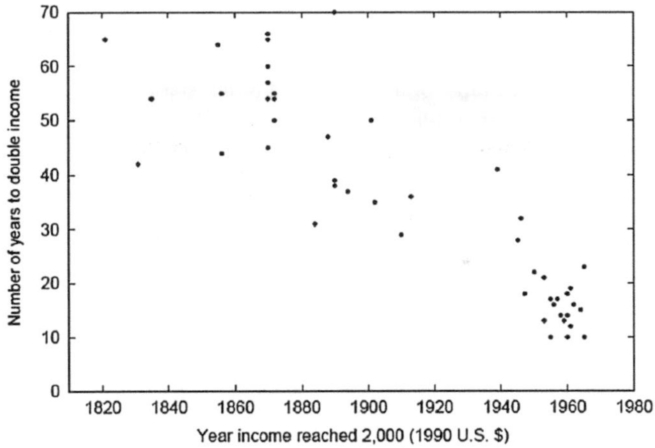

Fig. 4 Number of years to double GDP
Source: http://www.cema.edu.ar/publicaciones/download/documentos/270.pdf

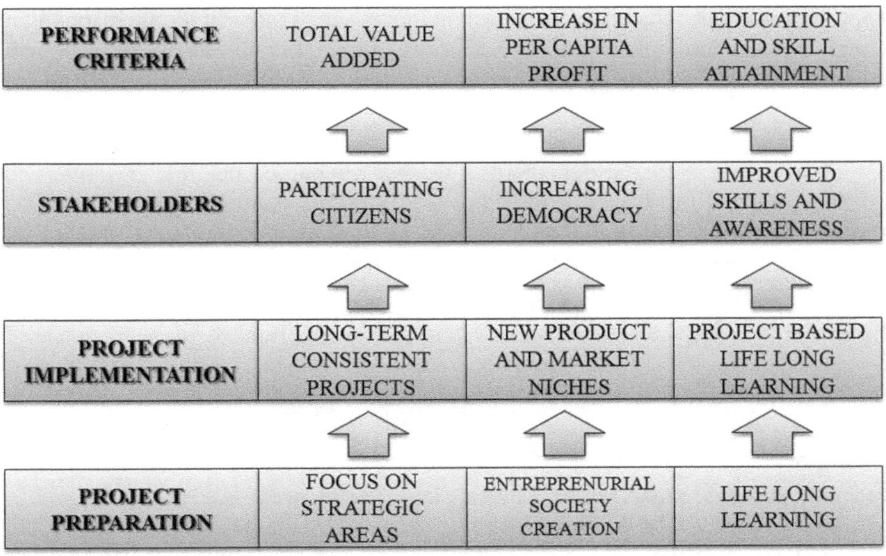

Fig. 5 The project strategy map

From top to bottom we see the demands of stakeholders from the project. The chart also shows each step to be taken in achieving the project goals. From the steps we see that strategic planning is an iterative process which relies on performance indicators. The following strategic moves further points out the structural format of handling economic problems related to degraded areas.

5 Sustainability Strategy

This strategy covers trade and production dynamics with the aim of maximizing farmland peoples incomes under the constraint of sustainability. Thus local, regional, national and international production and consumption dynamics should be considered at the phase of adaptation. The key question in the adaptation process will be the sustainability constraint. The strategic behavior relies on the regional assessment and prioritizing of the actions. This simply means that the same ecological properties could lead to different crop or service selection even among nearby vicinities, if the human education and skill dimension asks for different endowments. As stated earlier this approach would create a long term solution to a very regional issue if the threat (global warming) is global and diffused.

5.1 SWOT Analysis

SWOT analysis should be dynamic by nature looking at strengths, weaknesses, opportunities and threats with respect to other stakeholders (suppliers of other regions, consumption destinations and also the international dimension). This will enable us to see long term advantages and constraints that are locally and outside of the locality generated. This analysis will majorly help to the actions to be taken as well as providing the opportunities for the sustainability of the actions. In the externality discussions of our analysis we had stated the possible negative and positive externalities vaguely reflecting the direct accounting techniques. One should not forget that most issues related to the environment have little to do with direct costing vs. opportunity costing. This is the area where we see more complexity and omission of cost components which totally influences the optimal resource allocation.

5.2 Affordability Analysis

The affordability analysis aims to examine regions at an household level ability and intention of covering the project based costs of adaptation, considering their income, liabilities, potential of barrowing and available funds. This check simply assumes that, the sustainability of the adaptation project could not be financed by the third parties and beneficiary contributions, where the cost of the project will also increase the chance of success. Towards this end, the cost of adaptation that reflects the households should be pre-known before running an affordability analysis. As stated by Samuel Fankhausera, Sladjana Tepicb "There is also a sense that this affordability problem may become worse before it can get better." Improvements in the quality of infrastructure services for both rich and poor consumers – better access, more reliable supply, less wastage-are only possible if the underfunded energy and water industries are put back on a sound financial footing (Fankhausera and Tepicb 2007) have quoted, that, "The argument looks at the inner externalities of the project with respect to implementation. Consequently, the affordability analysis not only looks to the ability to pay but to the repercussions if the project is not implemented."

5.3 *Livelihood Analysis*

Looking at the literature we see that similar projects adapted in different parts of the world lead to different success or failure stories. It is easy to blame the implementers for the unexpected outcome. Experience shows us that one other cultural dimension of societies stems from how households perceive their livelihood objectives and what you are proposing – does it help them reach their own objectives (George et al. 2009), also considers the available resources in terms of land, labor, capital, entrepreneurial skills, along with the meaningful way of living given the available resources in terms of values, and provides the ways of achieving the preset goals and how the families would increase their available resources As well as the alternative methods of solving problems related to available resources? Further in the analysis, the available technology level, the level of absorption and diffusion of that technology is also investigated.

5.4 *Pest Analysis*

At the last stage of adaptation it is an inevitable fact that, the new production mix of the location will be sold out, distributed and consumed. This time another environment dimension will need more attention in terms of customers, shortly known as the marketing environment. Most agricultural production systems fail due to the lack of knowledge in marketing their output and services. PEST analysis stands for, the political, economical, socio-cultural and technological factors given below, which strongly influence the buying decisions of the societies. An academic group analyzed the different aspects of PEST analysis showing us the importance of marketing a product once it is produced (Marketing Teacher Ltd and Pest Analysis 2000).

5.4.1 Political Factors

The political arena has a huge influence upon the regulation of businesses, and the spending power of consumers and other businesses. Thus, one should consider issues such as:

1. How stable is the political environment?
2. Will government policy influence laws that regulate or tax your business?
3. What is the government's position on marketing ethics?
4. What is the government's policy on the economy?
5. Does the government have a view on culture and religion?
6. Is the government involved in trading agreements such as EU, NAFTA, ASEAN, or others?

5.4.2 Economic Factors

Marketers need to consider the state of a trading economy in the short and long-terms. This is especially true when planning for international marketing. You need to look at:

1. Interest rates.
2. The level of inflation Employment level per capita.
3. Long-term prospects for the economy Gross Domestic Product (GDP) per capita, and so on.

5.4.3 Socio-cultural Factors

The social and cultural influences on business vary from country to country. It is very important that such factors are considered. These factors include:

1. What is the dominant religion?
2. What are attitudes to foreign products and services?
3. Does language impact upon the diffusion of products onto markets?
4. How much time do consumers have for leisure?
5. What are the roles of men and women within the society?
6. What is the life expectancy of the population? Are the older generations wealthy?
7. Does the population have a strong/weak opinion on green issues?

5.4.4 Technological Factors

Technology is vital for competitive advantage, and is a major driver of globalization considering the following points:

1. Does technology allow for products and services to be made more cheaply and to a better standard of quality?
2. Do the technologies offer consumers and businesses more innovative products and services such as Internet banking, new generation mobile telephones, etc?

3. How is distribution changed by new technologies e.g., books via the Internet, flight tickets, auctions, etc?
4. Does technology offer companies a new way to communicate with consumers e.g., banners, customer relationship management (CRM), etc.

Pest analysis is highly vital for the economic sustainability of all economic attempts undertaken during and after the phase of adaptation.

5.4.5 Inter-Sectoral Strategy

Not only the regions, but the sectors give and take so many goods and services resources that, for the long term sustainability commodity and service flow dynamics should also be considered. Thus as most of us think, adaptation will not be as local as conceived. In the age of complex production, logistics and factor mobility are the key sources for competitiveness which strongly determines the per capita income. There is a common saying among doctors which states that there is no sickness as a headache. Although headache is the symptom, there might be dozens of factors causing the headache.

Economic analysis adapts the input–output matrix approach to analyze input giving sectors impact on final production units catching the full impact reflecting the expenditure on goods and services. An alternative approach could be the computable general equilibrium analysis technique adapted to measure sectoral sharing in terms of local, regional, domestic and international demand. Both rest on certain assumptions in terms of demand growth and resource utilization. An alternative and a more easy to use approach could be estimating the demand shifts with respect to income. Knowing that supply of goods and services without the demand dimension is not sustainable, inter-sectoral commodity and service flows should be endowed to the model to be able to capture the time dimension. Once more we will see that different income, education and cultural values in different strata of project implementation will be the area of highest difficulty in the cohesiveness of the project.

5.4.6 Education and Skills Strategy

Education and training towards new skills, leadership and change should be the first step towards adaptation in degraded areas. Education and training in general increases mobility and collaboration within a society. New roles, responsibilities and entrepreneurial capabilities depend on sound education and training. As stated earlier skill and knowledge formation should be towards future needs and trends. This in no way means that indigenous technology and the way of life can be omitted. On the contrary in search of adaptation should also have components of familiarity, mainly in areas where the cultures had been accumulating skills for lengthy time periods?

Education and training should accompany every stage of adaptation implementations. These efforts will increase the cohesiveness of implementation and understanding which helps to restructure societal values and sharing. One key factor behind all education and training efforts should be the learner centered education techniques being used. Program and learning outcomes that matches with the proposed study will give a wide chance of implementation and relevance. In this type of educational training programs all stakeholders should be active in setting program objectives. At the end of each program there should be assessment of education and training programs at each level and findings should be recorded and used for benchmarking.

To sum up, it seems that the project based adaptation fails for several reasons when compared to a holistic and strategic approach. Formerly, not only the indicating trends but long term effects and interactions should be emphasized given major priorities of the adaptation program. As more than ever having very restricted amounts of financial and natural resources, priorities should lead ever action taken. The second dimension of the proposed approach differing than the existing methodologies being used are the expected interactions among commodity groups and sectoral linkages which strongly effect the long term sustainability of adaptation projects.

Now we can turn back to the economic and social impact of the degraded Anthroscape.

6 The Degregation Issue

Oldeman, defines soil degradation as "a process that describes human-induced phenomena which lower the current or future capacity of soil to support human life" (Oldeman et al. 1991).

Land degradation is a more comprehensive term, showing the decrease in the capability of land yield under a given farm use. There are also studies concentrating on qualitative and quantitative assessment of soil and land degradation each having its own advantages (Van Lynden et al. 2004). But before we look into the soil and land degradation issue, may be it would be more consistent talking about the ecosystem to have a better picture of the threat. The total value of an ecosystem can be best described in the following figure.

The above figure illustrates the value of the good or the service in two main aspects, namely the use value and non-use value. The use value covers the direct (consumptive or non-consumptive) and indirect use and the optional value which covers what has been inherited from the past and the choice of today. In calculating

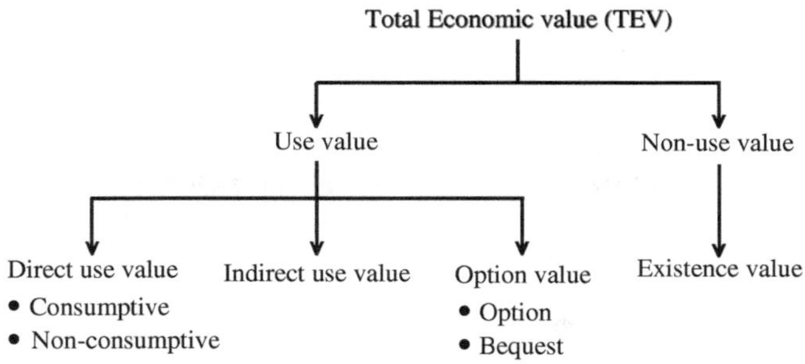

Fig. 6 Total economic value (Pagiola et al. 2004)

the total economic value the second component is the non-use value. This only reflects the existence value of the good or the service (Fig. 6).

An ecosystem creates benefits which strongly influences our daily life. The major contributions of the ecosystem can be summarized as; biodiversity conservation, downstream water services, recreation and extraction of forest products (World Bank Research Team 1997). This analysis shows us that land degradation negatively influences the benefit-generated ecosystem. The trend is more towards assessing the quantitative impacts of degradation vs. quantitative aspects. But in all cases if development can be defined as an ongoing improvement in human welfare and welfare is defined as increase in ones endowment enabling one to consume more goods and services, the optimization of this fascinating process is a need mainly due to the resource constraints around us. This in no way discriminates the developed or the developing world. It is a personal belief that the overall improvement in living standards over the past two decades created a confidence among nations that the pace of development has no constraints ant it is limitless. Van Lynden et al. (2004) mainly focusing on salinization, nutrient decline and soil pollution, conclude that, "the changes in land quality status estimated by compare, that in the assessment process, selection of laboratory method, sampling technique and method description are the critical factors in the assessment of soil degradation" (Van Lynden et al. 2004). Even the proposed methodology shows the role of land and soil degradation in the overall ecosystem performance. One should not forget that the estimation of all benefits and costs of the environment is not desirable or feasible due to cost and time constraints, model approach using abstractions with major elements and their interaction that should be accepted as the appropriate methodology. Perceive the issue from a different perspective, mainly through a local order or low entropy, provided bio-systems within certain boundaries are concerned. They offer a thermodynamic equilibrium model defined as the increase of energy through flow and stored biomass. Their model is original in the sense that, ecosystem and development phases show different properties in terms of useful energy and stored energy.

7 The Cost of Degradation

We will focus on the cost of degradation from two different perspectives; degradation and socioeconomic linkages and costs involved in degradation.

Geist and Lambin's comprehensive study in 2004 shows that from 132 case studies involved, at approximately 95% of the cases, the degradation of land is very closely linked with agricultural management performance and agricultural land use. These observations stimulate us to look at the socio-economic motivations that farmers have been creating for such an outcome. Within the assessment process the major difficulty comes from case and location specific features. The figure below is an attempt by (Shiferaw and Holden 1998), in explaining factors influencing the farmer's decision making process (Fig. 7).

Uriel and Zafar analyzes the farmers rationale under six headings, namely as the policy drivers such as the top down policies in local communities being the major rationale for degradation (Safriel and Adeel 2005). Improving bottom top participation and democratization gives very positive results in terms of reducing land degradation in all aspects. The second driver is the demography; population growth triggers the pressure towards degradation and health problems. Migration is another source of pressure towards land degradation. Looking into the trend there is a positive correlation between population and environment problems. The third driver is the land tenure policies; problems related to the security of farm land due to different reasons such as property rights usually leads to a loss of efficiency

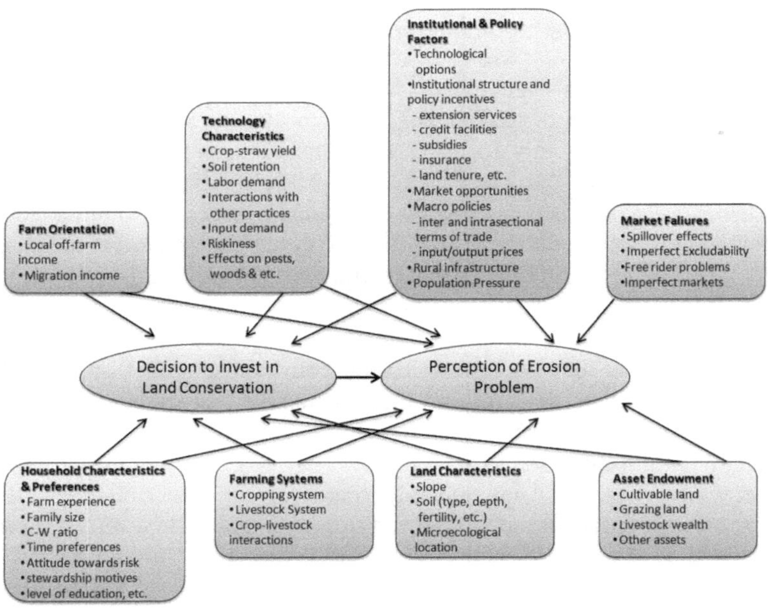

Fig. 7 Factors influencing farmers decision system

in terms of water availability, sedimentation, salinity or to inefficient use of the underground water. The fourth driver is the water policies; policies related to pricing, allocation, alternative uses of water other than agriculture and especially irrigation policies could be the main source of land degradation. The fifth driver is the governance approach; although large scale governmental policies can improve the livelihood of farmland improving sustainability, inappropriately managed governmental policies had negatively influenced land degradation. The last driver is the economic factor; the economic globalization trends of the recent decades decreased the focus largely from agricultural land, putting the emphasis on high tech manufacturing and to the service sector. The market trends and prices leading the economy as is due to very high levels of externalities, affect desertification at an increasing trend.

All these six major drivers coincide with another event, such as global warming, makes degradation the major agenda. From a farmers perspective, due to low farmers income, high cost of production and high opportunity cost of forgone output in terms of income also diminishes the cost of desertification. In technical terms marginal utility of today's income foregone seems to be higher than that of the future cost of degradation, especially under large externalities. That is why these inefficient market indicators need to be corrected by government intervention for the ecosystem sustainability. As shown in our previous economic analysis. Externalities leading to negative income distribution turn back and create another negative externality in term of inefficient land use and desertification.

An alternative approach offered by Osgood and Lipper approaches the issue using GIS and socio-economic data to analyze causal relationships (Osgood and Lipper 2009). Accepting that the quality of soil is a capital stock for famers, analysis aims to measure the outcome of farmers with one single data that is the quality of soil. In fact land degradation is nothing but loss of soil quality. Authors use the economic rationale and assuming that farmers maximize their income by the amount of yield generation which in turn will be determined by the soil quality. The major difficulty of the model seems to be stemming from land management under non-homogeneous land properties. It is none the less a very creative method comparing desertification by GIS and comparing the poverty levels with statistical data collected. Similar findings related to case specific property also holds in the research conducted by Anna Bolt at the World Bank or governmental level structural adjustment programs (SAP) concerning the impact on the environment. At a sector level an increase in product prices, leads to an increase in production thus having a negative impact on degradation. Farm practices totally depend on farmer decision making thus no meaningful generalization can be made in terms of SAP degradation relationship in terms of farm level practices. One last finding of the study is the higher agricultural income as a result of high prices enabling farmers to be engaged in productivity and conservation investments (Gueorguieva and Bolt 2003).

Turning back to the cost of degradation analysis, we will be restricting ourselves to three striking studies with alternative approaches. The first research is by Bolt, Ruta and Sarraf, which looks at the environment issue not from an ethical or sustainability dimension, but on the contrary whether individuals would like to spend their scarce resources (of course including their income) on environment, and at what amounts (Bolt et al. 2005). This is a very common method in economics because the price of a good simply shows the minimum amount that people are ready to pay to receive such a benefit. Because in its simplistic sense demand for a good has two prerequisites, firstly sufficient income holdings and secondly a willingness to pay given that income. This all relies on the basic assumption that, for all countries needs and wants are greater than the existing resources and making a choice at the cost of another good or service is a must. Thus this type of value measurement is independent from the intrinsic valuation of the environment, but instead the importance that the individual gives to the environmental component by the amount of money allocation. We economists deserve to be right when we consider the assumptions behind the model. In a case where individuals do not have the perfect information about the importance that he gives to an item and especially when there are costs born on an item but indirectly paid by some other person the model needs to be discussed twice. Research puts forward the following techniques in assessing alternative ways of costs borne as a result of environmental degradation. Authors developed a training manual which in brief can be summarized with the following figure (Bolt et al. 2005).

7.1 *Benefit Cost Analysis*

This calculation aims to calculate the sum of discounted benefit and costs over time. The Net Present value (NPV) is the most common technique to assess damage costs and benefits of a project. The biggest weakness of the model is in the selection of the appropriate interest rate. For practical purposes researchers use the barrowing interest rate in that sector. Predicting common years benefit and costs as well as the interest rate becomes more difficult when the economy shows sporadic fluctuations.

$$\mathrm{NPV} = \sum \mathrm{B_i} - \mathrm{C_i}/(1+\mathrm{r})^{\mathrm{i}} = \sum \mathrm{B_i}/(1+\mathrm{r})^{\mathrm{i}} - \sum \mathrm{C_i}/(1+\mathrm{r})^{\mathrm{i}},$$

where,

B_i = Benefits of the project in year i
C_i =Costs of the project in year i
R = discount rate
i = years

7.2 *Valuing Changes in Production*

When environmental conditions change the first impact will be on the quantity and quality produced. This will either decrease our output thus our revenue or will ask for additional inputs to make up for the loss in output. This automatically will be increasing our costs of production. Thus in terms of cost assessment first we have to find out the output decrease due to degradation. However, we should also keep in mind that loss in output is not only influenced by degradation. Some researchers use pretested impact coefficients related to other factors such as the weather conditions which may decrease the yield. In the second phase we have to find the market value for the loss. The key is catching cost and benefits when the production level varies. Again another difficulty comes from the reduction in the outputs revenue effect. However, due to the elasticity of the demand for agricultural products, decreases in output, increase the total revenue, ultimately becoming complicated to be accepted as a cost or a benefit.

7.3 *Averting and Mitigating Behavior*

In some cases the benefits and costs calculation differs depending on from which perspective you look at. If there is an epidemic due to non-clean water, health people calculate the cost of staying home sick as foregone earnings. Then the cost can be mitigated by preventive medicine, which in fact can decrease the negative effects of land degradation, i.e., the effect of the water lacking sanitation. Thus in such a case the cost of hazard can be declined as in the following,

$$S = S(P, D).$$

Where,
S = Environmental hazard
P = Level of environmental hazard
D = Level of mitigating behavior

Sometimes the measurement of mitigating behavior is not as simple as it seems. Mitigation has a cost, which decreases sick time (time lost) but also it creates comfort of not becoming sick for a lengthy time.

7.4 *Travel Cost Method*

In taking an environmental action not the action itself but the travel time cost (TCM) of commuting should be accepted as a direct cost. Natural locations have either no cost or very low cost for admission that should be considered as consumer surplus when the travel cost is also included into the visit cost. In the economic theory, TCM is a reveled preference approach for the valuation of economic actions. If we like to get the calculations more complicated travel cost will not be only travel cost because the real cost depends on the income, age, personal interests or other possible externalities.

The total methodology can be summarized under four steps; the First step is gathering explicit and implicit cost related to travel. This includes the opportunity cost of travel and the time cost of travel. We should also include the cost of time spent on the site. The second step is estimating the functional format of the trip given the data collected. The Third step is estimating the travel demand curve for individuals. We should not forget that each demand curve is estimated for each zone. The Forth step is driving the consumer surplus, below the demand curve which reflects the willingness to pay for the given service.

7.5 Hedonic Prices

The Hedonic prices assume that the demand for a good or a service is not a function of single variable. The Model assumes that bundle or mixes of variables are influencing the demand. Initially quality of land might seem the key determinant towards demand for natural resources. But other variables assessed in the demand function are also determining the willingness to pay.

In the application stage of the hedonic prices we have to follow four different stages. In the first stage we have to define which factors contribute to the price of the good.

$$P = P(X_1, X_2, X_3, \ldots\ldots X_N).$$

Where,

P = Price of the good or the service

X_i = Factors contributing to the price

In the second stage, we have to collect data related to the price of the good or the service and related to functional attributes (X_1, X_2, X_3, ... X_N). In the third stage we try to predict the parameters in front of X_1, X_2, X_3, ... X_N's which show the functional relationship of the contributing factors and the price. If the issue is finding land prices and factors influencing it, such as the environmental quality, given the parameters we can determine the demand curve for environmental quality. To be able to drive the environmental quality demand curve we have to have a second prediction related to the implicit price model.

7.6 Contingent Valuation Method

When we look at the market price, behind it, we assume that there is perfect available information about the product, its substitutes, and future expectations. But in reality information flow related to the market is not as perfect as expected. Thus an alternative method is encouraging individuals to reveal their preferences in terms of their willingness to pay. This method is known as contingent valuation.

This model enables even pricing of non-tradable goods. In the application stage, randomly selected individuals are asked to state how much these individuals are willing to pay for a change in the provision of a good or a service.

Contingent valuation analysis helps the establishment of a hypothetical product or service with well defined properties, then the method of payment should also be revealed to clarify the willing to pay price for the product. One other alternative could be using the auction market for the good or the service by phone, face to face, interviewing or some other method. The aim is assessing the highest willing to pay price. The Third stage is to test the willing to pay or willing to accept averages. Statistically speaking a small standard error is preferable for more accurate pricing techniques. One other method is using an OLS to predict responses. Both techniques are frequently used in practice. Success of the contingent valuation method usually depends on the preparation and administration of the survey. One should not forget that face to face interviews are far more time consuming but more accurate in terms of interpreting results.

7.7 Quantifying Health Services

This criterion measures the valuation of health damages as a result of environmental degradation. Such health damages could be, physical, biological, social and psychological. The analysis takes place under three stages. Initially we have to identify the hazards as a result of environmental degradation. The second step aims to estimate the impact of that hazard on health. The severity is measured by a dose coefficient. The third stage aims to explore the diffusion of the hazard. How many people, how many females, youth etc. The measurement could be in different forms; animal toxicological studies (experiments on animals), human clinical studies (controlled experiments on humans) and epidemiological studies looking at real world problems.

7.8 Valuing Health Effects

Valuation of health services is calculated by peoples' willingness to pay. The reasons why people are willing to pay for their health can rise from different reasons

- Time lost during the phase of sickness
- Medical costs associated with sickness
- Precautionary expenditures and
- Discomfort of sickness

Valuing mortality with the human capital approach; this approach focuses on productivity and income losses as a result of mortality, otherwise named as the incomes foregone method.

Valuing mortality through the cost of illness approach; these costs cover direct and indirect cost of illness, such as wages foregone.

Willingness to pay; these calculations cover how much people are willing to pay to avoid sickness or mortality. Sarraf et al. (2004) in their work on cost of environmental degradation explain the calculation process under three headings. Compared to the previous model, assessment procedure that seems to be far more simplified but easier to calculate.

7.8.1 Quantification of Environmental Degradation

This phase covers quantifying, changes in clean air, pollution in the river and health problems faced as a result of environmental degradation.

7.8.2 Quantification of Consequences of Degradation

In stage two we measure the negative impact of environmental degradation. Loss in the productivity of soil, days lost as a result of health problems stemming from environmental degradation (Fig. 8).

7.8.3 A Monetary Valuation of the Consequences

In this phase we assess monetary value for soil recovery, cost recovery from a sickness, reduced recreational values etc.

In their study towards cost implications of agricultural land degradation, a number of scientists used the EMM model which integrates the negative effect of

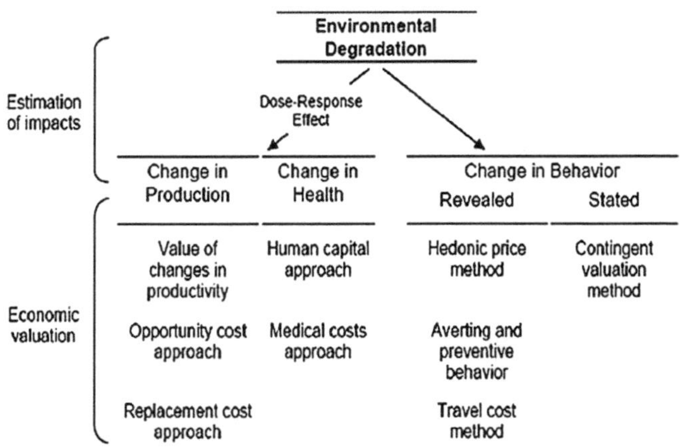

Fig. 8 Environmental degradation

the agricultural degradation to the rest of the economy. The Model asks for macro-micro integrated work linking agricultural degradation to soil loss, land productivity, farm income and poverty. EMM, in the assessment of a supply function, is measured under representative producer's response. On the contrary the demand function is disaggregated such as urban-nonurban looking at demand response at the per capita level. The model allows national and international substitution. The model seems to be more holistic but does not look at crop pattern mix as a result of supply and demand conditions (Diao and Sarpong 2007).

8 Conclusion

We have started this treatise, with the modest goal of exploring the social and economic impact of degradation and its possible international, national and local causes. Recent experience show us that the trend of degradation seems to be increasing, especially due to the externalities avoiding full costs to be seen and paid. The increasing impact of global warming warns us that mitigation policies on their own will not be an effective tool for the degraded land but policies of adaptation should be far more explored and improved. Statistical findings show us that land degradation and poverty usually goes hand to hand thus instead of solely income transfer policies, overall awareness and investment in human capital policies seems to solve the problem and watch over economic growth and environmental cognition.

Technological advances have helped to the developed and developing world improving the quality of life. But priorities towards economic growth without environmental considerations seem to be a short sighted view of welfare improvement. Strategic (long range) perspectives are needed to see the dynamic interaction of economic and social well being. Thus any segmented analysis solely focusing on project outcomes, income improvement or mitigating environmentally sound policies should be one more time questioned and holistic policies should be promoted.

References

Arrow KJ (1962) The Economic Implications of Learning by Doing. Review of Economic Studies, 29, 155–173

Blay D, Bonkoungou E, Chamshama SAO, Chikamai B (2004) Rehabilitation of Degraged Lands in Sub-Saharan Africa. Forestry Research Network for Sub-Saharan Africa, IUFRO

Bolt K, Ruta G, Sarraf M (2005) Estimating the Cost of Environmental Degradation. Environmental Department Papers, World Bank, No:106

Cass D (1965) Optimum Growth in an Aggregative Model of Capital Accumulation. Review of Economic Studies, 32, 233–240

Diao X, Sarpong DB (2007) Cost Implications of Agricultural Land Degradation in Ghana. International Food Policy Research Institute, IFPRI Discussion Paper 00698

Domar ED (1946) Capital Expansion, Rate of Growth, and Employment. Econometrica, 14, 137–147

Erk N, Çabuk, Ateş S (1998) Long-Run Growth and Physical Capital-Human Capital Concentration. METU International Economics Conference II, Sep. 9–12, Ankara

Fankhausera S, Tepicb S (2007) An Affordability Analysis for Transition Countries. Energy Policy, 35(2), 1038–1049

Fisher I (1930) The Teory of Interest. Macmillan, New York

George A, Pierret A, Boonsaner A, Valentin CO, Didier PO (2009) Potential and Limitations of Payments for Environmental Services (PES) as a Means to Manage Watershed Services in Mainland Southeast Asia. International Journal of the Commons, 3(1), 16–40

Griffin JM, Harry BS (1986) Energy Economics and Policy. 2nd edn. Academic Press College Division, Orlando, FL

Gueorguieva A, Bolt K (2003) A Critical Review of the Literature on Structural Adjustment and the Environment. The World Bank Environmental Department, Paper No 90

Harrod RF (1939) An Essay in Dynamic Theory. Economic Journal, 49, 14–33

Kapur S, Zdruli P, Akça E, Arnoldussen AH, Kapur B (2004) Anthroscapes of Turkey: Sites of Historic Sustainable Land Management (SLM). I: S. van Asselen, C. Boix-Fayos & A. Imeson (red), 'Briefing papers' of the second SCAPE workshop in Cinque Terre (Italy), 13–15 April 2004

Katyal JC, Vlek PLG (2000) Desertification-Concept Causes and Amelioration. ZEF-Discussion Papers on Development Policy, Bonn

Koopmans TC (1965) On the Concept of Optimal Economic Growth. The Econometric Approach to Development Planning, Amsterdam, North Holland

Lucas RE Jr (1988) On the Mechanics of Economic Development. Journal of Monetary Economics, 22(1), 3–42

Malthus TR (1798) An Essay on the Principle of Population. W. Pickering, London

Marketing Teacher Ltd, Pest Analysis, 2000–2009. http://www.marketingteacher.com/Lessons/lesson_pest.htm. Accessed 27.07.2009

Mikhail O, Swint J, Michael BMR, Moyé Lemuel A, Sabino M (1999) Technology Evolution. International Journal of Technology Assessment in Health Care, 15, 254–263

Oldeman LR, Hakkeling RTA, Sombroek WG (1991) World Map of the Status of Human-Induced Soil Degradation, 2nd revised edn. ISRIC/UNEP, Wageningen/Nairobi

Osgood, DL, Leslie L. Methodological Issues In Analyzing the Linkages Between Socio-Economic and Environmental Systems. FAO. http://www.fao.org/docrep/004/Y1796E/y1796e03.htm. Accessed 28.07.2009

Pagiola S, von Ritter K, Bishop J (2004) How Much Is an Ecosystem Worth? Assessing the Economic Value of Conservation. IUCN and The Nature Conservancy. The International Bank for Reconstruction and Development/The World Bank

Pomfret R (2002) State-Directed Diffusion of Technology: The Mechanization of Cotton Harvesting in Soviet Central Asia. The Journal of Economic History, 62(1), 170–188

Ramsey F (1928) A Mathematical Theory of Saving. Economic Journal, 38, 543–559

Ricardo D (1817) On the Principles of Political Economy and Taxation. Cambridge University Press, Cambridge

Rodriguez AM (2004) Learning Externalities and Economic Growth. http://ideas.repec.org/p/nbr/nberwo/11009.html. Accessed 28.07.2009

Romer PM (1986) Increasing Returns and long Run Growth. Journal of Political Economy, 94(5), 1002–1037

Romer PM (1987) Growth Based on Increasing Returns Due to Specialization. American Economic Review, 77(2), 56–62

Romer PM (1990) Endogeneous Technological Change. Journal of Political Economy, 98(5), part II, 71–102

Safriel U, Adeel Z (2005) Ecosystems and Human Well-being: Current State and Trends, Chapter 22, p. 648–650 www.millenniumassessment.org/documents/document.291.aspx.pdf

Sarraf M, Larsen B, Owaygen M (2004) Cost of Environmental Degradation. Environmental Economic Series No:97, Washington, DC

Sassenrath GF, Heilman P, Luschei E, Bennett GL, Fitzgerald G, Klesius P, Tracy W, Williford JR, Zimba PV (2008) Renewable Agriculture and Food Systems, vol 23. Cambridge University Press, Cambridge

Schumpeter JA (1934) The Teory of Economic Development. Harvard University Press, Cambridge, MA

Shiferaw B, Holden ST (1998) Resource Degradation and Adoption of Land Conservation Technologies in the Ethiopian Highlands: A Case Study in Andit Tid, North Shewa. Agricultural Economics, 18(3), 233–247

Smith A (1776) An Inquiry into the Nature and Causes of the Wealth of Nations. Random House, New York

Solow RM (1956) A Contribution to the Theory of Economic Growth. Quarterly Journal of Economics, 70(1), 65–94

Swan TW (1956) Economic Growth and Capital Accumulation. Economic Record, 32, 334–361

Turner RK, Van Den Bergh JCM, Soderqvist T, Barendregt A, Van Der Straaten J, Maltby E, Van Ierland EC (2000) Ecological–economic analysis of wetlands: scientific integration for management and policy. Ecological Economics, 35, 7–23

United Nations Statistics Division (UNSD) (2006) Millennium Development Goals (MDG) Indicator Database. http://mdgs.un.org/unsd/mdg/Data.aspx

Van Lynden GWJ, Mantel S, Van Oostrum A (2004) Guiding Principles for the Quantitative Assessment of Soil Degradation. FAO, Rome

World Bank Research Team (WBRT) (1997) Expanding the Measure of Wealth. Environmentally Sustainable Development Studies No. 17. World Bank, Washington, DC, p. 14

Young A (1928) Increasing Returns and Economic Progress. Economic Journal, 38, 527–542

Index

A
Adana, 2, 4, 13–17, 20, 24–26, 32, 35, 36, 38, 44, 138, 259–284, 314–316, 319
Aeolian
  sediments, 181
  transport, 181
Afforestation, 60–61, 65, 71–73, 78, 79, 89, 93–98, 101, 111, 115, 118–122, 357
Aggregate measures, 103–110
Agro
  environmental schemes, 171
  forestry, 135, 137, 244, 255
Agro-silvo-pastoral production system, 244
Alghero, 179–191
Amanos, 267
American civil war, 274–277
Analysis
  affordability, 394–395
  livelihood, 395
  PEST, 395, 397
Anatolia, 25, 64, 65, 135, 259, 267, 315
Anazarbos, 269, 271
Andosols, 134
Anthropic Regosols, 171, 173
Anthropogenic soils
  anthrosols, 179–181, 184, 187, 190
  processes, 181, 183, 184, 186, 187, 189, 190
Anthroscape
  Asian, 205–238
  Atriplex, 244, 251–254
  cotton, 272–273
  degraded, 146, 193–202
  economic and social impact of, 379–407
  historical, 259–284
  hydrogeologic, 314
  land quality classes, 2, 33–34, 41–43
  of the middle east, 57–123
  of the mountains, 260, 267, 270, 281, 315
  newly created, 250–253
  reshaping, 248
  soil formation, 193–202
  sustainability, 285–291
Anthrosols, 129–130, 135, 154, 164, 169, 179–180, 184, 187, 190, 207, 209, 215, 220, 223–225, 238
Approach
  holistic, 391, 398
  integrated, 44, 59
  multidisciplinary, 289
Apulia, 130, 132
Archaeological imprints, 155–156
Arenosol, 32, 136–138, 168, 169, 236, 335
Assyrians, 260, 262
Atriplex, 133, 244, 251–254
Averting and mitigating behavior, 403

B
Benefit cost analysis, 402
Byzantine empire, 265

C
Calcisol, 32, 35, 131–136, 139
Calcrete
  horizon and petrocalcic horizon, 135
  pedogenesis, 180–181
Calibration of model, 320–321, 329
Cambisols, 23, 32, 35, 128, 131, 134, 135, 137–139, 173, 174, 236
Carbon
  emission, 388
  sequestration, 129
  stocks, 218–221
Cartagena, 194, 294
Chalcolithic, 260
China, 205–238, 270, 351

Chinese soil taxonomy, 209, 215, 223, 226, 238
Cilicia, 259, 260, 265, 269–271, 273, 274, 278, 279
Clay Soil District, 163–178
Climate change
  projected impacts of, 311–330
  scenarios, 319, 320
Community measures, 103, 110–119
Convergence in economic growth, 385–390
Cotton, 22–24, 32, 132, 135, 136, 264, 269, 270, 272–278, 281, 282
Cydnos, 267

D
Degradation
  cost of, 127, 400–407
  of rangeland systems, 243–256
Desertification, 5, 25, 46, 57–123, 127, 140, 244, 253, 333, 334, 337, 391, 401
Dioscorides, 279
Driving forces, 4, 9–10, 46, 55
Durisols, 135–136

E
Early Bronze Age, 155, 156, 260
Egypt, 127, 132, 133, 140, 268, 274
Electrical resistivity (ER), 196, 197, 199, 296, 303, 304, 309
Eluviation, 212–215
Environmental
  decay, 386, 387
  degradation, 58, 127, 244, 247, 405, 406
  externalities, 386–390
  sustainability, 206
Epipedon
  anthrogleyic, 206, 225, 226, 238
  anthropic, 215–217
  anthrostagnic, 209, 217–218
  cumulic, 209, 216–217, 238
  fimic, 215, 217, 223
  siltigic/irragric, 209, 215, 216, 238
EU Soil thematic strategy, 51–56
Extra pedon, 181, 186, 187

F
Farmers decision system, 400
Fertile lands, 259–284
Floods, 267
Forest decay, 388
Fragmentation, 388
France, 131, 132, 134, 135, 137, 138, 274, 297

G
Geochemistry, 293
Geophysics, 293
Gleysols, 128, 132–133, 135, 138, 236
Greece, 129, 132, 134, 135, 137–139
Groundwater, 19, 22, 23, 128, 140, 181, 189, 211, 214, 215, 225, 229, 230, 312–314, 316–318, 320–330
Gypsisol, 132–136

H
Hadrian, 265, 270
Heavy metals, 150, 295, 298, 301, 302, 304
Hebron highlands, 63–64
Hedonic prices, 404
Herodotus, 260, 267
Histosols, 128–129, 168, 236, 288
Hittite, 27, 260
Horizon
  argic, 136, 137
  cambic/calcic, 136, 138, 235, 238
  duric, 136
  hydragric, 209, 218
  petroduric, 136
  salic, 132
Horse driven wells, 262–264
Human pressures, 125
Humret Es-Sahin watershed, 61–63, 82, 114
Hydrogeology, 311–330
Hydrology, 290, 319, 354

I
Illuviation, 137, 184, 210, 212–213, 223, 226, 230
  illuvial transport, 181
Impact analysis, 288–289
Impacts
  income inequality, 387, 388
  of misuse, 3, 9–10
  of winter resorts, 145–147
Indigenous technologies, 2, 45, 333–349, 397
Inheritance, 153–158
Intra pedon, 181, 186, 187
Irrigation, 7, 14, 16, 17, 19, 24, 30–32, 38, 118, 127, 130, 132–135, 137, 139, 140, 180, 184, 187, 206, 209–211, 213–217, 221, 223, 225, 229, 236, 238, 253, 261–264, 266, 270–272, 281, 314–321, 323–325, 327, 329, 330, 352–354, 356–362, 401
Israel, 59–61, 65–71, 73–75, 77–79, 81, 83, 92, 94–97, 99, 102, 105, 109, 112–117, 119, 131, 134, 137, 139

Italy, 9, 54, 56, 130–132, 134, 135, 137, 138, 149, 151, 155, 157, 179–202
Italian Alps, 143–160

J
Jordan, 60–63, 67, 70, 73, 74, 80–82, 84, 85, 96, 111, 119, 121, 126, 135, 137, 139
Justinian, 265, 266, 269, 270

K
Karst processes, 181
Kastanozem, 132, 134
Keli-calcrete surfaces, 283–284
Kizilöz Microcatchment, 64–65, 75–77, 100, 103
Kizzuwatna, 260

L
Land
capability, 2, 26, 288, 290
leveling, 210–211
restoration, 55, 56, 255
Land management
position, 41, 252
region, 125–141, 337
Landscape, 5, 52–55, 60, 63, 66, 120, 127, 129, 131, 132, 135, 139, 143–145, 151–155, 164–171, 173–175, 179, 180, 190, 206, 211, 212, 219, 223, 228, 230, 238, 244, 245, 247, 251, 252, 255, 313, 334–338, 352, 356, 364
Land use change, 5, 68, 165, 171, 206, 312
Lebanon, 131, 134, 137, 139
Legislation, 10, 19, 24, 25, 53, 302, 338
Leptosols, 8, 35, 128, 130–131, 134, 135, 137, 139
Linen, 268–270, 272, 281
Linthitic horizon, 215
Luvisols, 35, 127, 128, 131, 134, 135, 137–139, 168, 169, 171, 172, 236

M
Malaysia, 285–291
Management, 2, 5, 10, 18, 24, 27, 38, 39, 45, 48, 55, 59, 60, 62–66, 77–79, 82, 98, 125–141, 151, 163–165, 169–178, 184, 202, 209, 215, 221, 223, 244, 250, 251, 254, 270, 281, 288, 291, 313, 314, 330, 333–349, 351–362, 364, 385, 397, 400, 401
Marrakech, 133, 244–247, 255
Mediterranean, 2, 5, 7, 10, 23, 25, 26, 32, 37, 55, 59, 62, 63, 89, 114–120, 125–141, 181, 184, 187, 194, 244, 245, 250, 269, 278, 279, 281, 294, 311–330, 374, 377
Mersin, 44, 261
Micronorphology, 199, 229, 250
microstructure, 196, 199
Migration, 3, 10–19, 23, 38, 45, 47, 214, 286, 357, 400
Mining
dumpings, 145, 149, 150
mine tailings, 199, 294
plant establishment in mine tailings, 200–201
ponds, 294–297
Mismanagement, 20
Mitamashi, 351
Morocco, 127, 131–135, 137, 243–256
Murcia, 194, 293–309

N
Natural resources, 3, 18, 20, 25, 48, 127, 175, 351, 358, 398, 404
Neolithic, 53, 259, 261
Net primary production (NPP), 2, 39–43, 45–47
Niger, 333–349
Norway, 163–178

O
Oldman saltbush, 244
Organic matter, 52, 62, 67, 91, 98–100, 120, 129, 134, 139, 165, 170, 172, 173, 177, 194, 195, 199–202, 211–212, 214, 224, 225, 236, 253, 267, 336, 366–368, 370, 372, 377
Ottoman Empire, 260, 271, 274
Overgrazing, 19, 90, 139, 148, 149, 157–159, 245
Ownership, 10, 247, 255, 288

P
Paddy
cultivation culture, 351
history, 213
paddification, 213–214
soil chronosequences, 221–223
soils, 206, 207, 210–215, 217–221, 223, 225–238
Palaeo
environment, 253
micro-topography, 186
morphology, 183, 186
Palaeolithic, 155–156

Palestine, 60, 61, 63, 65–67, 80, 82, 85–88, 92, 98, 99, 101, 103, 106, 108, 111, 114, 115, 121, 260
Pastureland, 146, 148, 157–158, 247
Pedogenic
  features, 235
  iron, 221–223
  manganese, 221–223
Percogenic horizon, 215, 238
Petrocalcic Palexeralfs, 190
Phaeozem, 132, 134, 135, 139, 171, 237
Phoenicians, 278–279
Phosphate extractable organic nitrogen (PEON), 367–368, 371, 372, 377
Piri Reis, 266
Post-cultivation succession, 363–377
Project
  implementation, 397
  preparation, 140, 353
  strategy map, 392, 393
Projected impact, 311–330
Pyramos, 259, 261, 267, 279

Q
Quantification of carbon, 38–41
Quantification of degradation, 406

R
Rainfall, throughfall and stemflow, 71
Reconstructing the past, 363–377
Redoximorphic features, 205, 218, 226–227, 229, 230, 236, 238
Regenerating biodiversity, 363–377
Regosol, 128, 137–139, 171, 173, 174, 236
Resources
  non-renewable, 24, 287, 289
Rhizosphere, 365–367, 370–373, 375
Risk assessment, 293–309
Roman
  age, 155–156
  mining, 293–309
  terraces, 131
Root zone, 121, 201, 365–368, 372–375, 377
Runoff and sediment discharge, 65, 72, 86

S
Sahel, 333–349
Saline intrusion, 133, 314
Sarawak, 285–291
Sardinia, 179–191, 260
Saros, 259, 261, 267
Sassari, 244
Satoyama, 48, 354–361
Satsneferru, 268–269
Settlements
  civil, 145, 149–150, 159
  industrial, 145, 149–150, 159
Seyhan basin, 2, 3, 5–18, 22, 23, 38, 40–43, 45–47, 313
Seyhan dam, 267, 314, 319
Shifting agriculture, 285–291
Silk, 270, 272
Socio-economy, 10
Soil
  classification, 207–209, 233, 234, 236
  cover, runoff and erosion, 81–82, 91, 98–100
  degradation, 36, 51, 52, 139–140, 154, 163, 164, 398, 399
  description, 208
  development, 128, 163, 164, 223, 232
  erosion, 55, 57–123, 134, 154, 165, 173, 177, 180, 244, 247, 253, 342
  organic matter, 67, 91, 98–100, 120, 194, 200, 201, 211–212, 253, 367–368
  organic nitrogen, 365
  protection, 11, 51–53, 55, 56, 101, 122, 139–141
  quality, 164, 168, 253, 363–377, 401
  resources, 140, 154
  scape, 51
Soil units (WRB)
  Calcic, 134
  Chromic, 134
  Gleyic, 134
  Haplic, 131
  Leptic, 134
  Luvic, 134
  Mollic, 131
  Pachic, 134
  Siltic, 134
  Skeletic, 134
  Solonchak, 132–134
  Solonetz, 132–134
  Vertic, 134
Spain, 129, 131–139, 171, 193–202, 293–309
Spatial data, 289–291
Spinnable fibres, 269–270
Stewardship, 288, 291
Stone quarrying, 151–152
Sub-microscopy, 367
Suido, 352–361

Sustainability strategy, 392–398
Sustainable
 soil, 140, 333–349
 use, 20, 43, 44, 48, 125–141, 357
 water management, 333–349, 360
Swot analysis, 10, 20, 394
Sylvi-agro-pastoral watershed, 64–65
Syria, 131, 134, 135, 137, 139

T
Taiwan, 205–238
Taurus mountains, 260, 267, 270, 281, 315
Terracing mountain, 153–155, 159
Terra rossa, 127, 128, 137
Tomatoes, 132, 267, 282, 283
Traditional economic growth theory, 382–385
Turkey, 2, 5, 10, 19, 44, 60, 64–65, 68, 70, 73, 75–77, 103, 104, 119, 122, 129, 131–139, 274, 277, 280, 282, 313, 314, 318, 320, 329, 366, 388

U
Umbrisol, 134–135, 168, 169, 171

V
Vertisols, 8, 32, 128, 131–132, 135, 136, 138
Vineyard, 20, 21, 35, 55, 136, 137, 154, 155, 180, 184, 185, 187, 189, 190

W
Wakf, 266
Waste disposal, 145, 152–153, 159
Watersheds, 59–61, 64–66, 68, 71–73, 75, 76, 78, 79, 103, 106, 119, 121, 122, 171, 174, 175
Water stable aggregates (WSA), 366, 367, 370–377
Weed rhizospheres, 365, 370–373
Weeds, 287, 336, 364–369, 372
Western Balkans, 131, 134
World Reference Base (WRB), 127, 128, 154, 179, 180, 187, 209, 215, 236, 237
WSA. *See* Water stable aggregates

Y
Yatir forest, 60–61, 71, 81, 94, 96, 97, 99, 101–104, 106

FSC
www.fsc.org
MIX
Papier | Fördert gute Waldnutzung
FSC® C083411

Kolibri 360 GmbH
Ferdinand-Jühlke-Straße 7
99095 Erfurt, Deutschland
produktsicherheit@kolibri360.de